S0-AYZ-020

BIOLOGICAL EXPLORATIONS
A Human Approach
Fourth Edition

STANLEY E. GUNSTREAM

Pasadena City College
Pasadena, California

PRENTICE HALL
Upper Saddle River, New Jersey 07458

Library of Congress Cataloging-in-Publication Data is available upon request

Acquisitions Editor: Karen Horton
Editor in Chief: Sheri L. Snavely
Special Projects Manager: Barbara A. Murray
Project Management and Composition: NK Graphics
Manufacturing Manager: Trudy Pisciotti
Text Illustrations: Mary Dersch, CMI/The Professional Edge
Cover Designer: Jayne Conte
Cover Image: Doug Menuez, PhotoDisc, Inc.

© 2001, 1997 by Prentice-Hall, Inc.
Pearson Education
Upper Saddle River, New Jersey 07458

All rights reserved. No part of this book may be reproduced, in any form or by
any means, without permission in writing from the publisher.

Earlier edition, entitled *Human Biology: Laboratory Explorations*, copyright ©
1986 by Macmillan Publishing Company; copyright © 1994 by Macmillan
Publishing Company.

Color Insert Credits appear on page vii, which constitutes a continuation
of the copyright page.

Printed in the United States of America

10   9   8   7   5   4   3   2   1

ISBN 0-13-089446-X

Prentice-Hall International (UK) Limited, *London*
Prentice-Hall of Australia Pty. Limited, *Sydney*
Prentice-Hall Canada, Inc., *Toronto*
Prentice-Hall Hispanoamericana, S.A., *Mexico*
Prentice-Hall of India Private Limited, *New Delhi*
Prentice-Hall of Japan, Inc., *Tokyo*
Pearson Education Asia Pte Ltd., *Singapore*
Editora Prentice-Hall do Brasil, Ltda., *Rio de Janeiro*

CONTENTS

PREFACE

Biological Explorations: A Human Approach is a laboratory manual specifically designed for the laboratory component of courses in: (1) general biology where the human organism is emphasized, and (2) human biology. It is compatible with any modern textbook that emphasizes the human organism. The exercises are appropriate for three-hour laboratory sessions, but they are also adaptable for a two-hour laboratory format.

This laboratory manual is designed not only to enhance learning by students but to simplify the work of instructors. Its design assumes little if any prior experience by students in a biology laboratory.

MAJOR FEATURES

1. The **thirty-three exercises** provide a wide range of options for the instructor, and the range of activities within an exercise further increases the available options.
2. Each exercise is basically **self-directing**, which allows students to work independently without direct assistance by the instructor.
3. Each exercise, and its major subunits, are **self-contained** so that the instructor may arrange the sequence of exercises, or the activities within an exercise, to suit his or her preferences.
4. Each exercise begins with a list of **objectives** that outlines the minimal learning responsibilities of the student. The objectives also inform the student of the emphasis and scope of the activities to be accomplished and the minimal learning responsibilities.
5. Following the objectives, each exercise starts with a brief discussion of **background information** that is necessary to (a) understand the subject of the exercise, and (b) prepare the student for the activities that follow. The inclusion of the background information minimizes the need for introductory explanations by the in-

structor and assures that all lab sections receive the same introductory information.

6. Before beginning the laboratory activities, students are directed to demonstrate their understanding of the background information by labeling illustrations and completing the portion of the Laboratory Report that covers this material.
7. Over 200 **illustrations** are provided to enhance students' understanding. Students are asked to color-code significant structures in many illustrations as a means to aid learning. Eight plates of color photos are included in the organismic diversity section to assist students with the structure of difficult-to-observe organisms.
8. New **key terms** are in bold print for easy recognition by students, and they are defined when first used.
9. Required **materials** are listed for each activity in an exercise. The list is divided into materials needed by each student or student group and materials that are to be available in the laboratory. This list helps the student to obtain the needed materials and guides the laboratory technician in setting up the laboratory. The exercises utilize standard equipment and materials that are available in most biology departments.
10. Activities to be performed by students are identified by an **assignment** heading that clearly distinguishes activities to be performed from the background information. The assignment sections are numbered sequentially within each exercise and on the Laboratory Report to facilitate identification and discussion.
11. Clear **laboratory procedures** guide students through each activity, so that minimal assistance is needed from the instructor.
12. A **Laboratory Report** is provided for each exercise to guide and reinforce students' learning. The laboratory reports not only provide a place for students to record observations, collected data, and conclusions, but also provide a convenient means for the instructor to assess

student understanding. The separate laboratory reports may be removed from the manual without removing key information contained in the exercises and needed by students for study.

Where appropriate, laboratory reports contain a mini-practicum section where students are asked to identify structures or organisms set up by the instructor. This provides a suitable stimulus for students' study, as well as giving students some experience in lab practicums before taking the big one.

13. An **Instructor's Manual** accompanies the manual and contains (a) composite lists of equipment and supplies, (b) sources of supplies, (c) special techniques, (d) operational suggestions, and (e) answer keys for the laboratory reports.

ACKNOWLEDGMENTS

A number of users have provided helpful suggestions for improving the fourth edition of this book.

Special recognition is due those whose critical reviews contributed to the present revision:

Todd Rinkus, *Marymount University*
Theresa Page, *Texas Women's University*
Henry Patthey, *Nashville State Technical Institute*
Sallie M. Noel, *Austin Peay State University*
Deborah Whiting, *Indian River Community College*
Karen R. Zagula, *Wake Technical Community College*

I thank the members of the book team for their vital contributions to this revision. The support and contributions of Sheri Snavely, Editor in Chief and Karen Horton, Editorial Project Manager are gratefully acknowledged. It was a pleasure to work with Tonnya Norwood of NK Graphics, who skillfully guided the production process.

Adopters are encouraged to communicate to the author any suggestions that will improve the usefulness of this manual for their courses.

S.E.G.

Color Insert Credits

Plate 1: P1.lA-© David M. Phillips/Visuals Unlimited; P1.1B-© CNRI/Science Photo Library/Photo Researchers, Inc.; P1.1C-© Charles W. Stratton/Visuals Unlimited; P1.2-© Dr. Tony Brain/Sci. Photo Lib./Custom Medical Stock Photo; P1.3-© John D. Cunningham/Visuals Unlimited; P1.4-© T.E. Adams/Visuals Unlimited; **Plate 2:** P2.1-© M.I. Walker/Photo Researchers, Inc.; P2.2-© M. Abbey/Visuals Unlimited; P2.3-© M.I. Walker/Photo Researchers, Inc.; P2.4-© George J. Wilder/Visuals Unlimited; P2.5-© M. Abbey/Visuals Unlimited; **Plate 3:** P3.1-© Manfred Kage/Peter Arnold, Inc.; P3.2-© Manfred Kage/Peter Arnold, Inc.; P3.3-© M.I. Walker/Photo Researchers, Inc.; P3.4-© Eric V. Grave/Photo Researchers, Inc.; P3.5-© G.W. Willis/BPS; **Plate 4:** P4.1-© T.E. Adams/Visuals Unlimited; P4.2-© Cabisco/Visuals Unlimited; P4.3-© David M. Phillips/Visuals Unlimited; P4.4-© PHOTO-TAKE; P4.5-© Biophoto Ass./S.S./Photo Researchers, Inc.; **Plate 5:** P5.1-© Carolina Biological Supply Co./PHOTOTAKE; P5.2-© Carolina Biological Supply Co./PHOTOTAKE; P5.3-© Runk/Schoenberger/Grant Heilman Photography; P5.4-© Cabisco/Visuals Unlimited; P5.5-© CNRI/Sci. Source/Custom Medical Stock Photo; P5.6-© Bruce Iverson/Visuals Unlimited; **Plate 6:** P6.1A-© Hans Reinhard/Bruce Coleman Inc.; P6.1B-© Visuals Unlimited; P6.2-© S. Flegler/Visuals Unlimited; P6.3-© S. Flegler/Visuals Unlimited; P6.4-© Bill Keogh/Visuals Unlimited; **Plate 7:** P7.1-© M.J. Walker/Photo Researchers, Inc.; P7.2-© Manfred Kage/Peter Arnold, Inc.; P7.3-© M.I. Walker/Photo Researchers, Inc.; P7.4-© Cabisco/Visuals Unlimited; P7.6-© John D. Cunningham/Visuals Unlimited; **Plate 8:** P8.1-© Farrell Grehan/Photo Researchers, Inc.; P8.2-© Joyce Photographs/Photo Researchers, Inc.; P8.3-© Jeff L. Rotman; P8.4-© Phillip Sze/Visuals Unlimited; P8.5-© Daniel Gotshall/Visuals Unlimited.

Part 1   Cell Biology

1 Orientation

OBJECTIVES

After completion of the laboratory session, you should be able to:
1. Describe how to prepare for a laboratory session.
2. Describe laboratory safety procedures.
3. State the meaning of common prefixes and suffixes.
4. Use the metric system of measurement.
5. List the steps of the scientific method.
6. Define all terms in bold print.

Laboratory study is an important part of a course in biology. It provides opportunities for you to observe and study biological organisms and processes and to correlate your findings with the textbook and lectures. It allows the conduction of experiments, the collection of data, and the analysis of data to form conclusions. In this way, you experience the process of science, and it is this process that distinguishes science from other disciplines.

PROCEDURES TO FOLLOW

Your success in the laboratory depends on how you prepare for and carry out the laboratory ac-tivities. The procedures that follow are expected to be used by all students.

Preparation

Before coming to the laboratory, complete the following activities to prepare yourself for the laboratory session.

1. Read the assigned exercise to understand (a) the objectives, (b) the meaning and spelling of new terms, (c) the introductory background information, and (d) the procedures to be followed.
2. Label and color-code illustrations as directed in the manual, and complete the items on the laboratory report related to the background information.
3. Bring your textbook to the laboratory for use as a reference.

Working in the Laboratory

The following guidelines will save time and increase your chances of success in the laboratory.

1. Remove the laboratory report from the manual so that you can complete it without flipping pages. Laboratory reports are three-hole punched so you can keep completed reports in a binder.

2. Follow the directions explicitly and in sequence unless directed otherwise.
3. Work carefully and thoughtfully. You will not have to rush if you are well prepared.
4. Discuss your procedures and observations with other students. If you become confused, ask your instructor for help.
5. Answer the questions on the laboratory report thoughtfully and completely. They are provided to guide the learning process. Just filling in the blanks is not acceptable.

Laboratory Safety and Housekeeping

1. Use equipment with care. Report any problems to your instructor.
2. Inform your instructor of any breakage or spills, and ask for assistance in proper cleanup and disposal procedures.
3. Immediately report any injuries, even minor ones, to your instructor.
4. Tie back long hair and roll up loose sleeves when using open flames.
5. Clean glassware and equipment before and after use. Return each item to its proper location at the end of the session, and clean your workstation.
6. Do not smoke, eat, drink, or apply cosmetics in the laboratory.

Assignment 1

Complete item 1 on Laboratory Report 1 that begins on page 309.

BIOLOGICAL TERMS

One of the major difficulties encountered by beginning students is learning biological terminology. Each exercise has new terms emphasized in bold print so that you do not overlook them. Be sure to know their meanings prior to the laboratory session.

Biological terms are composed of a root word and either a prefix or a suffix, or both. The **root word** provides the main meaning of the term. It may occur at the beginning or end of the term, or it may be sandwiched between a **prefix** and a **suffix**. Both the prefix and suffix modify the meaning of the root word. The parts of a term

are often joined by adding *combining vowels* that make the term easier to pronounce. The following examples illustrate the structure of biological terms.

1. The term *endocranial* becomes *endo/crani/al* when separated into its components. *Endo* is a prefix meaning within; *crani* is the root word meaning skull; *al* is a suffix meaning pertaining to. Therefore, the literal meaning of endocranial is "pertaining to within the skull."
2. The term *leukocyte* becomes *leuko/cyte* when separated into its components. *Leuko* is a prefix meaning white, and *cyte* is a root word meaning cell. Thus, the literal meaning of leukocyte is "white cell."

Once you understand the structure of biological terms, learning the terminology becomes much easier. *Appendix A contains the meaning of common prefixes, suffixes, and root words*. Use it frequently to help you master new terms.

Assignment 2

Using Appendix A, complete item 2 on the laboratory report.

UNITS OF MEASUREMENT

Scientists use the **International System of Units (SI)**, commonly called the **metric system**, for making measurements. The basic reference units for these measurements are: **meter** for length, **gram** for mass, **liter** for volume, and **degree Celsius** for temperature. Each basic unit may be preceded by a prefix that modifies the value of the unit by one or more powers of ten as shown in Table 1.1 and summarized below.

$$
\begin{aligned}
\text{kilo-} &= \text{unit} \times 1{,}000 \\
\text{(none)} &= \text{unit} \\
\text{deci-} &= \text{unit} \div 10 \\
\text{centi-} &= \text{unit} \div 100 \\
\text{milli-} &= \text{unit} \div 1{,}000 \\
\text{micro-} &= \text{unit} \div 1{,}000{,}000
\end{aligned}
$$

TABLE 1.1
Common Metric System Units

Category	Symbol	Unit	Value	English Equivalent
Length	km	kilometer	1,000 m	0.62 mi
	m	meter*	1 m	39.37 in.
	dm	decimeter	0.1 m	3.94 in.
	cm	centimeter	0.01 m	0.39 in.
	mm	millimeter	0.001 m	0.04 in.
	μm	micrometer	0.000001 m	0.00004 in.
Mass	kg	kilogram	1,000 g	2.2 lb
	g	gram*	1 g	0.04 oz
	dg	decigram	0.1 g	0.004 oz
	cg	centigram	0.01 g	0.0004 oz
	mg	milligram	0.001 g	
	μg	microgram	0.000001 g	
Volume	l	liter*	1 l	1.06 qt
	ml	milliliter	0.001 l	0.03 oz
	μl	microliter	0.000001 l	

*Denotes the base unit.

Some English equivalents are included in Table 1.1 for comparison and to allow conversions between the English and the metric systems.

The major advantage of the metric system is that it allows easy conversion from one unit to another within a category by multiplying or dividing by the correct power of ten. The following examples show how this is done.

1. To convert 5.75 meters into millimeters, the first step is to determine how many millimeters are in a meter. Table 1.1 shows that 1 mm = 0.001 m or 1/1,000 of a meter. Thus, there are 1,000 mm in 1 m. Since you are converting meters to millimeters, you multiply 5.75 m by a fraction expressing that there are 1,000 mm in 1 m and cancel identical units. *You use the unit that you want to convert to as the numerator of the fraction.*

$$\frac{5.75 \text{ m}}{1} \times \frac{1,000 \text{ mm}}{1 \text{ m}} = 5,750 \text{ mm}$$

Note that this equation multiplies 5.75 by 1,000, which moves the decimal 3 places to the right and changes the units from meters to millimeters. Moving the decimal and

changing the unit is a quick way to do such problems.

2. To convert 125 centimeters into meters, the first step is to determine how many centimeters are in a meter. Table 1.1 shows that 1 cm = 0.01 m or 1/100 of a meter. Thus, there are 100 cm in 1 m. Since you are converting centimeters into meters, you multiply 125 cm by a fraction expressing that 1 m contains 100 cm and cancel identical units. Do you know why 1 m is the numerator of the fraction?

$$\frac{125 \text{ cm}}{1} \times \frac{1\text{m}}{100 \text{ cm}} = 1.25 \text{ m}$$

Note that this equation divides 125 cm by 100, which moves the decimal 2 places to the left and changes the units from centimeters to meters.

Length

Length is the measurement of a line, either real or imaginary, extending between two points. Your height, the distance between cities, and the size of a football field involve length. The basic unit of length is the meter.

Materials

Per student group
Metric ruler, clear plastic
Penny

Assignment 3

1. **Complete item 3a on the laboratory report. Perform the following procedures and record your data and calculations in item 3 on the laboratory report.**
2. Measure the diameter of a penny in millimeters. Convert your answer to centimeters and meters.
3. Measure the length of your little finger as shown in Figure 1.1. Record your answer on the laboratory report and in the class tabulation chart on the chalkboard.
4. Biologists usually analyze data statistically to determine if the findings are significant. We won't do that here, but you will look at certain characteristics of the data that have been collected.
5. **Record the class data in item 3d on the laboratory report.**
6. From the class data, determine the *range* of finger lengths and the *average* finger length.

(The average is calculated by dividing the sum of the finger lengths by the number of fingers measured.) **Complete item 3e on the laboratory report.**

7. Plot the little-finger lengths against the number of students that have each finger length on the graph shown in item 3f on the laboratory report. This is done for each finger length by placing a dot where an imaginary vertical line extending from the horizontal axis at a particular finger length intersects with an imaginary horizontal line extending from the vertical axis at the number of students having that finger length.
8. **Complete item 3 on the laboratory report.**

Mass

Mass is the characteristic that gives an object inertia, the resistance to a change in motion. It is the quantity of matter in an object. Mass is not the same as weight because weight is dependent upon the force of gravity acting on the object. The mass of a given object is the same on earth as on the moon, but its weight is much less on the moon because the moon's force of gravity is less than the earth's. As a nonscientist, you probably can get by using mass and weight interchangeably, although this is technically incorrect.

In this section, you will use a triple-beam balance to measure the mass of objects. Obtain a balance and locate the parts labeled in Figure 1.2. Note that each beam is marked with graduations. The beam closest to you has 0.1-g and 1.0-g graduations; the middle beam has 100-g

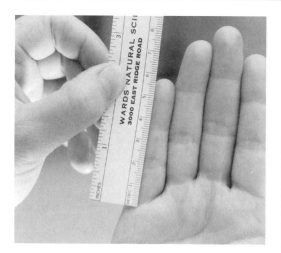

Figure 1.1. Measuring the little finger. Align the edge of the ruler along the midline of the finger.

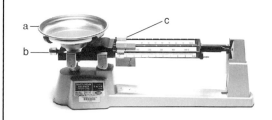

Figure 1.2. A triple-beam balance. (a) Pan. (b) Adjustment knob. (c) Movable mass.

graduations; and the farthest beam has 10-g graduations. There is a movable mass attached to each beam. When the pan is empty and clean and the movable masses are moved to zero—as far to the left as possible—the balance mark on the right end of the beam should align with the balance mark on the upright post at the right. If not, rotate the adjustment knob under the pan at the left until it does. Now the balance is ready to use.

Procedure for Measuring Mass

All three beams are used if you are measuring an object with a mass over 100 g; the first and third beams are used when measuring an object with a mass between 10 and 100 g; the first beam only is used when measuring an object with a mass of 10 g or less. Here is how to do it.

Place the object to be measured in the center of the pan. Move the movable mass on the middle (100 g) beam to the right, one notch at a time, until the right end of the beam drops below the balance mark. Then move the mass back to the left one notch. Now slide the movable mass on the third (10 g) beam to the right, one notch at a time, until the right end of the beam drops below the balance mark. Then move the mass back to the left one notch. Finally, slide the movable mass on the first (1 g) beam slowly to the right until the right end of the beam aligns with the balance mark. The mass of the object is then determined as the sum of the masses indicated on the three beams.

Materials

Per student group
Triple-beam balance
Wood blocks with different densities, numbered 1 to 3

Assignment 4

1. Measure the mass of each of the three numbered blocks of wood and record your data.
2. **Complete item 4 on the laboratory report.**

Volume

Volume is the space occupied by an object or a fluid. The liter is the basic unit of volume, but milliliters are the common units used for measuring and describing small volumes. Graduated cylinders and pipettes are used to measure fluids. Length may also be involved in volume determinations, because 1 cubic centimeter (cm^3 or cc) equals 1 ml.

Materials

Per student group
Beaker, 250 ml
Graduated cylinders, 10 and 50 ml
Medicine dropper
Pipettes, 1, 5, and 10 ml
Test tube
Test tube rack
Triple-beam balance
Wood blocks with different densities, numbered 1 to 3

Assignment 5

1. Measure the length, width, and depth of the wood blocks in centimeters. Then determine the volume of each block. Volume = length × width × depth.
2. The density of an object is defined as mass per unit of volume. Using the mass of each wood block from item 4a on the laboratory report, calculate the density of each block.

$$\text{Density} = \frac{\text{Mass}}{\text{Volume}}$$

3. **Complete items 5a to 5c on the laboratory report.**
4. Examine the pipettes and graduated cylinders to determine the meaning of the graduations marked on each.
5. Fill the beaker about half full with tap water. Pour a little water into the 10-ml graduated cylinder. Look at the top of the water column from the side and note that it is curved rather than flat. This curvature is known as the **meniscus**, and it results because water molecules tend to "creep up" and stick to the side of the cylinder. When measuring the volume of fluid in a cylinder or

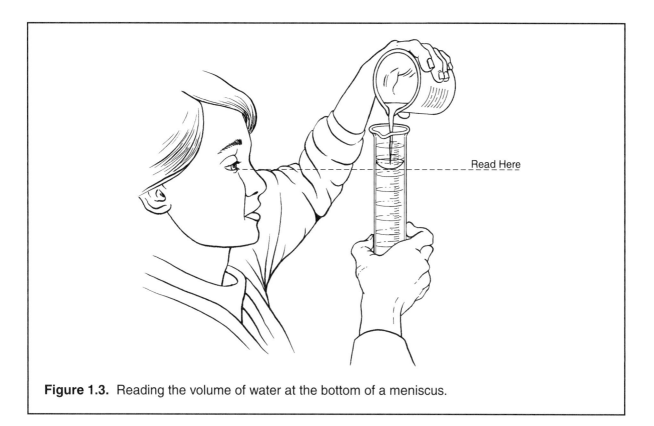

Figure 1.3. Reading the volume of water at the bottom of a meniscus.

pipette, you must read the volume at the *bottom of the meniscus*, as shown in Figure 1.3.

6. While observing the meniscus from the side, add water to the 10-ml graduated cylinder and fill it to the 8-ml mark. Repeat until you can do this with ease.

7. Determine the volume of the test tube by filling it with water and then pouring the water into the 50-ml graduated cylinder.

8. Empty the 10-ml cylinder and shake out as much water as possible. Determine the number of drops in 1 ml of water by adding water, drop by drop, to the 10-ml graduated cylinder with the medicine dropper (dropper pipette).

9. In later exercises, you will be asked to dispense a *dropper* of fluid. This means as much fluid as a medicine dropper will hold when the rubber bulb is fully depressed and then released after the tip has been submerged in the fluid. Determine the volume of a *dropper*.

10. Determine the mass of 30 ml of water using the 50-ml graduated cylinder and a triple-beam balance.

11. **Complete item 5 on the laboratory report.**

Temperature

Scientists use a Celsius thermometer to measure temperature, while nonscientists usually use a Fahrenheit thermometer. The differences in the scales of the two systems of temperature measurement are shown below.

	°C	°F
Boiling point of water	100	212
Freezing point of water	0	32

You can convert from one system to the other by using these formulas.

Celsius to Fahrenheit

$$°F = \frac{9 \times °C}{5} + 32$$

Fahrenheit to Celsius

$$°C = \frac{5 \times (°F - 32)}{9}$$

Materials

Per student group
Beaker, 250 ml
Celsius thermometer

Assignment 6

1. Examine the graduations on the Celsius thermometer.
2. Measure the temperature (°C) of air in the room and the temperature of cold tap water. Convert the temperatures to °F.
3. **Complete item 6 on the laboratory report.**

SCIENTIFIC METHOD

Many of the laboratory activities in this manual allow you to play the role of a biologist as you perform experiments, collect data, and make conclusions. You will use the **scientific method** of inquiry, a process that tests, in a systematic manner, possible answers to questions raised about nature. It also allows other scientists to duplicate the testing process you use.

The scientific method involves several sequential steps that are diagramed in Figure 1.4 and summarized briefly below. *Hypothetical examples* are provided for each step.

1. The process begins with making careful, thoughtful **observations** of nature directly or indirectly.

   Example: Observations suggest that a daily supplement of vitamin C may reduce the risk of cancer.

2. Observations raise questions in the biologist's mind that lead the biologist to state the **key question** to be answered. This key question is sometimes called "the problem."

   Example: Does a daily supplement of 50 mg of vitamin C reduce the risk of cancer?

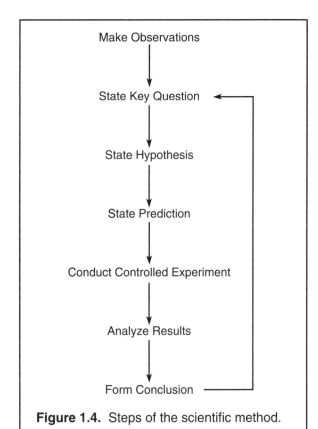

Figure 1.4. Steps of the scientific method.

3. A **hypothesis** is stated. This is a statement of the anticipated answer to the key question.

   Example: A normal diet plus a daily supplement of 50 mg of vitamin C reduces the risk of cancer.

4. A **prediction** is made, based on the hypothesis. This prediction is usually phrased in an "if ... then" manner. This tells the biologist what to expect in specific situations.

   Example: If daily supplements of 50 mg of vitamin C reduce the risk of cancer, then mice receiving this vitamin C supplement will develop fewer cancers than mice not receiving the supplement when both groups of mice are exposed to a carcinogen (cancer-causing substance).

5. A **controlled experiment** is designed and conducted to test the hypothesis if the hypothesis is testable by experiment. If it is not, additional observations are made to test the hypothesis.

Example: Daily supplements of 50 mg of vitamin C reduce the frequency of cancer in mice when they are exposed to a carcinogen.

Testing the hypothesis by a controlled experiment is highly preferable. There are three kinds of variables (conditions) in a controlled experiment. The **independent variable** (in our example, vitamin C) is the condition that is being evaluated for its effect on the **dependent variable** (in our example, cancer). **Controlled variables** are all other conditions that could affect the results but do not because they are controlled by being kept constant. A controlled experiment consists of two parts. The *experimental group* is exposed to the independent variable, but the *control group* is not. All other variables are controlled (kept constant).

Example: From a group of laboratory mice that have been bred to have identical hereditary compositions, 50 randomly selected mice are placed in the experimental group and 50 randomly selected mice are placed in the control group. The experimental group receives a supplement of 50 mg of vitamin C each day, but the control group does not. All other variables are kept identical for each group, i.e., exposure to a carcinogen, normal diet, temperature, humidity, light-dark cycles, water availability, and so forth. After 120 days, the mice are examined for cancers.

6. **Results** are collected and analyzed.

   Example: Among the experimental group, 5 mice developed cancers. Among the control group, 20 mice developed cancers.

7. A **conclusion** about whether the hypothesis is accepted or rejected is made, based on the results of the experiment. A conclusion often leads to the formation of a new hypothesis and additional experiments.

   Example: The hypothesis in this experiment is supported. The frequency of cancer was reduced in mice receiving a daily supplement of 50 mg of vitamin C.

Assignment 7

Complete item 7 on the laboratory report.

2

The Microscope

OBJECTIVES

After completion of the laboratory session, you should be able to:
1. Identify the parts of compound and dissecting microscopes and state the function of each.
2. Describe and demonstrate the correct way to:
   a. Carry a microscope.
   b. Clean the lenses.
   c. Prepare a wet-mount slide.
   d. Focus with each objective.
   e. Determine the total magnification.
   f. Estimate the size of objects.
3. Define all terms in bold print.

A **microscope** is a precision instrument and an essential tool in the study of cells, tissues, and minute organisms. It must be handled and used carefully at all times. Most of the microscopic observations in this course will be made with a **compound microscope**, but a **dissecting microscope** will be used occasionally. A microscope consists of a lens system, a controllable light source, and a mechanism for adjusting the distance between the lens system and the object to be observed.

To make the observations required in this course, you must know how to use a microscope effectively. This exercise provides an opportunity for you to develop skills in microscopy.

THE COMPOUND MICROSCOPE

The major parts of the compound microscope are shown in Figure 2.1. Refer to this figure as you read this text, and *label the parts* indicated on the figure. Your microscope may be somewhat different from the one illustrated.

The **base** of the microscope rests on the table and, in most microscopes, contains a built-in **light source** and a **light switch**. Some microscopes have a light-intensity (voltage) control knob on the base, usually associated with the light switch. The **arm** rises from the base and supports the stage, lens system, and control mechanisms of the microscope. The **stage** is the flat surface on which microscope slides are placed for viewing. **Stage clips**, or a **mechanical stage**, hold the slide in place.

Most microscopes have a **condenser** located below the stage. It concentrates the light on the object being viewed, and may be raised or lowered by the **condenser control knob**. Usually, the condenser should be raised to its highest position. An **iris diaphragm** is built into the base

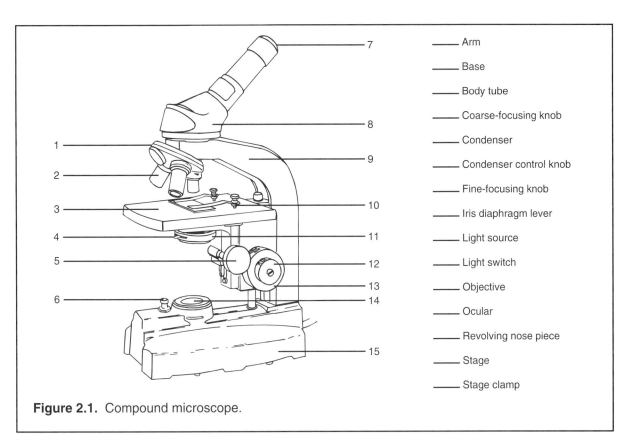

Figure 2.1. Compound microscope.

Labels at right:
— Arm
— Base
— Body tube
— Coarse-focusing knob
— Condenser
— Condenser control knob
— Fine-focusing knob
— Iris diaphragm lever
— Light source
— Light switch
— Objective
— Ocular
— Revolving nose piece
— Stage
— Stage clamp

of the condenser. The **iris diaphragm control lever** (a rotatable wheel in some microscopes) varies the amount of light entering the condenser and the lens system.

The **body tube** is supported by the arm of the microscope, and has an **ocular lens** at its upper end and a **revolving nosepiece** with the attached **objective lenses** at its lower end. The nosepiece is rotated to bring different objectives into viewing position. The objectives usually click into viewing position.

Student microscopes usually have three objective lenses. The shortest is the **scanning objective**, which has a magnification of 4×. The **low-power objective**, with a 10× magnification, is intermediate in length. The **high-power (high-dry) objective** is the longest, and usually has a magnification of 40×, but may have a 43× or 45× magnification in some microscopes. In this manual, the high-power objective is often called the 40× objective. Some microscopes have an **oil-immersion objective** (100× magnification) that is a bit longer than the high-power objective.

There are two focusing knobs. The **coarse-focusing knob** has the larger diameter and is used to bring objects into rough focus when using the 4× and 10× objectives. The **fine-focusing knob** has a smaller diameter and is used to bring objects into fine focus. It is the *only* focusing knob used with the high-power and oil-immersion objectives.

Magnification

The magnification of each lens is fixed and inscribed on the lens. The ocular lens usually has a 10× magnification. The powers of the objectives may vary, but are usually 4×, 10×, and 40×. The **total magnification** is calculated by multiplying the power of the ocular lens by the power of the objective lens.

Resolving Power

The quality of a microscope depends on its ability to **resolve** (distinguish) objects. Magnification without resolving power is of no value. Modern microscopes increase both magnification and resolution by a careful matching of their light source and precision lenses. Most microscopes have a blue light filter located in either the condenser or the light source, since resolving power increases as the wavelength of light decreases. Blue light has a short wavelength. Student microscopes can usually resolve objects that are 0.5 μm or more apart. The best light microscopes can resolve objects that are 0.1 μm or more apart.

Contrast

Sufficient **contrast** must be present among the parts of an object for the parts to be distinguishable. Contrast results from the differential absorption of light by the parts of the object. Sometimes, stains must be added to a specimen to increase the contrast. A reduction in the amount of light improves contrast when viewing unstained specimens.

Focusing

A microscope is focused by increasing or decreasing the distance between the specimen on the slide and the objective lens. The focusing procedure used depends on whether your microscope has a **movable stage** or **movable body tube** (see Figure 2.2). Both kinds of focusing procedures are described below. Use the one appropriate for your microscope. As a general rule, you should start focusing with the low-power (10×) objective unless you are viewing an object whose large size requires starting with the 4× objective.

Focusing with a Movable Body Tube

1. Rotate the 10× objective into viewing position.
2. While using the coarse-focusing knob and

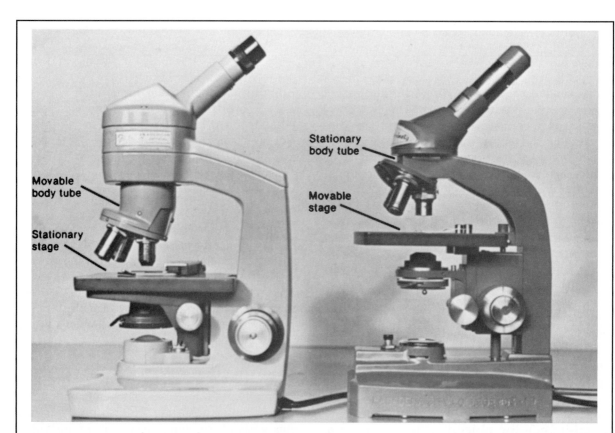

Figure 2.2. Comparison of two microscope designs.

looking from the side (not through the ocular), lower the body tube until it stops or until the objective is about 3 mm from the slide.

3. While looking through the ocular, slowly raise the body tube by turning the coarse-focusing knob toward you until the object you are seeking becomes visible. Use the fine-focusing knob to bring the object into sharp focus.

Focusing with a Movable Stage

1. Rotate the 10× objective into viewing position.
2. While using the coarse-focusing knob and *looking from the side* (not through the ocular), raise the stage to its highest position or until the slide is about 3 mm from the objective.
3. While looking through the ocular, slowly lower the stage by turning the coarse-focusing knob away from you until the object comes into focus. Use the fine-focusing knob to bring the object into sharp focus.

Switching Objectives

Your microscope is **parcentric** and **parfocal**. This means that if an object is centered and in sharp focus with one objective, it will be centered and in focus when another objective is rotated into the viewing position. However, slight adjustments to recenter the object and refocus the microscope (with the fine-focusing knob) may be necessary. As you switch objectives from 4× to 10× to 40× to increase magnification, the (1) working distance, (2) diameter of the field, and (3) light intensity are *reduced* as magnification increases. Note this relationship in Figure 2.3.

Slide Preparation

Specimens to be viewed with a compound microscope are placed on a **microscope slide** and are usually covered with a **cover glass**. Specimens may be mounted on slides in two different ways. A **prepared slide** (permanent slide) has a permanently attached cover glass, and the specimen is usually stained. A **wet-mount slide** (temporary slide) has the specimen mounted in a liquid, usually water, and covered with a cover glass. In this course, you will observe commercially prepared permanent slides and wet-mount

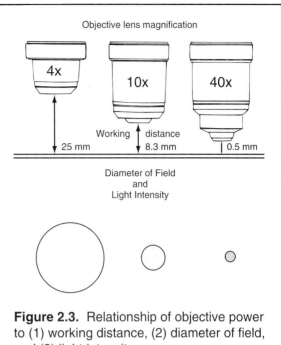

Figure 2.3. Relationship of objective power to (1) working distance, (2) diameter of field, and (3) light intensity.

slides that you will make. Wet-mount slides are prepared as shown in Figure 2.4.

Care of the Microscope

You should carry a microscope upright and in front of you, not at your side. Use one hand to support the base of the microscope and the other to grasp the arm. See Figure 2.5. Develop the habit of cleaning the lenses prior to using the microscope. Use only special lint-free lens paper. If the lens paper does not clean the lenses, inform you instructor. If any liquid gets on the lenses during use, wipe it off immediately and clean the lenses with lens paper.

When you are finished using the microscope, perform these steps:

1. Remove the slide. Clean and dry the stage.
2. Clean the lenses with lens paper.
3. Rotate the nosepiece so that none of the objective lenses projects beyond the front of the stage.
4. Raise the stage to its highest position *or* lower the body tube to its lowest position in accordance with the type of microscope you are using.

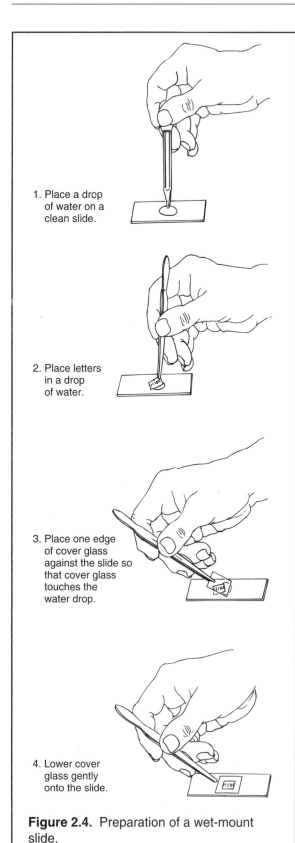

1. Place a drop of water on a clean slide.

2. Place letters in a drop of water.

3. Place one edge of cover glass against the slide so that cover glass touches the water drop.

4. Lower cover glass gently onto the slide.

Figure 2.4. Preparation of a wet-mount slide.

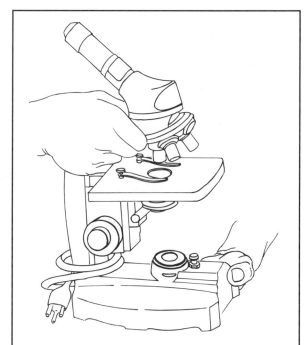

Figure 2.5. How to carry the microscope. Grasp the arm of the microscope with one hand and support the base with the other.

5. Unplug the light cord and loosely wrap it around the arm, below the stage. Add a dust-cover, if the microscope has one.
6. Return the microscope to the correct cabinet cubicle.

Assignment 1

1. Label Figure 2.1.
2. Obtain the microscope assigned to you. Carry it as described above and place it on the table in front of you. Locate the parts shown in Figure 2.1. Clean the lenses with lens paper. Try the knobs and levers to see how they work.
3. Raise the condenser to its highest position and keep it there. Plug in the light cord and turn on the light. If your microscope has a voltage-control knob, adjust it to an intermediate position to prolong the life of the bulb.
4. Rotate the 4× objective into viewing position and look through the ocular. The circle of light that you see is called the **field of view** or simply the **field**.
5. While looking through the ocular, open and

close the iris diaphragm and note the change in light intensity. Repeat for each objective and note that the light intensity decreases as the power of the objective increases. Therefore, you will need to adjust the light intensity when you switch objectives. *Remember to use reduced light intensity when you are viewing unstained and rather transparent specimens.*

6. If your microscope has a voltage-control knob, repeat item 5 while leaving the iris diaphragm open but changing the light intensity by altering the voltage.
7. **Complete item 1 on Laboratory Report 2 that begins on page 313.**

Developing Microscopy Skills

The following microscopic observations are designed to help you develop skill in using a compound microscope.

Materials

Per student
Compound microscope
Dissecting instruments
Medicine dropper
Metric ruler, clear plastic
Water in dropping bottle

Per lab
Kimwipes
Lens paper
Microscope slides and cover glasses
Newspaper
Pond water culture
Prepared slides of fly wing

Assignment 2

1. Obtain a microscope slide and cover glass. If they are not clean, wash them with soap and water, rinse, and dry them. Use a paper towel to dry the slide, but use Kimwipes to blot the water from the fragile cover glass.
2. Use scissors to cut three sequential letters from a newspaper, with the letter *i* as the middle letter.
3. Prepare a wet-mount slide of the letters as shown in Figure 2.4. Use a paper towel to soak up any excess water. If too little water is present, add a drop at the edge of the

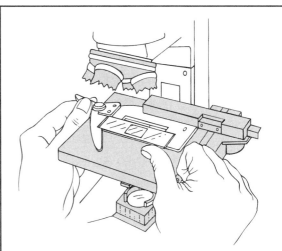

Figure 2.6. Placing the slide in the mechanical stage. The retainer lever is pulled back to place the slide against the stationary arm. Then, the retainer lever is released to secure the slide.

cover glass and it will flow under the cover glass.
4. Place the slide on the microscope stage, with the letters over the **stage aperture**, the circular opening in the stage. Secure the slide with either a mechanical stage or stage clamps. See Figure 2.6. The slide should be parallel to the edge of the stage nearest you with the letters oriented so that they may be read with the naked eye.
5. Rotate the 10× objective into viewing position and bring the letters into focus, using the focusing procedure described above that is appropriate for your microscope.
6. Rotate the 4× objective into the viewing position. Center the letter *i* and bring it into sharp focus. Can you see all of the *i*? Can you see the other letters? What is different about the orientation of the letters when viewed with the microscope instead of the naked eye?
7. Move the slide to the left while looking through the ocular. Which way does the image move? Practice moving the slide while viewing through the ocular until you can quickly place a given letter in the center of the field.
8. Center the *i* and bring it into sharp focus. Rotate the 10× objective into position. Is the

i centered and in focus? If not, center it and bring it into sharp focus. How much of the *i* can you see?

9. Rotate the 40× objective into position. Is the *i* centered and in focus? All that you can see at this magnification are "ink blotches" that compose the *i*. If you do not see this, center the *i* and bring it into focus with the *fine-focusing knob. Never use the coarse-focusing knob with the high-power objective*.

10. Practice steps 1 through 4 until you can quickly center the dot of the letter *i* and bring it into focus with each objective. **Remember**, you are *never* to start observations with the high-power objective. Instead, start at a lower power and work up to the 40× objective.

11. **Complete item 2 on the laboratory report**.

12. Remove the slide and set it aside for later use.

Depth of Field

When you view objects with a microscope, you obviously are viewing the objects from above. The vertical distance within which structures are in sharp focus is called the **depth of field**, and it decreases as magnification increases. You will learn more about depth of field by performing the observations that follow.

Assignment 3

1. Obtain a prepared slide of a fly wing. Observe the tiny spines on the wing membrane with each objective, starting with the 4× objective. Can you see all of a spine at each magnification?

2. Using the 4× objective, locate a large spine at the base of the wing where the veins converge, and center it in the field. Can you see all of it?

3. Rotate the 10× objective into position and observe the spine. Can you see all of it? Practice focusing up and down the length of the spine, and note that you can see only a portion of the spine at each focusing position.

4. Rotate the 40× objective into position and observe the spine. At each focusing position, you can see only a thin "slice" of the spine. To de-

termine the spine's shape, you have to focus up and down the spine using the *fine-focusing knob*.

The preceding observations demonstrate that when viewing objects that have a greater depth (thickness) than the depth of field, you see only a two-dimensional plane that is "optically cut" through the object in a horizontal manner. To discern an object's three-dimensional shape, a series of these two-dimensional images must be "stacked up" in your mind as you focus through the depth of the object.

5. **Complete item 3 on the laboratory report**.

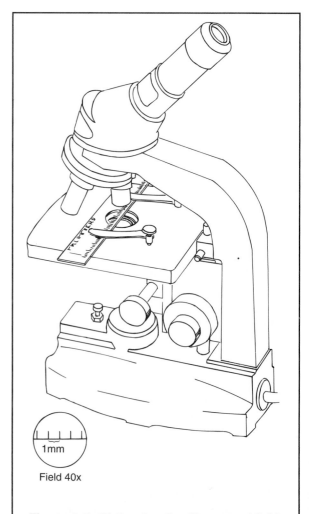

1mm

Field 40x

Figure 2.7. Estimating the diameter of field.

Diameter of Field

When using each objective, you must know the diameter of the field of view that it provides in order to estimate the size of observed objects. Estimate the diameter of this field for each magnification of your microscope as described in the section that follows.

Assignment 4

1. Place the clear, plastic ruler on the microscope stage as shown in Figure 2.7. The edge of the ruler should extend across the diameter of the field. Focus on the metric scale of the ruler with the 4× objective, and adjust the ruler so that one of the millimeter marks is at the left edge of the field. Estimate and record the diameter of the field at 40× by counting the spaces between the millimeter marks and adding any fraction of a millimeter that remains at the end.
2. Use these equations to calculate the diameter of field at 100× and 400×:

$$\frac{\text{Diam. (mm)}}{\text{at } 100\times} = \frac{40\times}{100\times} \times \frac{\text{diam. (mm) at } 40\times}{1}$$

$$\frac{\text{Diam. (mm)}}{\text{at } 400\times} = \frac{40\times}{400\times} \times \frac{\text{diam. (mm) at } 40\times}{1}$$

3. Return to your slide of newspaper letters, and estimate the diameter of the dot of the letter *i* and the length of the letter *i* including the dot.
4. **Complete item 4 on the laboratory report.**

Application of Microscopy Skills

In this section you will use the skills and knowledge gained in the preceding portions of the exercise.

Assignment 5

1. Prepare a wet-mount slide of two crossed hairs, one blond and the other brunette. Ob-

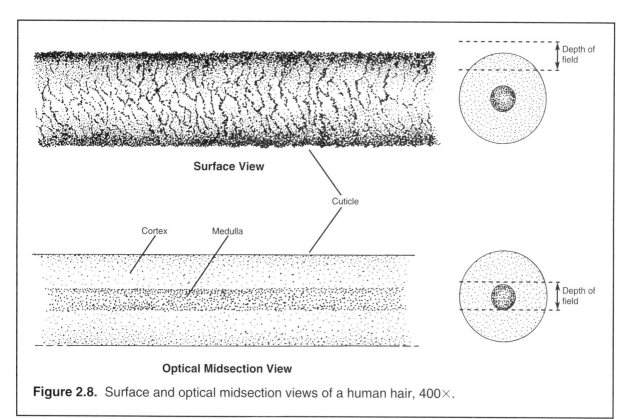

Surface View

Cuticle

Cortex   Medulla

Depth of field

Depth of field

Optical Midsection View

Figure 2.8. Surface and optical midsection views of a human hair, 400×.

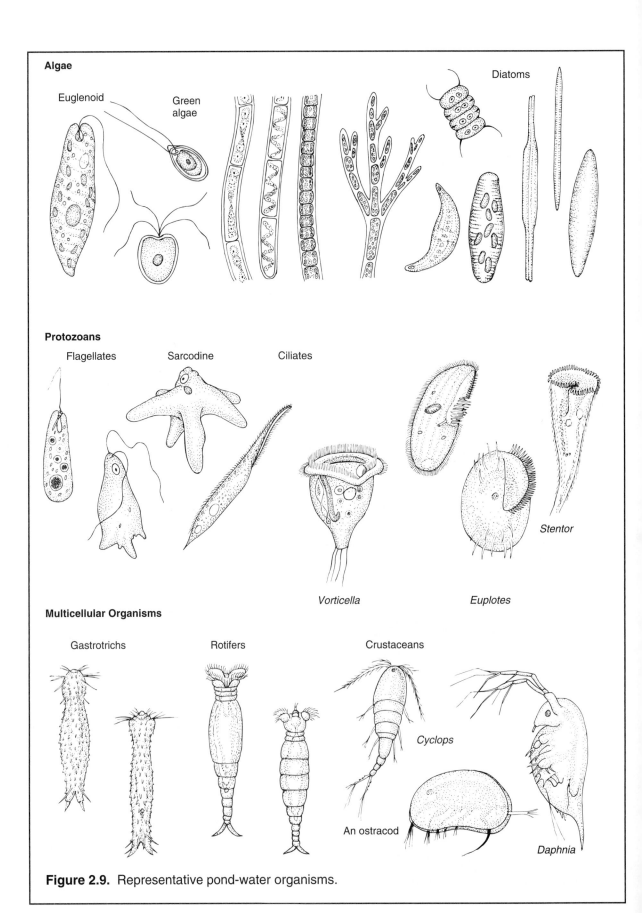

Algae

Euglenoid

Green algae

Diatoms

Protozoans

Flagellates

Sarcodine

Ciliates

Vorticella

Euplotes

Stentor

Multicellular Organisms

Gastrotrichs

Rotifers

Crustaceans

Cyclops

An ostracod

Daphnia

Figure 2.9. Representative pond-water organisms.

tain 1-cm lengths of hair from cooperative classmates.

2. Using the 4× objective, center the crossing point of the hairs in the field and observe. Are both hairs in sharp focus?

3. Examine the crossed hairs at 100× magnification. Are both hairs in sharp focus? Determine which hair is on top by using focusing technique. If you have trouble with this, see your instructor.

4. Examine the crossed hairs at 400× magnification. Are both hairs in focus? Move the crossing point to one side and focus on the blond hair. Using focusing technique, observe surface and optical midsection views of the hair as shown in Figure 2.8. At 400× magnification, the depth of field is less than half of the diameter of the hair.

5. **Complete items 5a to 5d on the laboratory report.**

6. Make wet-mount slides of pond-water samples and examine them microscopically. The object of this is to sharpen your microscopy skills rather than to identify the organisms in the water. However, you may see organisms like those shown in Figure 2.9. Note the size, color, shape, and motility of the organisms. **Draw a few of the organisms in the space for item 5e on the laboratory report**.

7. Prepare your microscope for return to the cabinet as described previously, and return it.

THE DISSECTING MICROSCOPE

A **dissecting microscope** is used to view objects that are too large or too opaque to observe with a compound microscope. The two oculars of the dissecting microscope enable stereoscopic observations, and are usually 10× in magnification. Most student models of dissecting microscopes have two objectives that provide 2× and 4× magnification, so that the total magnification is 20× and 40×. Some models have a zoom feature that enables observations to be made at intermediate magnifications. Objects are usually viewed with reflected light instead of transmitted light, although some dissecting microscopes provide both types of light sources.

The parts of a dissecting microscope are shown in Figure 2.10. Note the single focusing knob and the two oculars. The oculars may be moved inward or outward to adjust for the distance between the pupils of your eyes. One ocular has a focusing ring that may be adjusted to accommodate for differences in visual acuity in your eyes.

Materials

Per student
Coin
Desk lamp
Dissecting microscope
Metric ruler, clear plastic

Assignment 6

1. Obtain a dissecting microscope from the cabinet, and locate the parts shown in Figure 2.10.

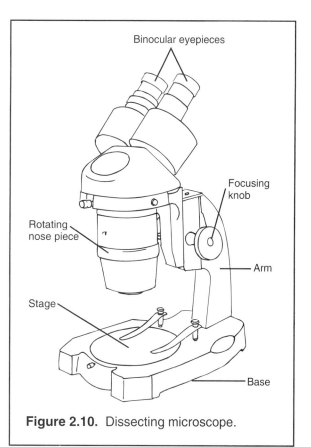

Figure 2.10. Dissecting microscope.

2. Place a coin or other object on the stage, illuminate it with a desk lamp, and examine it with both objectives. To accommodate for differences in acuity in your eyes, focus first with the focusing knob while viewing through the ocular that *cannot* be individually adjusted. Then use the **focusing ring** on the other ocular to bring the object into sharp focus for that eye.

3. Practice focusing until you are good at it. Move the coin on the stage, noting the direction in which the image moves. Remove the coin.

4. Place your finger tips, nails down, at the center of the stage. Focus on the ridges and grooves that produce your unique fingerprint.

5. Determine the diameter of field at each magnification.

6. **Complete items 6 and 7 on the laboratory report.**

3

The Cell

OBJECTIVES

After completion of the laboratory session, you should be able to:
1. Describe the basic characteristics of prokaryotic and eukaryotic cells.
2. Identify the basic parts of eukaryotic cells when viewed with a microscope.
3. Describe the major functions of the parts of a eukaryotic cell.
4. Distinguish animal and plant cells.
5. Define all terms in bold print.

Living organisms exhibit two fundamental characteristics that are absent in nonliving things: **self-maintenance** and **self-replication**. The smallest unit of life that exhibits these characteristics is a single living cell. Thus, a single cell is the *structural and functional unit of life*:

1. All organisms are composed of cells.
2. All cells arise from preexisting cells.
3. All hereditary components of organisms occur in cells.

Two different types of cells occur in the biotic world: prokaryotic cells and eukaryotic cells. Prokaryotic cells are rather primitive and occur only in bacteria and cyanobacteria. All other organisms consist of eukaryotic cells. Your study in this exercise will emphasize the structure of eukaryotic cells.

PROKARYOTIC CELLS

Prokaryotic cells lack a nucleus and membrane-bound organelles. An **organelle** is a specific cellular structure that performs a specific function. In a prokaryotic cell, a single circular chromosome is located in an irregular **DNA (deoxyribonucleic acid) region** in the interior of the cell. A small amount of **cytoplasm** containing **ribosomes** lies between the DNA region and the **cell membrane**. A supportive and protective **cell wall** is secreted just exterior to the cell membrane. See Figure 3.1.

Materials

Per student
Compound microscope

Per lab
Kimwipes
Lens paper
Prepared slides of bacteria and cyanobacteria

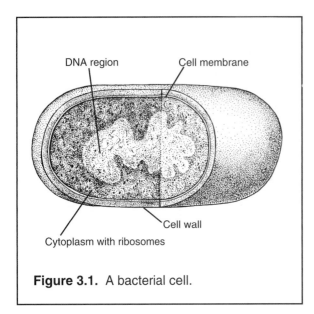

DNA region
Cell membrane
Cell wall
Cytoplasm with ribosomes

Figure 3.1. A bacterial cell.

Assignment 1

1. Examine the prepared slides of bacteria and cyanobacteria at $400\times$ total magnification. Can you detect the cellular components?
2. **Complete item 1 on Laboratory Report 3 that begins on page 317.**

EUKARYOTIC CELLS

Eukaryotic cells are considerably larger than prokaryotic cells, and they possess a nucleus and other membrane-bound organelles. As you read the following descriptions of cell anatomy, locate and label the organelles on Figures 3.2 and 3.3 and note their functions in Table 3.1. The structures of the cellular organelles are shown as viewed with an electron microscope; most organelles are too small to be seen with your microscope.

The Animal Cell

Animal cells are surrounded by a **plasma (cell) membrane** that is composed of two back-to-back phospholipid layers and associated proteins. Intracellular membranes have this same structure. The bulk of the cell consists of **cytoplasm**, the semifluid or gel-like substance in which the nucleus and other organelles are embedded. The cytoplasm is supported by a lattice formed of very fine **microfilaments** (label 13) and slightly larger **microtubules** (label 15) that form the cytoskeleton.

The **nucleus** is a large, spherical organelle that contains the **chromosomes**. In nondividing cells, the chromosomes are uncoiled and elongated so that only bits and pieces of them may be seen as **chromatin granules** (label 4) at each focal plane when viewing the nucleus with a microscope. In dividing cells, the chromosomes coil tightly and appear as dark-staining, rod-shaped structures. The spherical, dark-staining structure in the nucleus is the **nucleolus**, which is composed of ribonucleic acid (RNA) and protein. The nucleus is surrounded by a **nuclear envelope** that is composed of two adjacent membranes perforated by pores. The pores enable materials to move between the nucleus and cytoplasm.

The outer membrane of the nuclear envelope is continuous with the **endoplasmic reticulum (ER)**, a series of folded membranes that permeate the cytoplasm. **Smooth ER** (label 3) lacks ribosomes; **rough ER** is studded with ribosomes (shown as dots in the figures). **Ribosomes** (label 14) are tiny organelles consisting of RNA and protein that may occur singly, in clusters, or in chains. They are located either free in the cytoplasm or on the rough ER. The **Golgi complex** (label 2) is a stack of membranes associated with the ER, usually near the nucleus.

A pair of **centrioles**, short cylindrical bodies composed of microtubules, are oriented perpendicular to each other near the nucleus. **Mitochondria** (label 17) are elongated structures formed of a larger, folded, inner membrane surrounded by a smaller, nonfolded membrane. Mitochondria are more abundant in cells with a high metabolic rate. **Vacuoles** (label 9), small fluid-filled spaces enveloped by a membrane, may be scattered in the cytoplasm. **Lysosomes** (label 8) are smaller than vacuoles and contain powerful digestive enzymes. **Secretory vesicles** (label 1), formed by the Golgi complex are tiny sacs that carry materials to the cell membrane for export.

Some animal cells possess **cilia** (label 10), hair-like projections that move particles over the cell surfaces in multicellular forms. Sperm cells have a **flagellum**, a whip-like structure, that provides cell movement.

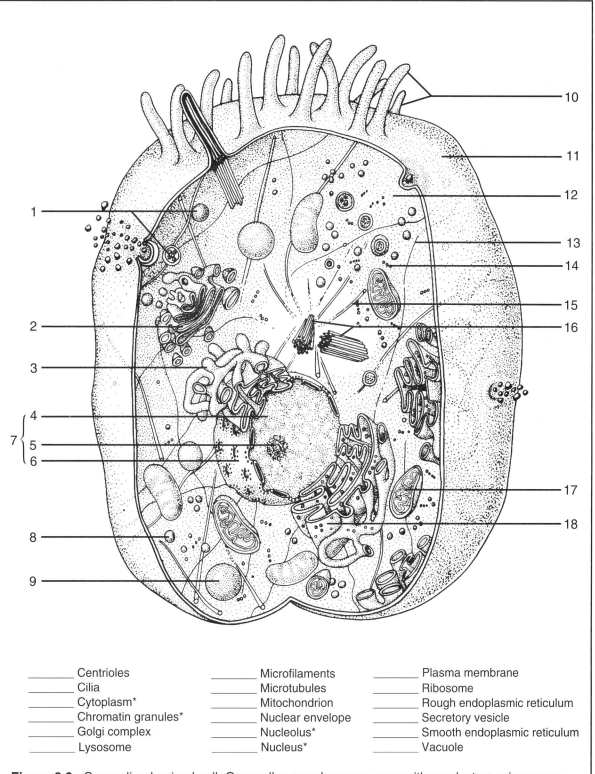

Figure 3.2. Generalized animal cell. Organelles are shown as seen with an electron microscope. Organelles visible with your microscope are noted with asterisks.

_____ Centrioles
_____ Cilia
_____ Cytoplasm*
_____ Chromatin granules*
_____ Golgi complex
_____ Lysosome

_____ Microfilaments
_____ Microtubules
_____ Mitochondrion
_____ Nuclear envelope
_____ Nucleolus*
_____ Nucleus*

_____ Plasma membrane
_____ Ribosome
_____ Rough endoplasmic reticulum
_____ Secretory vesicle
_____ Smooth endoplasmic reticulum
_____ Vacuole

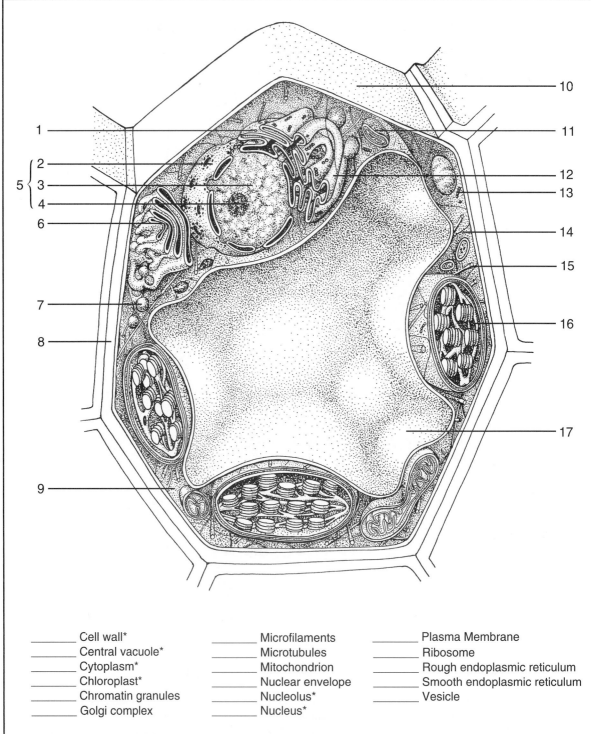

_____ Cell wall*	_____ Microfilaments	_____ Plasma Membrane
_____ Central vacuole*	_____ Microtubules	_____ Ribosome
_____ Cytoplasm*	_____ Mitochondrion	_____ Rough endoplasmic reticulum
_____ Chloroplast*	_____ Nuclear envelope	_____ Smooth endoplasmic reticulum
_____ Chromatin granules	_____ Nucleolus*	_____ Vesicle
_____ Golgi complex	_____ Nucleus*	

Figure 3.3. Generalized plant cell. Organelles are shown as seen with an electron microscope. Organelles visible with your microscope are noted with asterisks.

TABLE 3.1
Function of Cell Structures

Cell Structure	*Function*
Cell surface	
Cell (plasma) membrane	Controls the passage of materials into and out of the cell; maintains the integrity of the cell.
Cell wall	Protects and supports the cell. Absent in animal cells.
Cilia and flagella	Move fluid over cell surface, or move the cell through fluid. Cilia absent in plant cells. Flagella present on reproductive cells of some plants.
Hereditary material organization	
Nuclear envelope	Controls the passage of materials into and out of the nucleus.
Nucleus	Acts as control center of the cell; contains the hereditary material.
Chromosomes	Contain DNA, the determiner of inheritance, which controls cellular processes.
Nucleolus	Assembles proteins and RNA that will form ribosomes.
Cytoplasmic structures	
Centrioles	Form microtubule systems, including spindle fibers in dividing animal cells. Absent in plant cells.
Endoplasmic reticulum	Membranous channels for the movement of materials from place to place within a cell; provide membranous surfaces for chemical reactions.
Golgi apparatus	Stores, modifies, and packages materials for export from the cell.
Lysosomes	Contain enzymes to digest worn out or damaged cells or cell parts. Absent in plant cells.
Microfilaments	Provide support for the cell; involved in movement of organelles and entire cells.
Microtubules	Provide support for the cell; involved in movement of organelles and entire cells.
Mitochondria	Sites of aerobic cellular respiration, which releases energy from nutrients and forms ATP.
Plastids	Chloroplasts are sites of photosynthesis. Chromoplasts contain pigments giving color to flowers and fruits. Leukoplasts are often sites of starch storage. Absent in animal cells.
Ribosomes	Sites of protein synthesis.
Vacuole	Contains water and solutes, wastes, or nutrients.
Vacuole, central	Contains water and solutes; provides hydrostatic pressure, which helps support cell. Provides lysosome function in plant cells. Absent in animal cells.
Vesicles	Carry substances from place to place within the cell. Secretory vesicles carry substances to the plasma membrane for export from the cell.

The Plant Cell

Plant cells contain all of the organelles found in animal cells, except centrioles and lysosomes. Mature plant cells are characterized by the presence of cell walls, plastids, and a central vacuole.

A rigid **cell wall**, formed of cellulose, is located just exterior to the cell membrane.

Plastids are enveloped by a double membrane and are classified according to the pigments that they contain. **Chloroplasts** contain chlorophyll and carotenes and are the only plastids shown in Figure 3.3. **Chromoplasts** (not shown) contain various red, orange, or yellow pigments. **Leukoplasts** (not shown) contain no pigments and are colorless.

Immature plant cells have numerous small vacuoles, but in mature cells these vacuoles combine to form a large **central vacuole** that constitutes much of the cell volume.

Assignment 2

1. Label Figures 3.2 and 3.3 and color-code the organelles.
2. **Complete item 2 on the laboratory report.**

MICROSCOPIC STUDY

In this section, you will prepare wet-mount slides for the study of cell structure. You will start with plant cells, because they do not require special preparation and their cellular components are easier to observe.

Materials

Per student
Compound microscope
Dissecting instruments

Per lab
Dropping bottles of:
    iodine solution (IKI)
    sodium chloride, 0.9%
Kimwipes
Lens paper
Microscope slides and cover glasses
Toothpicks, flat
Medicine droppers
Amoeba proteus culture
Elodea shoots
Onion bulb scales, red

Onion Epidermal Cells

Epidermal cells of an onion scale show many of the features found in nongreen plant cells and are excellent subjects to use in beginning your study of cells.

Assignment 3

1. Label Figure 3.4 using information from the previous section.
2. Prepare a wet-mount slide of the inner epidermis of an onion scale as shown in Figure 3.5. Use a drop of iodine solution as the mounting fluid and add a cover glass.
3. Examine the cells at total magnifications of 100× and 400×. Note the arrangement of the cells. Locate the parts shown in Figure 3.4.
4. Prepare a wet-mount slide of the red epidermis from the outer surface of the onion scale. The color is due to **anthocyanin**, a water-soluble pigment in the central vacuole. Note the size of the vacuole and the location of the cytoplasm and nucleus.
5. **Complete item 3 on the laboratory report.**

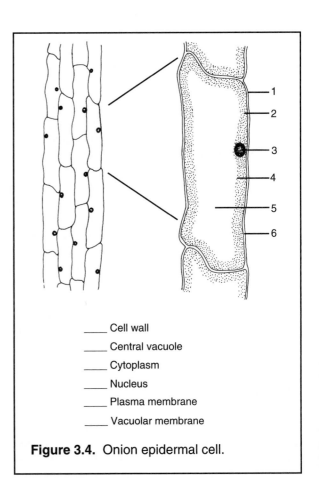

_____ Cell wall
_____ Central vacuole
_____ Cytoplasm
_____ Nucleus
_____ Plasma membrane
_____ Vacuolar membrane

Figure 3.4. Onion epidermal cell.

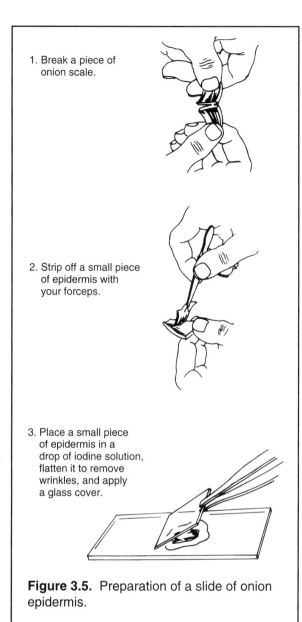

1. Break a piece of onion scale.

2. Strip off a small piece of epidermis with your forceps.

3. Place a small piece of epidermis in a drop of iodine solution, flatten it to remove wrinkles, and apply a glass cover.

Figure 3.5. Preparation of a slide of onion epidermis.

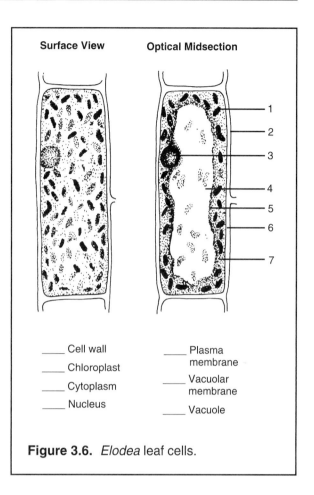

Surface View   **Optical Midsection**

1
2
3
4
5
6
7

_____ Cell wall

_____ Chloroplast

_____ Cytoplasm

_____ Nucleus

_____ Plasma membrane

_____ Vacuolar membrane

_____ Vacuole

Figure 3.6. *Elodea* leaf cells.

Elodea Leaf Cells

Elodea is a water plant with simple leaves composed of cells that are easy to observe and that exhibit the characteristics of green plant cells.

Assignment 4

1. Label Figure 3.6.
2. Prepare a wet-mount slide of an *Elodea* leaf in this manner:

   a. Use forceps to remove a young leaf from near the tip of the shoot and mount it in a drop of water.
   b. Add a cover glass and observe at total magnifications of 40× and 100×.
3. Note the arrangement of the cells and the "spine" cells along the edge of the leaf. Locate a light green area for study. The thickness of the leaf is composed of more than one layer of cells. Switch to the 40× objective and focus through the thickness of the leaf. Determine the number of cell layers present.
4. Using the 40× objective, focus through the depth of a cell and locate as many parts shown in Figure 3.6 as possible. The nucleus is spherical and slightly darker than the cytoplasm.
5. Focus carefully to observe surface and optical midsection views. See Figure 3.6. Note the basic shape of a cell.
6. Examine a spine cell at the edge of the leaf,

using reduced illumination to locate the nucleus, vacuole, and cytoplasm uncluttered with chloroplasts.

7. **Complete item 4 on the laboratory report.**

Human Epithelial Cells

The epithelial cells lining the inside of your mouth are easily obtained for study, and they exhibit some of the characteristics of animal cells.

Assignment 5

1. Label Figure 3.7.
2. Prepare a wet-mount slide of human epithelial cells as shown in Figure 3.8, using 0.9% sodium chloride (NaCl) as the mounting fluid.
3. Add a cover glass and observe at total magnifications of 100× and 400×. What cell structures can you see? Note the differences between these cells and the plant cells.
4. **Complete item 5 on the laboratory report.**

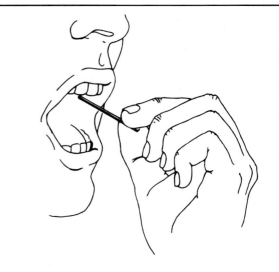

1. Gently scrape inside of mouth with a toothpick to obtain epithelial cells.

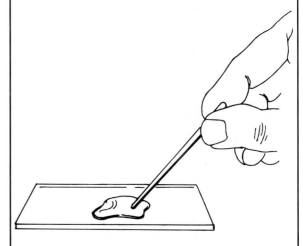

2. Swirl the toothpick in a drop of 0.9% NaCl on a clean slide.

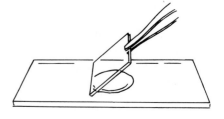

3. Gently apply a cover glass.

Figure 3.8. Preparation of a slide of human epithelial cells.

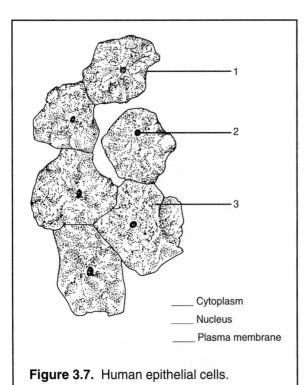

_____ Cytoplasm

_____ Nucleus

_____ Plasma membrane

Figure 3.7. Human epithelial cells.

The Amoeba

The common freshwater protozoan *Amoeba* exhibits many characteristics of animal cells, and it is large enough for you to see the cellular structure rather well. The nucleus, cytoplasm, vacuoles, and cytoplasmic granules are readily visible. The flowing movement allows an amoeba to capture minute organisms that are digested in **food vacuoles**. An amoeba also has **contractile vacuoles** that maintain its water balance by collecting and pumping out excess water.

Ectoplasm, the clear outer portion of the cytoplasm, is located just interior to the plasma membrane. Most of the cytoplasm consists of the granular **endoplasm**. This portion of the cytoplasm contains the organelles, and may be either gel-like, the **plasmagel**, or fluid, the **plasmasol**. Reversible changes between plasmagel and plasmasol result in the flowing **amoeboid movement** of an amoeba. Your white blood cells also exhibit amoeboid movement as they slip through capillary walls and wander among the body's tissues, engulfing disease-causing organisms and cellular debris.

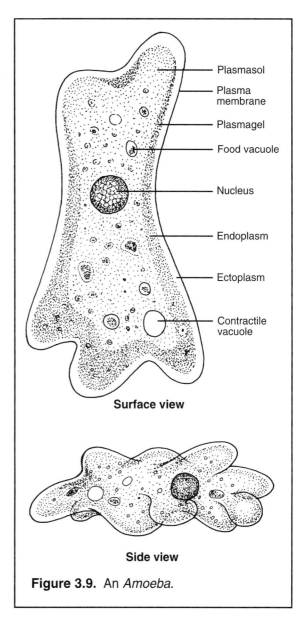

Surface view

Side view

Figure 3.9. An *Amoeba*.

Assignment 6

1. Use a medicine dropper to obtain fluid *from the bottom* of the culture jar and place a drop on a clean slide. Do *not* add a cover glass.
2. Use the 10× objective to locate and observe an *Amoeba*. Note the manner of movement and locate the nucleus, cytoplasm, and vacuoles. The nucleus is spherical and a bit darker than the granular cytoplasm. Contractile vacuoles appear as spherical bubbles in the cytoplasm, while food vacuoles contain darker food particles within the vacuolar fluid. Compare your specimen with that shown in Figure 3.9 and Plate 2.5. Observe the way an amoeba moves, while remembering that some of your white blood cells move about in a similar way.
3. Note the appearance of the cytoplasm. Does the appearance of the inner and outer portions differ? Observe the nucleus and vacuoles. Gently add a cover glass to your slide and observe the amoeba at 400×. When the pressure of the cover glass becomes too great, the cell membrane will rupture, and the contents of the cell will spill out, killing the amoeba.
4. **Complete item 6 on the laboratory report.**

Assignment 7

Review the differences between animal and plant cells by completing item 7 on the laboratory report.

Your instructor has set up several microscopes as a mini-practicum so that you can check your understanding of cell types and cell structure.

Your task is to identify the type of cells observed and the cell structures identified by the microscope pointers. **Complete item 8 on the laboratory report.**

4

Chemistry of Cells

After completion of the laboratory session, you should be able to:
1. Determine the number of protons, neutrons, and electrons in an atom of an element from data shown in a periodic table of the elements.
2. Diagram a shell model of a simple atom.
3. Describe the formulation of ionic and covalent bonds in accordance with the octet rule.
4. Describe and perform simple tests for carbohydrates, fats, and proteins.
5. Define all terms in bold print.

Life at its most fundamental level consists of complex chemical reactions. Therefore, having some understanding of the chemicals in living organisms and the nature of chemical reactions is important in your study of biology.

Chemical substances are classified into two groups: elements and compounds. An **element** is a substance that cannot be broken down by chemical means into any simpler substance. Table 4.1 lists the most common elements found in living organisms. An **atom** is the smallest unit of an element that retains the properties (characteristics) of the element.

Two or more elements may combine to form a **compound**. Water (H_2O) and table salt (NaCl) are simple compounds; carbohydrates, fats, and proteins are complex compounds. The smallest unit of a compound that retains the properties of a compound is a **molecule**. A molecule is formed of two or more atoms joined by **chemical bonds**.

TABLE 4.1
Common Elements in the Human Body

Element	Symbol	Percentage*
Oxygen	O	65
Carbon	C	18
Hydrogen	H	10
Nitrogen	N	3
Calcium	Ca	2
Phosphorus	P	1
Potassium	K	0.35
Sulfur	S	0.25
Sodium	Na	0.15
Chlorine	Cl	0.15
Magnesium	Mg	0.05
Iron	Fe	0.004
Iodine	I	0.0004

*By weight.

ATOMIC STRUCTURE

Atoms are composed of three basic components: **protons**, **neutrons**, and **electrons**. Protons and neutrons have a mass of one atomic unit and are located in the central region of an atom, the nucleus. Protons possess a positive electrical charge of one (+1), but neutrons have no charge. Electrons have a negative electrical charge of one (−1), almost no mass, and orbit at near the speed of light around the atomic nucleus. The attraction between positive and negative charges is what keeps the electrons spinning about the nucleus.

Atoms of a given element differ from those of all other elements in the number and arrangement of their protons, neutrons, and electrons, and these differences determine the properties of each element. Chemists have used these differences to construct a **periodic table of the elements**.

Table 4.2 is a simplified periodic table of the first 20 elements. Each element is identified by its **symbol**. Note the location of the **atomic number**, which indicates the number of protons in each atom. Since a neutral atom possesses the same number of protons as electrons, the atomic number of an element also indicates the number of electrons in each atom of the element. Note that the elements are sequentially arranged in the table according to their atomic numbers.

The atom's **mass number** (sometimes called atomic weight) equals the sum of protons plus neutrons in each atom. Because the number of neutrons may vary slightly in atoms of the same element, the atomic mass reflects the *average*

TABLE 4.2
Simplified Periodic Table of the Elements—1 Through 20[*]

I	II	III	IV	V	VI	VII	VIII
1 H hydrogen 1.0							2 He helium 4.0
3 Li lithium 7.0	4 Be beryllium 9.0	5 B boron 11.0	6 C carbon 12.0	7 N nitrogen 14.0	8 O oxygen 16.0	9 F fluorine 19.0	10 Ne neon 20.0
11 Na sodium 23.0	12 Mg magnesium 24.3	13 Al aluminum 27.0	14 Si silicon 28.1	15 P phosphorus 31.0	16 S sulfur 32.1	17 Cl chlorine 35.5	18 Ar argon 40.0
19 K potassium 39.1	20 Ca calcium 40.1						

Atomic Number — Atomic Symbol — Atomic Mass

[*] The dark line separates metals, on the left, from nonmetals.

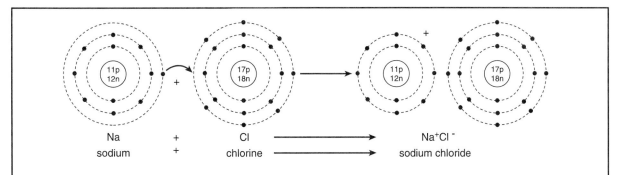

Figure 4.1. The formation of sodium chloride, an ionic reaction. Chlorine receives an electron from sodium, with the result that each atom has eight electrons in the outer shell.

mass of the atoms. Atoms of an element that vary slightly in mass due to a difference in the number of neutrons are called **isotopes**.

The **shell model** is a useful, although not quite accurate, representation of atomic structure. The energy levels occupied by electrons are represented by circles (shells) around the nucleus. Electrons are shown as dots on the circles.

Examine Figure 4.1. Note the arrangement of the electrons in sodium (Na) and chlorine (Cl). The innermost shell can contain a maximum of only two electrons. Atoms with two or more shells can contain a maximum of eight electrons in the outermost shell.

Assignment 1

1. Use the simplified periodic table of the elements (Table 4.2) to determine the number of protons, electrons, and neutrons in the first 20 elements.
2. Compare the number of electrons in the outer shell of the elements shown in Table 4.2 with their position in the table.
3. **Complete item 1 on Laboratory Report 4 that begins on page 321.**

REACTIONS BETWEEN ATOMS

The **octet rule** states that atoms react with each other to achieve eight electrons in their outermost shells. An exception to this is that the first shell is filled by only two electrons.

The octet rule is achieved by (1) an atom's losing or gaining electrons or (2) an atom's sharing electrons, with another atom. Atoms with one, two, or three electrons in their outer shell donate (lose) these electrons to other atoms so that the adjacent shell, filled with electrons, becomes the outer shell. Atoms with six or seven electrons in the outer shell receive (gain) electrons from adjacent atoms to fill their outer shells. Other atoms usually attain an octet by sharing electrons.

Ionic Bonds

Typically, an atom has a net charge of zero, because it has the same number of protons (positive charges) and electrons (negative charges). If an atom gains electrons, it becomes negatively charged, since the number of protons in the atom is constant. Similarly, the loss of electrons causes an atom to become positively charged. An atom with an electrical charge is called an **ion**.

Ionic bonds are formed between two ions by the attraction between opposite charges that have resulted from gaining or losing electrons. Examine Figure 4.1. Sodium and chlorine unite by an ionic bond to form sodium chloride (NaCl). Note that when the single electron in the outer shell of sodium is transferred to chlorine, both the sodium and the chlorine ions have eight electrons in their outer shell. This transfer causes sodium to have a positive charge ($+1$) and chlorine to have a negative charge (-1). The opposite charges hold the two ions together to form the NaCl molecule. This type of bonding occurs between **metals** (electron donors) and **nonmetals** (electron recipients).

Covalent Bonds

When nonmetals react with nonmetals, electrons are shared instead of being passed from one atom to another. This sharing of electrons usually forms a **covalent bond**. The electrons are shared in pairs, with one member of each pair coming from each atom. The electrons spend some time in the outer shell of each atom, so that the outer shell of each atom meets the octet rule.

Examine Figure 4.2. Note how one atom of oxygen (O) and two atoms of hydrogen (H) share their electrons to form one molecule of water (H_2O). Since the atom of oxygen is sharing a single pair of electrons with each atom of hydrogen, each atom of hydrogen is joined with the atom of oxygen by a single covalent bond.

Table 4.3 shows the covalent bonding capacity of the four elements that form about 96% of the human body. The bonding pattern shows each potential covalent bond, which is represented by a single line. A common practice is to draw structural formulas of molecules that have covalent bonds by using lines to represent the bonds. One line indicates a single bond (one pair of

TABLE 4.3
Bonding Capacities

Element	Number of Bonds	Potential Bonds
Carbon	4	—C—
Nitrogen	3	N
Oxygen	2	—O—
Hydrogen	1	H—

shared electrons), two lines indicate a double bond (two pairs of shared electrons), and so forth.

Assignment 2

1. Study Figures 4.1 and 4.2 to understand the characteristics of ionic and covalent bonds. Compare the shell model and structural formula of a water molecule in Figure 4.2.
2. **Complete item 2 on the laboratory report.**

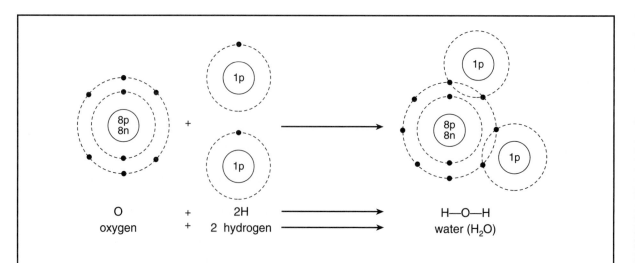

Figure 4.2. The formation of water, a covalent reaction. A pair of electrons is shared between each hydrogen atom and the oxygen atom. Shared electrons are counted as belonging to each atom.

ACIDS, BASES, AND PH

An **acid** is a substance that releases **hydrogen ions** (H^+) when it is dissolved in water. The greater the number of (H^+) released, the greater is the strength of the acid. Hydrogen chloride is a strong acid.

$$HCl \longrightarrow H^+ + Cl^-$$

A **base** is a substance that releases **hydroxide ions** (OH^-) when dissolved in water. The greater the number of (OH^-) released, the stronger is the base. Sodium hydroxide is a strong base.

$$NaOH \longrightarrow Na^+ + OH^-$$

Chemists use a **pH scale** to indicate the strength of acids and bases. Figure 4.3 shows the pH scale, which ranges from 0 to 14 and indicates the proportionate concentration of hydrogen and hydroxide ions at the various pH values. When the concentration of hydrogen ions increases, the concentration of hydroxide ions decreases, and vice versa. A change of one (1.0) in the pH number indicates a 10-fold change in the concentration of hydrogen ion's because the pH scale is a logarithmic scale.

Pure water has a pH of 7, the point at which the concentration's of hydrogen and hydroxide ions are equal. Observe this in Figure 4.3. Acids have a pH less than 7; bases have a pH greater than 7. The greater the concentration of hydro-gen ions, the stronger the acid and the *lower* the pH number. The greater the concentration of hydroxide ions, the stronger the base and the *higher* the pH number.

Living organisms are sensitive to the concentrations of hydrogen and hydroxide ions, and must maintain the pH of their cells within narrow limits. For example, your blood is kept very close to a pH of 7.4. Organisms control the pH of cellular and body fluids by using buffers. A **buffer** is a compound or a combination of compounds that can combine with or release hydrogen ions to keep the pH of a solution relatively constant.

Materials

Per student group
Beakers, 100 ml, 2
Dropping bottles of:
  bromthymol blue
  hydrochloric acid (HCl), 1.0%
Graduated cylinder, 50 ml
pH test papers, wide an narrow ranges

Per lab
Buffer solution, pH 7, 1,000 ml
Distilled water, pH 7, 1,000 ml
Dropping bottles of:
  Alka-Seltzer® solution
  detergent solution
  household ammonia
  lemon juice
  mouthwash
  white vinegar
  unknowns

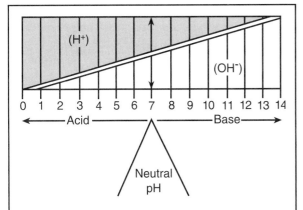

Figure 4.3. The pH scale. The diagonal line indicates the proportionate amount of hydrogen (H^+) and hydroxide (OH^-) ions at each pH value.

Assignment 3

1. **Complete items 3a and 3b on the laboratory report.**
2. Use pH test papers to determine the pH of the solutions provided by your instructor. Place 1 drop of the solution to be tested on a small strip of pH test paper while holding it over a paper towel. Compare the color of the pH paper with the color code on the dispenser to determine the pH. Use the wide-range test paper first to determine the approximate pH. Then, use a narrow-range test paper for the final

determination. **Record your pH determinations in item 3c on the laboratory report.**

3. Determine the effect of a buffer on pH as follows.

   a. Place 25 ml of (*i*) distilled water at pH 7, and (*ii*) buffer solution at pH 7, in separate, labeled beakers. Determine and record the pH of each.

   b. Add 6 drops of bromthymol blue to each beaker. Mix by swirling the liquid. The fluid in each beaker should be a pale blue. Bromthymol blue is a pH indicator that is blue at pH 7.6 and yellow at pH 6.0.

   c. Place the beaker of distilled water and bromthymol blue solution on a piece of white paper. Add 1.0% HCl drop by drop to the solution and mix thoroughly by swirling the liquid between the addition of each drop. Count and record the number of drops required to turn the solution yellow, which indicates a pH of 6.

   d. Use the preceding method to determine the number of drops of 1.0% HCl required to lower the pH of the buffer solution from 7 to 6.

3. **Complete item 3d on the laboratory report.**

IDENTIFYING BIOLOGICAL MOLECULES

Carbohydrates, lipids, and proteins are complex organic compounds found in living organisms. Carbon atoms form the basic framework of the molecules of these types of substances. In this section, you will analyze plant and animal materials for the presence of carbohydrate, lipid, and protein compounds. Work in groups of two to four students. If time is limited, your instructor will assign certain portions to separate groups, with all groups sharing the data.

Carbohydrates

Carbohydrates are formed of carbon, hydrogen, and oxygen. Their molecules are characterized by H—C—OH grouping, in which the ratio of hydrogen to oxygen is 2:1. **Monosaccharides** are simple sugars containing three to seven carbon atoms, and they serve as building blocks for more complex carbohydrates. The combination of two monosaccharides forms a **disaccharide** sugar, and the union of many monosaccharides

forms a **polysaccharide**. Six-carbon sugars, such as glucose, are primary energy sources for living organisms. Starch in plants and glycogen in animals are polysaccharides used for nutrient storage. See Figure 4.4.

Use the iodine test for starch and Benedict's test for reducing sugar (most 6-carbon and some 12-carbon sugars) to identify the presence of these carbohydrates. See Figure 4.5.

Iodine Test for Starch

1. Place 1 dropper (1 ml) of the liquid to be tested in a clean test tube. (A "dropper" means *one dropper full* of liquid.)
2. Add 3 drops of iodine solution to the liquid in the test tube and shake gently to mix.
3. A gray to blue-black coloration of the liquid indicates the presence of starch, with the order from gray to blue indicating an increasing starch concentration.

Benedict's Test for Reducing Sugars

1. Place 1 dropper (1 ml) of the liquid to be tested in a clean test tube.
2. Add 3 drops of Benedict's solution to the liquid in the test tube and shake gently to mix.
3. Heat the tube to near boiling in a water bath for 2-3 min. (A water bath consists of a beaker that is half-full of water, heated on a hot plate or over a burner. The test tube is placed in the beaker so that it is heated by the water in the beaker rather than directly by the heat source.)
4. A light green, yellow, orange, or brick-red coloration of the liquid indicates the presence of reducing sugars, with this progression of colors indicating a progressively increasing sugar concentration.

Lipids

Lipids include an array of oily and waxy substances, but the most familiar are the neutral fats (triglycerides). Lipids are composed of carbon, oxygen, and hydrogen, and have fewer oxygen atoms than carbohydrates. A fat molecule consists of three long fatty acid chains joined to a single glycerol molecule. See Figure 4.6. Fats serve as an important means of nutrient storage in organisms and are not soluble in water.

The presence of lipids may be determined by a

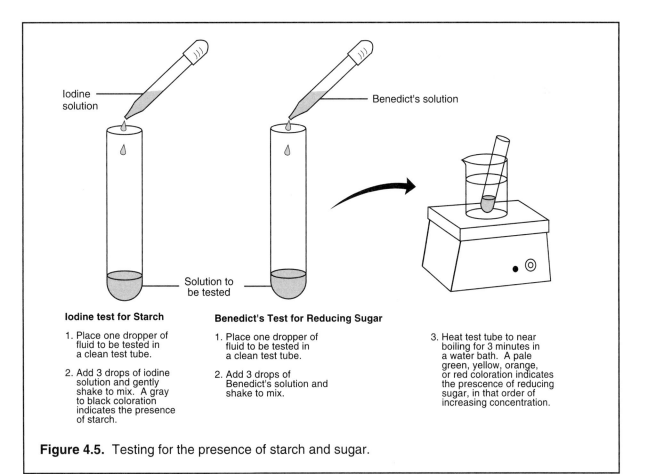

$C_6H_{12}O_6$
a. Glucose

$C_{12}H_{22}O_{11}$
b. Maltose

c. Starch

Figure 4.4. Carbohydrates. (a) Glucose is a monosaccharide. (b) Maltose is a disaccharide composed of two glucose units. (c) Starch is a polysaccharide composed of many glucose units.

Iodine solution

Benedict's solution

Solution to be tested

Iodine test for Starch

1. Place one dropper of fluid to be tested in a clean test tube.

2. Add 3 drops of iodine solution and gently shake to mix. A gray to black coloration indicates the presence of starch.

Benedict's Test for Reducing Sugar

1. Place one dropper of fluid to be tested in a clean test tube.

2. Add 3 drops of Benedict's solution and shake to mix.

3. Heat test tube to near boiling for 3 minutes in a water bath. A pale green, yellow, orange, or red coloration indicates the prescence of reducing sugar, in that order of increasing concentration.

Figure 4.5. Testing for the presence of starch and sugar.

Figure 4.6. A triglyceride is composed of one glycerol molecule combined with three fatty acids. Note that a fatty acid consists mostly of a string of carbon atoms with attached hydrogen atoms.

paper spot test and Sudan IV, a dye that selectively stains lipids.

Paper Spot Test

1. Place a drop of the liquid to be tested on a piece of paper. Blot it with a paper towel and let it dry. If the substance to be tested is a solid, it can be rubbed on the paper. Remove the excess with a paper towel. Animal samples may need to be gently heated.
2. After the spot has dried, hold the paper up to the light and look for a permanent translucent spot on the paper, indicating the presence of lipids.

Sudan IV Test

1. Place a dropper of distilled water in a clean test tube and add 3 drops of Sudan IV. Shake the tube from side to side to mix well.
2. Add 10 drops of the substance to be tested and mix again. Let the mixture stand for 5 min.
3. Lipids will be stained red and will rise to the top of the water, or will be suspended in the water in the form of tiny globules.

Proteins

Proteins are usually large, complex molecules composed of many **amino acids** joined together by **peptide bonds**. The molecules of proteins

contain nitrogen in addition to carbon, oxygen, and hydrogen. Proteins play important roles by forming structural components of cells and tissues and by serving as enzymes. See Figure 4.7.

Proteins may be identified by a biuret test that is specific for peptide bonds.

Biuret Test

1. Place a dropper of the solution to be tested in a clean test tube.
2. Add a dropper of 10% sodium hydroxide (NaOH) and mix by shaking the tube from side to side. *Caution:* NaOH is a caustic substance that can cause burns. If you get it on your skin or clothing, wash it off *immediately* with lots of water.
3. Add 5 drops of 0.5% copper sulfate ($CuSO_4$) to the tube and shake from side to side to mix. Observe after 3 min.
4. A violet color indicates the presence of proteins.

Materials

Per student group
Beaker, 250 ml
Boiling chips
Brown paper, 1 sheet
Dissecting instruments
Hot plate

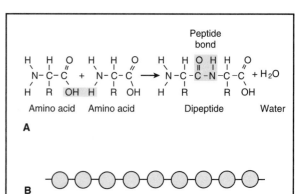

Figure 4.7. Proteins are formed of amino acids joined together by peptide bonds. **A.** Formation of a peptide bond. **B.** Amino acids composing a tiny part of a protein are arranged somewhat like beads on a string.

Medicine droppers, 3
Mortar and pestle
Petri dish, plastic
Test tubes, 12
Test-tube holder
Test-tube rack
Dropping bottles of:
    Copper sulfate, 0.5%
    Benedict's solution
    Iodine solution (IKI)
    Sodium hydroxide, 15%
    Sudan IV dye

Per lab
Apple
Onion
Potato
Dropping bottles of:
    albumin, 0.5%
    corn oil
    egg white, 0.5%
    glucose, 0.5%
    soluble starch, 0.5%

Assignment 4

1. Be sure that you understand the tests to be performed.
2. Prepare a set of standards that yield positive and negative results for each test that you will use. The standards will help you interpret your results when testing substances with unknown organic compounds. Prepare the standards as follows.
   a. Obtain 8 test tubes and number them 1A through 4A and 1B through 4B and place them in a test tube rack. Add the following substances to the test tubes. You will use *droppers* (about 1 ml) of the test materials.
      1A: 1 dropper of 0.5% starch solution
      1B: 1 dropper of water

2A: 1 dropper of 0.5% glucose solution
2B: 1 dropper of water
3A: 1 dropper of water plus 5 drops corn oil
3B: 1 dropper of water
4A: 1 dropper of 0.5% albumin solution
4B: 1 dropper of water
   b. Add the reagents to the test tubes and mix the contents by shaking the tubes from side to side.
      1A and 1B: Add 3 drops iodine solution to each tube.
      2A and 2B: Add 3 drops Benedict's solution to each tube and heat both tubes in a boiling water bath for 3 min.
      3A and 3B: Add 3 drops of Sudan IV to each tube. Read your results after 5 min.
      4A and 4B: Add 1 dropper of 15% sodium hydroxide and drops of 0.5% copper sulfate to each tube. (**Caution:** NaOH can cause burns. If spilled on skin or clothing, wash it off *immediately* with lots of water.)
   c. For the paper spot test of lipids, apply a drop of corn oil and a drop of water to different parts of a brown paper. Blot with a paper towel and write the name of each substance near its spot. Read your results after the spots have dried.
   d. **Record the results of your set of standards in item 4a on the laboratory report.**
3. Perform each of the tests on the "unknown" substances provided by your instructor.
   a. It will be necessary to macerate (grind to a pulp) solid materials to obtain a liquid for testing. Place a 1-cm cube of the substance to be tested in a mortar. Add 1 dropper of water. Use a pestle to crush and grind up the cube.
   b. Remove the liquid with a dropper for testing.
   c. **Record your results in item 4c on the laboratory report.**

5   Enzymes

OBJECTIVES

After completion of the laboratory session, you should be able to:
1. Explain the role of enzymes in living cells.
2. Describe the action of catalase.
3. State conclusions based on the experiments.
4. Define all terms in bold print.

The chemical reactions in cells would not occur fast enough to support life without the action of enzymes. **Enzymes** are organic catalysts that greatly accelerate the rate of chemical reactions in cells by reducing the required activation energy. All chemical reactions require a certain amount of activation energy to start. For example, energy (heat) is added by a lighted match to start a fire.

Figure 5.1 contrasts the required activation energies needed to start a chemical reaction with and without an enzyme. By lowering the required activation energy, an enzyme greatly increases the rate of a chemical reaction—by up to a million times a second in some cases.

Enzymes are proteins; therefore, each enzyme consists of a specific sequence of amino acids. Weak hydrogen bonds that form between some of the amino acids help to determine the three-dimensional shape of the enzyme, and it is this shape that allows the enzyme to fit onto a specific **substrate molecule** (the substance the enzyme acts upon). The enzyme and the substrate molecule must fit together like a lock and key.

The interaction of an enzyme and substrate is shown in Figure 5.2. In this reaction, the substrate is split into two products. Another way to express an enzymatic reaction is as follows:

$$E + S \rightarrow ES \rightarrow E + P$$

The *enzyme* (E) combines with the *substrate molecule* (S) to form a temporary *enzyme-substrate complex* (ES), in which the specific reaction occurs. Then the *product molecules* (P) separate from the enzyme, and the unchanged enzyme is recycled to combine with another substrate molecule. Note that the enzyme is not altered in the reaction, which means that a few enzyme molecules can catalyze a great number of reactions.

An enzyme is inactivated by a change in shape, and its shape is altered by anything that disrupts the enzyme's pattern of hydrogen bonding. For example, many enzymes function best within rather narrow temperature and pH ranges, because substantial changes in temperature or pH disrupt their hydrogen bonds and al-

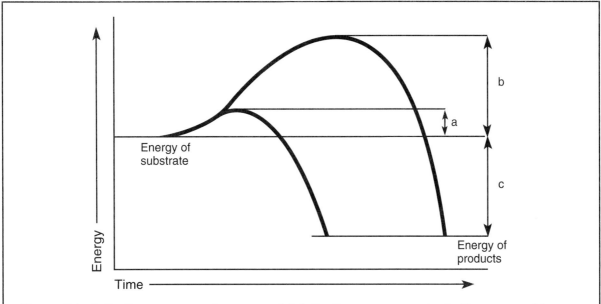

Figure 5.1. Activation energy and enzymes. (a) Activation energy required with enzyme. (b) Activation energy required without enzyme. (c) Net energy released by the reaction.

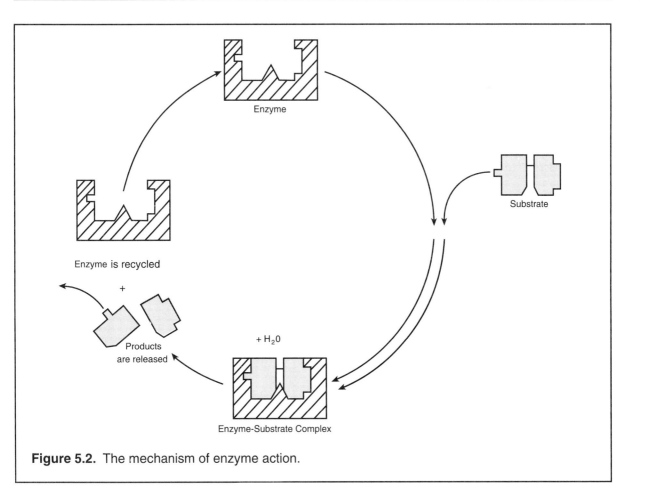

Figure 5.2. The mechanism of enzyme action.

ter their shapes. However, some enzymes function well over rather broad temperature and pH ranges because their hydrogen bonds are not easily disrupted. It is the unique hydrogen bonding pattern of each enzyme that determines the sensitivity of the enzyme to changes in temperature and pH.

Action of Catalase

All organisms using molecular oxygen produce hydrogen peroxide (H_2O_2) as a harmful by-product of some cellular reactions. Hydrogen peroxide is a strong oxidizing agent that can cause serious damage to cells. Fortunately, cells have an enzyme, **catalase**, that quickly breaks down hydrogen peroxide into water and oxygen, preventing cellular damage.

$$\text{Hydrogen peroxide} \xrightarrow{\text{(catalase)}} \text{Water} + \text{Oxygen}$$
$$2H_2O_2 \qquad\qquad 2H_2O \quad\; O_2$$

In a cell, oxygen released by the above reaction is used for other cellular processes, but when the reaction occurs in a test tube, oxygen gas bubbles to the surface, producing a layer of foam on the surface of the hydrogen peroxide solution. The amount of foam produced and the speed with which it forms are measures of catalase activity. In the following experiments, you will determine the rate of catalase activity by measuring the thickness of the foam layer.

Materials

Per student group
Beakers, 250 ml, 3
Celsius thermometer
Glass-marking pen
Hot plate
Medicine dropper
Metric ruler
Mortar and pestle
Scalpel or knife
Test tubes, 6
Test-tube rack
Dropping bottles of:
    buffer solutions, pH 4, 6, 8, 10
    hydrogen peroxide, 3%
    tap water

Per lab
Water bath, 70 °C
Crushed ice
Apple
Liver, $\frac{1}{8}$ lb
Onion
Potato
Ground beef, $\frac{1}{8}$ lb

Assignment 1

Complete items 1a to 1e on Laboratory Report 5 that begins on page 325.

Assignment 2

Do all cells contain an equal amount of catalase? Answer this question by doing the following experiment.

Hypothesis. All cells of aerobic organisms contain the same quantity of catalase.

Prediction: If the cells contain the same quantity of catalase, then catalase extracted from cells of different aerobic organisms will breakdown hydrogen peroxide at the same rate.

1. Place 1 dropper of hydrogen peroxide in each of 6 numbered test tubes.
2. Use a mortar and pestle to macerate (grind to a pulp) a small amount (0.5 cm^3) of beef liver in 2 droppers of tap water.
3. Use a medicine dropper to add 1 drop of tap water to Tube 1. Use a different medicine dropper to add 1 drop of liver extract to Tube 2. (The drop should land directly in the peroxide rather than running down the side of the tube.) What happens?
4. Measure the thickness of the foam layer (mm) in each tube after 1 min, and **record its thickness in item 2a on the laboratory report.**
5. Use the same procedure for ground beef, apple, onion, and potato. Measure the thickness of the foam layers after 1 min. *Wash the dropper and mortar and pestle thoroughly after macerating each tissue.*
6. **Complete item 2 on the laboratory report.**

Assignment 3

Is liver catalase active within a narrow or broad temperature range? Find out by doing the following experiment.

Hypothesis: Catalase is active within a narrow temperature range.

Prediction: If catalase is active within a narrow temperature range, then catalase will not breakdown hydrogen peroxide at all temperatures when exposed to a broad range of temperatures.

1. Macerate a cube ($0.5\ cm^3$) of liver in 2 droppers of tap water using a mortar and pestle as before. Place 1 drop of the liver extract and 1 dropper of tap water in each of 4 test tubes numbered 1A, 2A, 3A, and 4A.
2. Place 1 dropper of hydrogen peroxide in each of 4 test tubes numbered 1B, 2B, 3B, and 4B.
3. Place all of the tubes for 5 min as follows.
   Tubes 1A and 1B: in a beaker of crushed ice.
   Tubes 2A and 2B: in a test tube rack at room temperature.
   Tubes 3A and 3B: in a water bath at 70 °C.
   Tubes 4A and 4B: in a boiling water bath.
   Measure and record the temperature of exposure for each tube.
4. After 5 min, pour the peroxide from Tubes 1B, 2B, 3B, and 4B into corresponding Tubes 1A, 2A, 3A, and 4A. Measure the thickness of the foam layer after 1 min.
5. **Complete item 3 on the laboratory report.**

Assignment 4

Is liver catalase active within a narrow or broad pH range? Find out by doing the following experiment.

Hypothesis: Catalase is active within a narrow pH range.

Prediction: If catalase is active within a narrow pH range, then catalase will not breakdown hydrogen peroxide at all pH values when exposed to a broad range of pH values.

1. Macerate a cube ($0.5\ cm^3$) of liver in 2 droppers of tap water using a mortar and pestle before. Place 1 drop of the liver extract in each of 4 test tubes numbered 1, 2, 3, and 4.
2. Add to these tubes 1 dropper of the following buffers.
   Tube 1: pH 4
   Tube 2: pH 6
   Tube 3: pH 8
   Tube 4: pH 10
   Place the tubes in a test tube rack for 5 min at room temperature.
3. After 5 min, add 1 dropper of hydrogen peroxide to each tube. Measure the thickness of the foam layer after 1 min.
4. **Complete item 4 on the laboratory report.**

Assignment 5

Complete item 5 on the laboratory report.

6

Diffusion and Osmosis

After completion of the laboratory session, you should be able to:
1. Describe and explain the cause of Brownian movement, diffusion, and osmosis.
2. Determine the effects of temperature, molecular weight, and concentration gradient on the rate of diffusion and osmosis.
3. Use the scientific method in your laboratory investigations.
4. Define all terms in bold print.

Materials constantly move into and out of cells. The **selective permeability** of the cell membrane permits some materials to pass through it, but prevents others from doing so. Furthermore, the materials that can pass through the membrane may change from moment to moment.

Materials move through a cell membrane by two different processes. **Passive transport**, commonly called **diffusion**, results from the normal, random motion of molecules, and does not require an expenditure of energy by the cell. **Active transport** requires the use of energy by the cell, does not depend on molecular motion, and may move materials either with or against a

concentration gradient. In a similar way, materials that would normally diffuse out of a cell may be prevented from doing so by **active retention**.

You may wish to refresh your understanding of the scientific method in Exercise 1, since you will be using it in this exercise. A general hypothesis, prediction, and null hypothesis are provided for each experiment to help you think through the steps of the scientific method.

BROWNIAN MOVEMENT

Molecules of liquids and gases are in constant, random motion. A molecule moves in a straight-line path until it bumps into another molecule, and then it bounces off into a different straight-line course. Since this motion is temperature dependent, the higher the temperature the greater the molecular movement. This movement cannot be observed directly. However, tiny particles suspended in a liquid may be observed with a microscope as they are moved in a random fashion as the result of bombardment by molecules composing the liquid. This vibratory movement is indirect evidence of molecular motion, and is called **Brownian movement** after Robert Brown, who first described it in 1827.

Materials

Per student
Compound microscope

Per lab
Dissecting needles
Microscope slides and cover glasses
Powdered carmine dye
Dropping bottles of detergent-water solution

Assignment 1

1. Place a drop of water-detergent solution on a clean slide.
2. Dip a *dry* tip of a dissecting needle into powdered carmine, and, while holding the tip above the drop of fluid on the slide, tap the needle with your finger to shake only a few dry particles into the drop. Add a cover glass.
3. Examine the dye particles at 400× to observe Brownian movement.
4. **Complete item 1 on Laboratory Report 6 that begins on page 327.**

DIFFUSION

The constant, random motion of molecules is what enables diffusion to occur. **Diffusion** is the net movement of the same kind of molecules from an area of their higher concentration to an area of their lower concentration. Thus, the molecules move down a **concentration gradient**. Molecules that are initially unequally distributed in a liquid or gas tend to move by diffusion throughout the medium until they are equally distributed. Diffusion is an important means of distributing materials within cells and of passively moving substances through cell membranes.

Diffusion and Temperature

Hypothesis: The rate of diffusion is affected by temperature.

Prediction: If temperature affects the rate of diffusion, then the same water-soluble substance placed in water at different temperatures will diffuse at different rates.

You are to test the hypothesis by performing a controlled experiment. The temperature of wa-

ter in each of two beakers will each serve as the control for the other beaker. There is only one independent variable, temperature. All other variables are controlled.

Materials

Per student group
Hot plate
Beakers, 250 ml, 2
Beaker tongs
Celsius thermometer

Per lab
Forceps
Granules of potassium permanganate
Ice water

Assignment 2

1. **Complete items 2a to 2c on the laboratory report.**
2. Fill a beaker two-thirds full with water and heat it on a hot plate to about 50 °C. Use beaker tongs to place the beaker on a pad of paper towels.
3. Place an equal amount of ice water (but no ice) in another beaker. Record the temperature of the water in each beaker.
4. Keeping both beakers motionless so that the water is also motionless, drop a granule of potassium permanganate into the water in each beaker. Record the time. Observe the rate of diffusion of the potassium permanganate molecules during a 15-min period. Is the rate identical in each beaker?
5. **Complete item 2 on the laboratory report.**

Diffusion and Molecular Mass

Hypothesis: The rate of diffusion of a substance is affected by its molecular mass.

Prediction: If molecular mass affects the rate of diffusion, then heavier molecules will diffuse more slowly than lighter molecules.

You are to test the hypothesis by measuring the rate of diffusion of potassium permanganate (mol. mass 158) and methylene blue (mol. mass 320) through an agar gel that is about 98% water.

Materials

Per student group
Petri dish of agar gel

Per lab
Forceps
Granules of:
    methylene blue
    potassium permanganate

Assignment 3

1. Place equal-sized granules of potassium permanganate and methylene blue about 5 cm apart on an agar plate. Gently press the granules into the agar with forceps to assure good contact. Record the time.
2. After 1 hr, determine the distance each substance has diffused by measuring the diameter of each colored circle on the agar plate.
3. **Complete item 3 on the laboratory report.**

Diffusion and Molecular Size

At times, materials seem to diffuse through cell membranes if their molecules are sufficiently small, while larger molecules are prevented from passing through. This selectivity based on molecular size can be observed in an experiment in which a cellulose membrane is used to simulate a cell membrane. The cellulose membrane has many microscopic pores scattered over its surface, and molecules smaller than the pores can pass through the membrane.

You will use the iodine test for starch and Benedict's test for reducing sugars to determine the results of the experiment. See Figure 4.5.

Starch Test

1. Place one full dropper of the substance to be tested in a clean test tube.
2. Add 3 drops of iodine solution to the test tube, and shake to mix. A gray to blue-black coloration indicates the presence of starch in this order of increasing starch concentration.

Benedict's Test

1. Place one full dropper of the substance to be tested in a clean test tube.

2. Add 5 drops of Benedict's solution to the test tube, and mix by shaking.
3. Heat to near boiling for 2–3 min in a boiling water bath. A water bath is set up by placing the test tube in a beaker, which is about half full of water, and by heating the beaker on a hot plate. Add 3 or 4 boiling chips to the beaker to prevent spatter of hot water. In this way the test tube is heated by the water in the beaker rather than directly by the hot plate. A light green, yellow, orange, or brick-red coloration of the liquid in the test tube indicates the presence of reducing sugars, in that same order of increasing concentration.

Materials

Per student group
Hot plate
Beaker, 250 ml
Boiling chips
Rubber band
Test tube, 24 × 200 mm
Test tubes, 14 × 150 mm, 2
Test-tube holder
Test-tube rack
Dropping bottle of:
    Benedict's solution
    iodine solution (IKI)

Per lab
Cellulose tubing, 1-in. width
Flasks or squeeze bottles of:
    glucose, 20%
    soluble starch, 0.1%

Assigment 4

1. Set up the experiment as shown in Figure 6.1. *Be certain* to rinse the outside of the sac before inserting it into the test tube.
2. Place the test tube in a beaker or rack for 20 min.
3. Examine your test tube and record any color change that has occurred.
4. Test the solution outside the sac to see if glucose has diffused from the sac. Place one dropper of the solution outside the sac in a clean test tube and perform Benedict's test.
5. **Complete item 4 on the laboratory report.**

1. Fill a large test tube two thirds full with water and add 4 droppers of iodine solution. Place test tube in a beaker.

2. Soak a 20-cm length of cellulose tubing in water. Then tie a knot in one end to form a sac.

3. Fill the sac half full with starch solution and add 5 droppers of glucose solution.

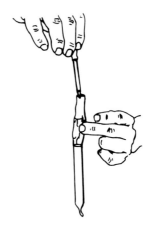

4. Hold sac closed and rinse outside of sac under the tap.

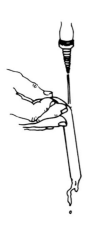

5. Insert the sac into the test tube.

6. Bend top of sac over the lip of test tube and secure it with a rubber band. Let stand for 20 min.

Figure 6.1. Preparation of an experiment to study the effect of molecular size on the diffusion of molecules through a cellulose membrane.

6. Clean the test tubes thoroughly, using a test-tube brush.

OSMOSIS

Water is essential for life. It is the **solvent** of living systems and is the aqueous medium in which the chemical reactions of life occur. Substances dissolved in a solvent are called **solutes**.

Water is the most abundant substance in cells, and contains both organic and inorganic solutes. A water molecule is a small molecule that diffuses freely in and out of cells in accordance with the concentration of water inside and outside the cells. **Osmosis** is the name given to the diffusion of water through a semipermeable or selectively permeable membrane. Like all substances, water diffuses down a **concentration gradient**, moving from an area of higher concentration to an area of lower concentration. Remember that the concentration of water increases as the concentration of solutes decreases. Thus, a 5% salt solution contains 95% water, while a 10% salt solution contains 90% water.

Osmosis and Concentration Gradients

When unequal concentrations of solutes in aqueous solutions are present on opposite sides of a semipermeable membrane, water always moves from the **hypotonic solution** (lower solute concentration) into the **hypertonic solution** (higher solute concentration), that is, down the concentration gradient of water. If the concentrations of solutes on the opposite sides of the membrane are equal, the solutions are said to be **isotonic solutions**. They are at **osmotic equilibrium**, and no net movement of water occurs through the membrane.

Hypothesis: The rate of osmosis is affected by the concentration gradient of water across a membrane.

Prediction: If the concentration gradient affects the rate of osmosis, then osmosis will occur more rapidly when the gradient is greater.

In this experiment, you will test the hypothesis by immersing sacs of 20% sucrose, 10% sucrose, and water in beakers of water to determine their gain in mass (weight). The percentage of mass increase that occurs within 1 hr for each sac is a measure of the rate of osmosis as water moves into the sacs. Water molecules can move freely through the pores in the sacs, but sucrose molecules are too large to pass through.

Materials

Per student group
Beakers, 400 ml, 3
Cellulose tubing, 1-in. diameter
Marking pen, felt-tip, waterproof
String

Per lab
Squeeze bottles of:
   20% sucrose, 4 per lab

10% sucrose, 4 per lab
Triple-beam balance

Assignment 5

1. **Complete items 5a to 5f on the laboratory report.**
2. Set up the experiment as shown in Figure 6.2. Using a felt-tip marker pen, mark the string of each sac to code for its contents. Blot the sacs and measure their mass to the nearest 0.1 g. **Record your measurements in item 5g on the laboratory report**.
3. Set the beakers with their sacs aside for 1 hr. Then blot the sacs dry and measure their mass again to the nearest 0.1 g.
4. **Complete item 5 on the laboratory report.**

Osmosis and Living Cells

In this section, you will observe the movement of water into and out of living cells that are exposed to hypertonic and hypotonic solutions.

Materials

Per student group
Compound microscope
Dropping bottles of:
   10% NaCl solution
   distilled water

Per lab
Microscope slides and cover glasses
Elodea shoots
Osmosis demonstration using celery sticks

Assignment 6

1. **Complete items 6a to 6c on the laboratory report.**
2. xamine the osmosis demonstration using celery sticks. Celery sticks were placed in (a) distilled water and (b) 10% salt solution at the beginning of the lab session. Evaluate the crispness and flexibility of the celery sticks. **Complete item 6d on the laboratory report**.
3. Mount an *Elodea* leaf in distilled water. Ob-

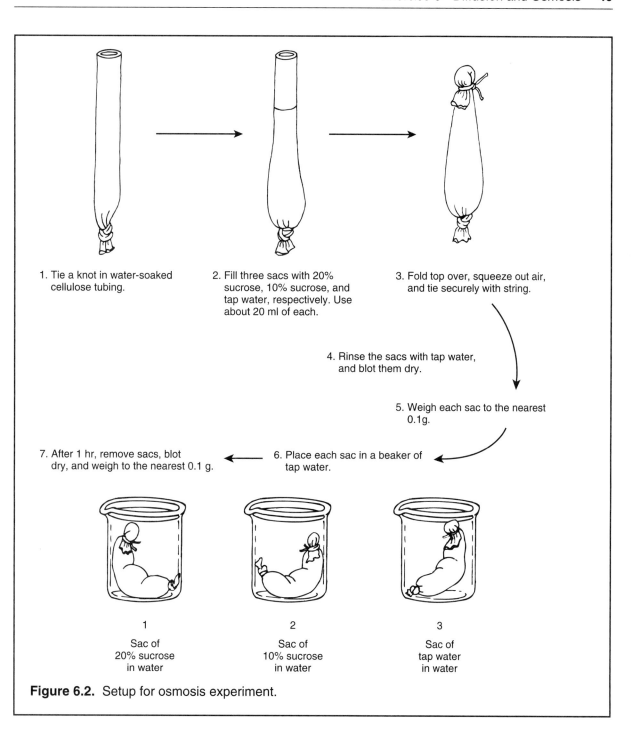

1. Tie a knot in water-soaked cellulose tubing.

2. Fill three sacs with 20% sucrose, 10% sucrose, and tap water, respectively. Use about 20 ml of each.

3. Fold top over, squeeze out air, and tie securely with string.

4. Rinse the sacs with tap water, and blot them dry.

5. Weigh each sac to the nearest 0.1g.

7. After 1 hr, remove sacs, blot dry, and weigh to the nearest 0.1 g.

6. Place each sac in a beaker of tap water.

1
Sac of
20% sucrose
in water

2
Sac of
10% sucrose
in water

3
Sac of
tap water
in water

Figure 6.2. Setup for osmosis experiment.

serve the leaf and its cells at 100×. Focus on an optical midsection of a cell at 400×. Note the distribution of the protoplasm, chloroplasts and central vacuole.

4. **In the space for item 6e on the laboratory report, draw an optical midsection view of a normal Elodea cell.** Show the distribution of the protoplasm and the central vacuole.

5. Place 1 drop of 10% salt solution at the edge of the cover glass, and draw it under the cover glass as shown in Figure 6.3. Repeat the process. Now the leaf is mounted in salt solution. What do you think will happen?

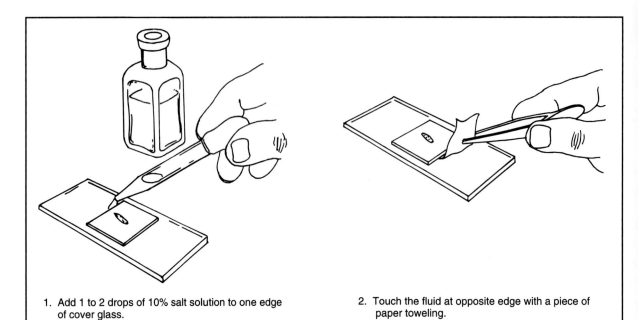

1. Add 1 to 2 drops of 10% salt solution to one edge of cover glass.

2. Touch the fluid at opposite edge with a piece of paper toweling.

Figure 6.3. How to draw a solution under a cover glass.

6. Quickly observe the leaf at 100× and watch what happens. Examine the distribution of protoplasm and chloroplasts at 400×.
7. **Make a drawing of a cell mounted in salt solution in the space for item 6e on the laboratory report.**
8. Remove the *Elodea* leaf from the salt solu-tion, rinse it, and remount it on a clean slide in distilled water. Quickly observe it with your microscope. What happens?
9. **Complete the laboratory report.**
10. Clean and dry the objectives and stage of your microscope to remove any salt water that may be present on them.

7 Photosynthesis

OBJECTIVES

After completion of the laboratory session, you should be able to:
1. Determine experimentally materials or conditions necessary for photosynthesis.
2. Write the summary equation for photosynthesis and identify the role of each reactant.
3. Separate and identify the chloroplast pigments.
4. Describe the effect of light quantity on the rate of photosynthesis.
5. Define all terms in bold print.

Living organisms require a constant supply of energy to power their metabolic (living) processes. Animals obtain energy by eating food. Organic nutrients (carbohydrates, proteins, and fats) in food contain chemical energy in their chemical bonds, and this energy is used to power cellular processes. Unlike animals, plants are able to synthesize their own organic nutrients. They use **photosynthesis** to convert light energy into the chemical energy of organic nutrients. The end product of photosynthesis is glucose ($C_6H_{12}O_6$), a disaccharide carbohydrate. Some glucose is immediately converted into starch, a polysaccharide, for storage within a photosynthesizing cell.

Leaves are the primary organs of photosynthesis in most plants. As you investigate factors essential for photosynthesis, you will use the presence of starch in leaf cells as evidence for the occurrence of photosynthesis.

Materials

Per student group
Alcohol
Beakers, 100 ml, 250 ml
Boiling chips
Hot plate
Petri dish
Scissors
Test tubes, 2
Test-tube rack
Dropping bottles of:
    Benedict's solution
    iodine (IKI) solution

Per lab
Setup for photosynthesis and CO_2 experiment
    (Figure 7.2)
    bell jars, 2
    glass funnels, 2
    fluorescent lamps, 2
    Petri dish
    rubber stoppers for bell jars, 1-hole

soda lime granules

fairy primrose (*Primula malacoides*), 4-in. pots

Setup for photosynthesis and light experiment

leaf shields

fluorescent lamps

fairy primrose (*Primula malacoides*), 4-in. pots

Setup for photosynthesis and chlorophyll experiment

fluorescent lamps

Coleus plants, green and white, 4-in. pots

CARBON DIOXIDE AND PHOTOSYNTHESIS

Plants absorb water and inorganic nutrients through their roots, and water (H_2O) serves as the source of hydrogen atoms in the carbohydrate end product of photosynthesis. But what is the source of the carbon atoms? Carbon compounds are not absorbed through the roots. Can the small amount (.03%) of carbon dioxide (CO_2) in the air be the source? The following experiment will answer this question.

Hypothesis: Carbon dioxide is necessary for photosynthesis.

Prediction: If carbon dioxide is necessary for photosynthesis, then photosynthesis will not occur in the absence of carbon dioxide.

Two healthy plants where placed in darkness for 48 hr and then placed under separate bell jars. See Figure 7.1. Plant A has been exposed to air containing CO_2. Plant B has been exposed to air from which CO_2 has been removed by soda lime. Both plants have been exposed to light for 24 hr prior to the laboratory session. You will test the hypothesis by determining the presence or absence of starch in a leaf from each plant, since the presence of starch is an indicator of photosynthesis.

Plant A is exposed to normal air.

Plant B is exposed to air lacking CO_2, because CO_2 has been removed by the soda lime.

Figure 7.1. Setup for the carbon dioxide and photosynthesis experiment.

1. Add boiling chips to the beaker and heat the water to boiling. Add the leaf and boil for 3-5 min to rupture cells.

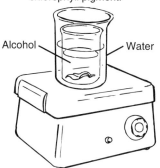

2. Add boiling chips to the beaker of alcohol. Boil the leaf in alcohol in water bath to extract chlorophyll pigment.

Alcohol   Water

3. Place leaf in a petri dish and completely cover it with iodine solution for 3 min. Rinse with water and examine.

Figure 7.2. Procedure for testing a leaf for the presence of starch.

Assignment 1

1. Remove a leaf from each plant. Cut off the petiole of the leaf from plant A so that you can distinguish the leaves.
2. Carefully follow the procedures in Figure 7.2 to test each leaf for the presence of starch. A gray to blue-black color indicates the presence of starch. **Caution:** *Alcohol is flammable; do not get it near an open flame.*
3. **Complete item 1 on Laboratory Report 7 that begins on page 331.**

LIGHT AND PHOTOSYNTHESIS

Leaves are exposed to sunlight only during daylight hours, but they carry on energy-powered living processes at night as well. Does photosynthesis occur only when plant leaves are exposed to light? Perform the following experiment to answer this question.

Hypothesis: Light is necessary for photosynthesis.

Prediction: If light is necessary for photosynthesis, then photosynthesis will not occur in the absence of light.

Healthy green plants were placed in darkness for 48 hr. Leaf shields, that screen part of a leaf from light, were placed on the leaves, and the plants have been exposed to light for 24 hr prior to the laboratory session. You will test the hypothesis by comparing the position of the light shield (area receiving no light) with the distribution of starch in the leaves.

Assignment 2

1. **Complete items 3a and 3b on the laboratory report.**
2. Remove a leaf from one of the plants. **Trace or sketch the leaf in item 3c, showing the position of the leaf shield.**
3. Remove the leaf shield and test the leaf for the presence of starch as shown in Figure 7.3.
4. Compare the distribution of starch and the position of the leaf shield.
5. **Complete item 2 on the laboratory report.**

CHLOROPHYLL AND PHOTOSYNTHESIS

Nearly all plants have green leaves because of the abundance of **chlorophyll** in their chloroplasts. Chlorophyll gives the color to green leaves because it reflects green wavelengths of light while absorbing other wavelengths. The widespread abundance of chlorophyll suggests that it plays a vital role in plants. Is chlorophyll necessary for the absorption of light that powers

1. Remove a leaf from a healthy plant exposed to light for 24 hr.

2. With scissors cut the leaf into small pieces.

3. Place leaf fragments in a test tube with 2 droppers of water. Place test tube in a beaker half filled with water and boil 5 min.

4. Pour off the fluid into another test tube.

5. Add 5 drops of Benedict's solution to the fluid.

6. Place test tube in boiling water for 5 min. Yellow or orange color is positive test for glucose.

Figure 7.3. Procedure for testing a leaf for the presence of glucose.

photosynthesis? Perform the following experiment to answer this question.

Hypothesis: Chlorophyll is necessary for photosynthesis.

Prediction: If chlorophyll is necessary for photosynthesis, then photosynthesis will not occur in the absence of chlorophyll.

Some plants have variegated leaves that have an unequal distribution of chlorophyll. Some parts of variegated leaves lack chlorophyll. Such leaves are good subjects for investigating whether chlorophyll is necessary for photosynthesis. You will test the hypothesis by comparing the distribution of chlorophyll and starch in variegated leaves of *Coleus* plants that have

been exposed to light for 24 hr prior to the laboratory session.

Assignment 3

1. Remove a variegated leaf from a *Coleus* plant on the stock table. Cut the leaf in half along the midrib. Sketch half of the leaf, showing the distribution of chlorophyll.
2. Test one half of the leaf for the presence of starch as shown in Figure 7.3, and compare the starch and chlorophyll distributions.
3. Use scissors to separate the green and non-green portions of the other half of the leaf, and test *each portion* separately for the pres-

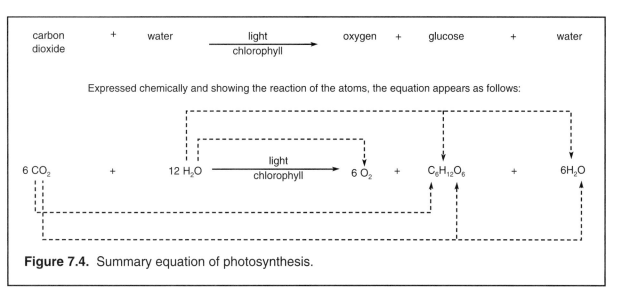

Figure 7.4. Summary equation of photosynthesis.

ence of sugar, as shown in Figure 7.4. A light green, yellow, or orange coloration indicates the presence of glucose. Compare your results with the distribution of chlorophyll.

4. **Complete item 3 on the laboratory report.**

SUMMARY OF PHOTOSYNTHESIS

You have discovered that water, carbon dioxide, light, and chlorophyll are necessary for photosynthesis to occur. Photosynthesis occurs within chloroplasts where light-absorbing **chlorophyll** and necessary enzymes are located. Study the summary equation of photosynthesis in Figure 7.4. Note the reactants and products, and trace the flow of atoms from reactants to products.

Photosynthesis is a complex series of reactions that may be divided into two stages: a light reaction and a dark reaction. In the **light reaction**, chlorophyll converts light energy into chemical energy, and water molecules are split, releasing oxygen molecules. In the **dark reaction**, chemical energy formed in the light reaction is used to join hydrogen atoms split from water with carbon dioxide molecules to form glucose. Although the dark reaction occurs during exposure to light, it does not *require* light energy, hence its name.

Assignment 4

Complete item 4 on the laboratory report.

CHLOROPLAST PIGMENTS

When white light is passed through a prism, the component wavelengths are separated to form a rainbow-like spectrum that we perceive as colors ranging from violet (wavelength of 380 nanometers, nm) through blue, green, yellow, and orange to red (wavelength of 760 nm). Light energy decreases as the wavelength of light increases.

When white light strikes a leaf, some wavelengths are absorbed and others are reflected or transmitted. The greatest absorption of light by chlorophyll is in the red, blue, and violet wavelengths. See Figure 7.5. Most leaves appear green because the green wavelengths of light are reflected rather than being absorbed by chlorophyll, which is usually the dominant pigment in the leaf. Most leaves also have other light-absorbing pigments in their chloroplasts, notably the yellow **carotenes** and the yellow-orange **xanthophylls**. These pigments absorb different wavelengths of light than chlorophyll and pass the captured energy to chlorophyll.

Chloroplast pigments may be separated by

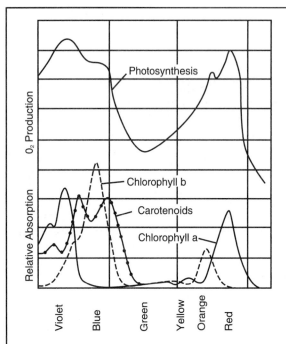

Figure 7.5. Absorption spectrum of chloroplast pigments and action spectrum of photosynthesis.

paper chromatography, a technique that takes advantage of slight differences in solubility of the pigments. The most soluble pigment will be carried farthest along the paper by the solvent used in this technique, and the least soluble pigment will be carried the shortest distance.

Your instructor has prepared a concentrated solution of chloroplast pigments for your use. **Caution:** *The pigment solution and the developing solvent are flammable and volatile. Keep them away from open flames.*

Materials

Per student group
Chromatography jar with cork stopper
Chromatography paper, 1-in. width
Paper clip
Scissors
Straight pin

Per lab
Chloroplast pigment solution
Chromatography developing solvent
Chromatography waste collecting jar
Paintbrushes, watercolor, fine-tipped
Pliers, pointed nose

Assignment 5

1. Obtain the jar, cork, straight pin, paper clip, and chromatography paper for the setup shown in Figure 7.6. Handle the paper by the edges, since oils from your fingers can adversely affect the results. Cut small notches about 2 cm from one end of the paper, and cut off the corners of this end.
2. Insert a pin into the cork stopper and bend it to form a hook. Attach the paper clip to the

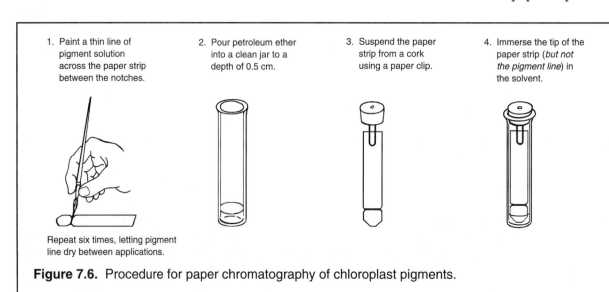

1. Paint a thin line of pigment solution across the paper strip between the notches.

Repeat six times, letting pigment line dry between applications.

2. Pour petroleum ether into a clean jar to a depth of 0.5 cm.

3. Suspend the paper strip from a cork using a paper clip.

4. Immerse the tip of the paper strip (*but not the pigment line*) in the solvent.

Figure 7.6. Procedure for paper chromatography of chloroplast pigments.

paper and suspend it in the jar *before adding the solvent*. The end of the paper should be *just above the bottom of the jar*. Adjust the paper clip or cut off the paper to achieve this length. Then remove the paper and clip for step 3.

3. Use a small paintbrush to paint a line of chloroplast pigment solution across the paper between the notches. Let it dry. Repeat this step six times to obtain a dark pigment line.

4. Pour the developing solvent into your jar *under a fume hood* to a depth of about 1 cm.

5. Return to your workstation, and insert the cork and suspended chromatography paper into the jar. The end of the paper, but not the pigment line, should be in the solvent. *Do not move the jar for 30 min.*

6. After 30 min, remove the paper and place it on a paper towel. Pour the developing solvent into the waste jar under the fume hood.

7. **Attach your chromatogram to your lab report** and label the separated pigments: bright yellow carotene at the top of the paper, 1 or 2 bands of yellow xanthophylls, blue-green chlorophyll a, and yellow-green chlorophyll b.

8. **Complete item 5 on the laboratory report.**

LIGHT QUANTITY AND THE RATE OF PHOTOSYNTHESIS

You have shown that light is necessary for photosynthesis. Plants are exposed to different quantities of light throughout the day and from season to season. In addition, some plants grow mostly in shade while others thrive in direct sunlight. Do you think that the quantity of light affects the rate of photosynthesis? Let's find out.

Hypothesis: Light quantity affects the rate of photosynthesis.

Prediction: If light quantity affects the rate of photosynthesis, then the rate of photosynthesis will be faster when a plant is exposed to more light and slower when it is exposed to less light.

As noted in the summary equation of photosynthesis (Figure 7.4), oxygen molecules are one of the products. For each molecule of oxygen produced, a molecule of carbon dioxide is used in the formation of glucose. Some of the oxygen is used by the plant cells, and the excess is released into the atmosphere. This excess oxygen forms the oxygen in the atmosphere.

Since the rate of oxygen production is an indicator of the rate of photosynthesis, you will test the hypothesis by measuring the rate of oxygen production by an *Elodea* shoot exposed to different quantities of light. Light quantity decreases with distance from a light source, so you will vary the light quantity by changing the distance between the *Elodea* shoot and the light source. See Figure 7.7. A good range of light quantities is obtained by placing the *Elodea* shoot at 25 cm, 50 cm, 75 cm, and 100 cm from a 150-watt spot lamp. If you use a fluorescent lamp, decrease the distance by half at each setting.

Materials

Per student group
Heat filter (rectangular glass container of water)
Light source: 150-watt spot lamp or fluorescent lamp
Meter stick
Pipette, 1 ml
Plastic tubing, 7 cm
Ring stand
Ring stand test-tube clamps
Scalpel or razor blade
Syringe and needle, 5 or 10 ml
Test tube
Tubing clamp, screw type

Per lab
Sodium bicarbonate solution, 2%
Elodea shoots

Assignment 6

1. **Complete items 6a and 6b on the laboratory report.**

2. Fill a test tube about three-fourths full of 2% sodium bicarbonate ($NaHCO_3$) solution, which will provide an adequate concentration of CO_2.

3. Make a diagonal cut through the stem of a leafy shoot of *Elodea* about 10–13 cm from its tip. Use a sharp scalpel, being careful not to crush the stem. Insert the *Elodea* shoot, tip down, into the test tube so that the cut end is 1–2 cm below the surface of the solution.

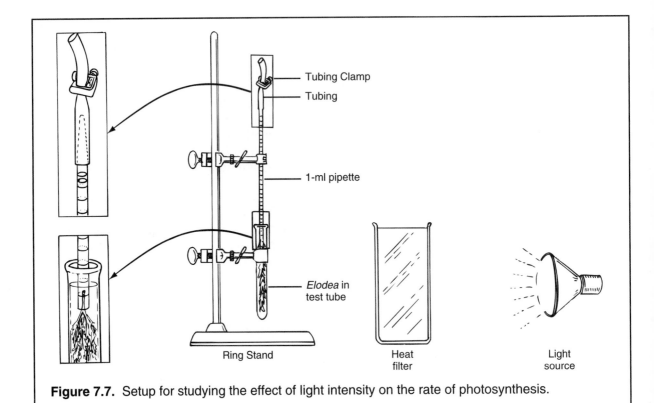

Figure 7.7. Setup for studying the effect of light intensity on the rate of photosynthesis.

Place the tube in a test-tube holder on a ring stand. If bubbles of oxygen are released too slowly from the cut end of the stem, recut the stem at an angle to obtain a good production of bubbles.

4. Place a short piece of plastic tubing snugly over the tip (pointed end) of a 1-ml pipette as shown in Figure 7.7. Place a screw-type tubing clamp on the tubing and tighten it until it is *almost closed.* Place the pipette in a ring-stand clamp with the tip up, and lower the base of the pipette into the test tube so that the cut end of the *Elodea* shoot is inserted into the pipette. Secure the pipette in this position. See Figure 7.7. Study the 0.01-ml graduations on the pipette to ensure that you known how to read them.

5. If you are using a spot incandescent light source, fill a rectangular glass container with water to serve as a heat filter. The heat filter must be placed between the *Elodea* shoot and the spot lamp, about 10 cm in front of the *Elodea.* A heat filter is not necessary if using a fluorescent lamp.

6. You will vary the light quantity by placing the light source at 25 cm, 50 cm, 75 cm, and 100 cm from the tube containing the *Elodea* shoot. Start at the 25-cm distance and arrange your setup as shown in Figure 7.7. Turn off the room lights.

7. After allowing 10 min for equilibration, tighten the screw clamp to close the tubing on the tip of the pipette. Insert the needle of a 5- or 10-ml syringe into the tubing and gently pull out the syringe plunger to raise the level of the fluid in the pipette to the 0.9-ml mark. Then remove the syringe.

8. Bubbles of oxygen should start forming at the cut end of the *Elodea* stem and rise into the pipette, displacing the solution.

9. Read and record the fluid level in the pipette, and record the time of the reading. Be sure to take your readings at the bottom of the meniscus, as shown in Figure 7.8. **Record your data in the chart in item 6c on the laboratory report.**

10. *Exactly* 3 min later, record another reading. Then determine the volume of oxygen produced within the 3-min interval. Repeat the process two more times to determine oxygen production during three separate 3-min intervals at this distance from the light source.

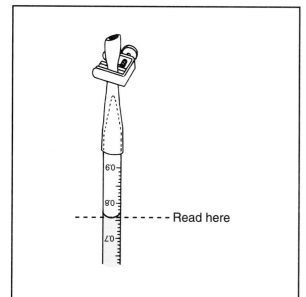

Figure 7.8. Reading the fluid level in the pipette at the meniscus.

Then calculate the average volume of oxygen produced per minute, as milliliters of oxygen per minute (ml O_2/min), and record this figure in the chart in item 6c on the laboratory report.

11. Use the same procedure as in 9 and 10 to determine the rate of oxygen production at 50 cm, 75 cm, and 100 cm. Move the light source to obtain the new distance, and change the water in the heat filter at each new distance. Allow 10 min for your test system to equilibrate at the new distance before starting your readings.

12. **Complete the laboratory report.**

8 Cellular Respiration

OBJECTIVES

After completion of the laboratory session, you should be able to:
1. Describe the relationship between cellular respiration, ADP, ATP, and cellular work.
2. Compare aerobic and anaerobic forms of respiration.
3. Determine experimentally the effect of temperature on aerobic cellular respiration in plants and animals.
4. Define all terms in bold print.

Living organisms are able to use the energy stored in chemical bonds of organic nutrients to supply their energy needs. The three major groups of organic nutrients are **carbohydrates**, **fats**, and **proteins**. In order to extract the energy stored in these nutrients, their chemical bonds must be broken by **cellular respiration**.

Cellular respiration is an *enzymatically controlled oxidation* that breaks bonds sequentially and releases energy in small amounts, so that this energy may be "captured" in high-energy phosphate bonds (~P). The captured energy (~P) combines with **adenosine diphosphate (ADP)** to form **adenosine triphosphate**

(**ATP**). ATP is the immediate source of energy for cellular work. It transfers ~P to power the chemical reactions within the cell. Usable energy is always transferred as ~P. Study Figure 8.1.

The two types of cellular respiration are aerobic and anaerobic. Compare the summary equation in Figure 8.2.

Most organisms depend on the **aerobic respiration** of organic nutrients in order to produce ATP. Aerobic respiration requires oxygen and yields a net of 36 ATP molecules for each molecule of glucose respired. About 40% of the released energy is transferred to ATP, while the remainder is "lost" as heat. Aerobic respiration is about twice as efficient as an automobile engine.

Anaerobic respiration does not require oxygen, but it produces a net of only 2 ATP molecules for each glucose molecule respired. Only a few bacteria and yeasts can survive on the low ATP output of anaerobic respiration, and some of these organisms are used commercially to produce alcohol and industrial solvents. *Fermentation* is a synonym for anaerobic respiration in bacteria and fungi.

Anaerobic respiration occurs in humans, but only for very brief periods, when the body's energy (ATP) need for muscle contraction exceeds the oxygen supply to the muscles, such as when

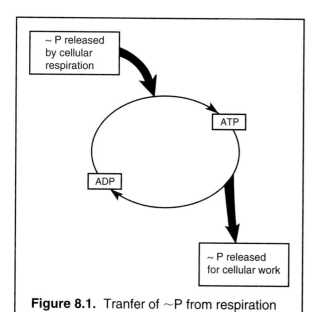

Figure 8.1. Tranfer of ~P from respiration to cellular work.

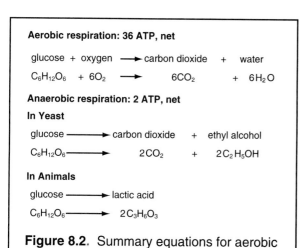

Figure 8.2. Summary equations for aerobic and anaerobic respiration.

This exercise focuses on aerobic cellular respiration. It will be detected by the production of CO_2 and measured by the consumption of O_2.

Assignment 1

Complete item 1 on Laboratory Report 8 that begins on page 335.

Materials

Per student group
Drinking straws, 2
Glass-marking pen
Glass tubing, 2-cm lengths, 9
Medicine dropper
Rubber-bulbed air syringe
Test tubes, 3
Test-tube rack
Bromthymol blue, 0.004%, in dropping bottle

Per lab
Crickets, live
Pea seeds, germinating
Respiration and heat demonstration
   Celsius thermometers, 3
   vacuum bottles, 3
   cotton plugs for vacuum bottles

The equation for aerobic respiration (Figure 8.2) indicates that CO_2 is a product of the reaction. Therefore, an accumulation of CO_2 may be used as an indicator of cellular respiration. A dilute solution of bromthymol blue, a pH indicator, may be used to detect an increase in CO_2 concentration. Carbon dioxide easily dissolves in water, and, as shown in Figure 8.3, it combines with water to form carbonic acid. Carbonic acid

a sprinter runs the 200-meter dash. Anaerobic respiration provides ATP for muscle contraction, but results in an accumulation of lactic acid in the muscles and blood. The rapid breathing and heart rate after such exertion is the body's way of providing an increased amount of oxygen to muscles and the liver to metabolize the accumulated lactic acid and to remove the excess carbon dioxide. Once this has been accomplished, both breathing and the heart rate return to normal.

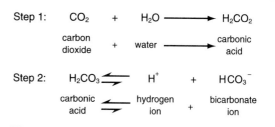

Figure 8.3. Reaction of carbon dioxide and water.

then dissociates, releasing hydrogen ions (H^+) that, in turn, react with bromthymol blue, causing it to turn yellow.

Assignment 2

Hypothesis: Carbon dioxide produced by cellular respiration is released by living organisms at concentrations greater than that in atmospheric air.

Prediction: If CO_2 is a product of cellular respiration, then respiratory gases released by living organisms should turn a bromthymol blue solution yellow.

1. Experiment 1 will determine if your exhaled breath contains a greater concentration of CO2 than the concentration present in atmospheric air.
   a. Place 3 drops of 0.004% bromthymol blue into each of two numbered test tubes.
   b. Exhale your breath through a drinking straw into test tube 1 for 3 min.
   c. Use a rubber-bulbed syringe to pump atmospheric air into test tube 2 for 3 min.
   d. **Record any color change in the bromthymol blue solutions in the table in item 2 on the laboratory report.**
2. Experiment 2 will determine if germinating pea seeds and crickets produce CO_2 concentrations greater than that in the air.
   a. Place 3 drops of 0.004% bromthymol blue into each of 3 numbered test tubes.
   b. Place several short segments of glass tubing into each tube so that the pea seeds and crickets will be kept out of the solution.
   c. Place 6–10 germinating pea seeds into tube 1, place 3–6 crickets into tube 2, and place nothing else in tube 3.
   d. Gently (to prevent a sudden increase in air pressure from injuring the crickets) insert a rubber stopper or cotton plug into each tube. Place the tubes in a test-tube rack for 20 min. Then observe any color change in the bromthymol blue.
   e. **Record your results on the laboratory report and complete item 2.**

RESPIRATION AND HEAT PRODUCTION

Only about 40% of the energy released by aerobic respiration is captured in high-energy phosphate bonds of ATP; the remainder is lost as heat. Heat produced by aerobic respiration maintains normal body temperatures in humans and other homeothermic animals.

Assignment 3

Do you think simpler organisms like germinating seeds and crickets produce heat by aerobic respiration? Perform the following experiment to find out.

Hypothesis: Heat produced by cellular respiration is released by living organisms.

Prediction: If heat produced by cellular respiration is released by living organisms, then the temperature in vacuum bottles containing living organisms should be higher than ambient temperature.

1. A few hours earlier, your instructor set up three vacuum bottles to test the null hypothesis. Bottle 1 contains germinating pea seeds. Bottle 2 contains live crickets. Bottle 3 contains air only. A thermometer has been inserted through the cotton stopper of each bottle to measure the temperature inside the bottle.
2. Read and record the temperatures in the bottles.
3. **Complete item 3 on the laboratory report.**

TEMPERATURE AND RESPIRATION RATE

Now you will consider the *rate* of aerobic respiration, i.e., the number of reactions per unit time. Do you think that the rate of cellular respiration is constant, or that, like other chemical reactions, it is affected by factors such as temperature? Let's find out.

Hypothesis: The rate of cellular respiration is affected by changes in temperature.

Prediction: If temperature affects the rate of cellular respiration, then this rate will change as temperature increases within a normal range.

Germinating Peas and Crickets

The rate of oxygen consumption is a good measure of the rate of aerobic respiration, because six molecules of oxygen are consumed for each glucose molecule respired. You will test the hypothesis by measuring the rate of oxygen consumption by germinating peas and crickets at three different temperatures. In order to allow comparisons among the organisms, it is necessary to determine the oxygen consumption per hour per gram (O_2/hr/g) of body mass for each organism.

These experiments are best done by groups of four students, with each group assigned to a particular temperature: 10 °C, room temperature, or 40 °C. If time is limited, your instructor may assign a different experiment to each group, with all groups sharing the data.

When your group has been formed, read through the experiments and establish a division of labor to cover the tasks. After the experiments have been set up, one person should keep the time, one or two persons should take the readings, and one person should record the data.

Materials

Per student group
Celsius thermometer
Dropping bottle of 10% NaCl, colored
Glass marking pen
Respirometer (Figure 8.4)
   beaker, 500 or 1,000 ml
   pipettes, 1 ml, 3
   rubber stoppers for test tubes, 1-hole, 3
   test tubes, 3
Test-tube rack

Per lab
Cotton, absorbent
Crushed ice
Soda lime
Triple-beam balance
Water bath, 40 °C
Crickets, live
Germinating pea seeds

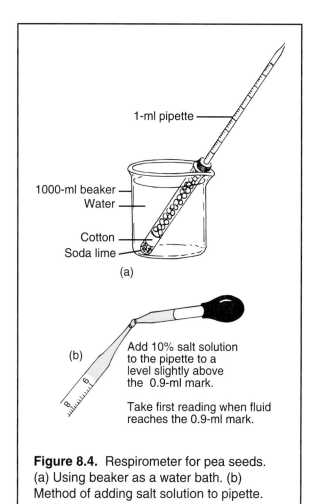

Figure 8.4. Respirometer for pea seeds. (a) Using beaker as a water bath. (b) Method of adding salt solution to pipette.

Within the figure:
1-ml pipette
1000-ml beaker
Water
Cotton
Soda lime
(a)
(b)
Add 10% salt solution to the pipette to a level slightly above the 0.9-ml mark.

Take first reading when fluid reaches the 0.9-ml mark.

Assignment 4

1. Set up three numbered test tubes at the assigned temperature. The interior of each tube must be dry. Place soda lime in each tube to a depth of about 2 cm. Then insert a loose pad of cotton to separate the specimens from the soda lime. Place the tubes in a test-tube rack.
2. Obtain 10–12 pea seeds and 3–6 crickets, which are to be placed in the test tubes. Blot any water from them and measure their mass to the nearest 0.1 g on the balance.
3. Add specimens to the tubes as follows:
   Tube 1: germinating pea seeds
   Tube 2: crickets
   Tube 3: nothing
4. Insert the base of a 1-ml pipette into the one-hole stopper so that it is flush with the inner

surface of the stopper. Insert the stopper into a test tube. Repeat for each tube.

5. Place the respirometer at the assigned temperature for 10 min to permit temperature equilibration. It is important that the respirometer is placed at an angle (not vertical), as shown in Figure 8.4 to prevent the salt solution from running down into the pipette. After 10 min, place a small drop of 10% sodium chloride solution at the tip of each pipette, and note how it is drawn into the pipettes of the experimental tubes. Since carbon dioxide is absorbed by the soda lime, oxygen consumption can be measured by the movement of the fluid toward the test chamber.

6. You are to determine the movement of the fluid for each tube during five 3-min test intervals. Record the reading on the pipette at the beginning and end of each 3-min test period. Be sure to take your readings at the front edge of the fluid. Discard the lowest and highest readings (Why?) and calculate the average of the remaining three readings. Subtract the average movement in the control tube (tube 3) from the average movement in the experimental tubes. Calculate the average respiration rate (ml O_2/hr/g) for each specimen as shown below. **Record your data in item 4a on the laboratory report.**

$$\frac{\text{ml } O_2}{3 \text{ min}} \times \frac{60 \text{ min}}{1 \text{ hr}} = \text{ml } O_2/\text{hr}$$

$$\frac{\text{ml } O_2/\text{hr}}{\text{mass in grams}} = \text{ml } O_2/\text{hr/g}$$

7. Clean the apparatus and your workstation.
8. Exchange data with groups doing the experiment at different temperatures.
9. **Complete item 4 on the laboratory report.**

Frog and Mouse

Now you will investigate the effect of temperature on the rate of cellular respiration in a frog, a poikilothermic animal whose body temperature varies directly with ambient temperature, and a mouse, a homeothermic animal whose body temperature is constant in spite of moder-

ate changes in ambient temperature. Do you think that exposure to temperatures of 10 °C, room temperature, and 40 °C will have the same effect on the rate of aerobic respiration in each animal? Let's find out.

Hypothesis: The rate of cellular respiration in poikilothermic and homeothermic animals increases with an increase in the temperature of exposure.

Prediction: If the rate of cellular respiration in poikilothermic and homeothermic animals increases with an increase in temperature, then the rate of cellular respiration in both a frog and a mouse will be slowest at 10 °C and fastest at 40 °C.

The hypothesis will be tested by measuring oxygen consumption at 10 °C, room temperature, and 40 °C. Your group will be assigned to do part of the experiment at one temperature by your lab instructor. Exchange results with student groups doing part of the experiment at other temperatures.

The manometer (U-tube) of the respirometer (Figure 8.5) connects the experimental and control chambers. A pressure change in one chamber will cause the movement of the manometer fluid toward the chamber with the lowest pressure. Since carbon dioxide is absorbed by the soda lime, oxygen consumption can be measured by the movement of fluid toward the experimental chamber. Note that any change in pressure in the control chamber is automatically reflected in the level of the fluid.

Perform the experiment as described below.

Materials

Per student group
Celsius thermometer
Respirometer (Figure 8.5)
    colored water for manometer
    glass jars, wide-mouth
    glass tubing
    pipette, 5 ml (manometer)
    plastic tubing
    rubber stoppers for glass jars, 2-hole
    ring stands with ring clamps, 2
    tubing clamps, pinch type
    water bath at assigned temperature
    wire screen

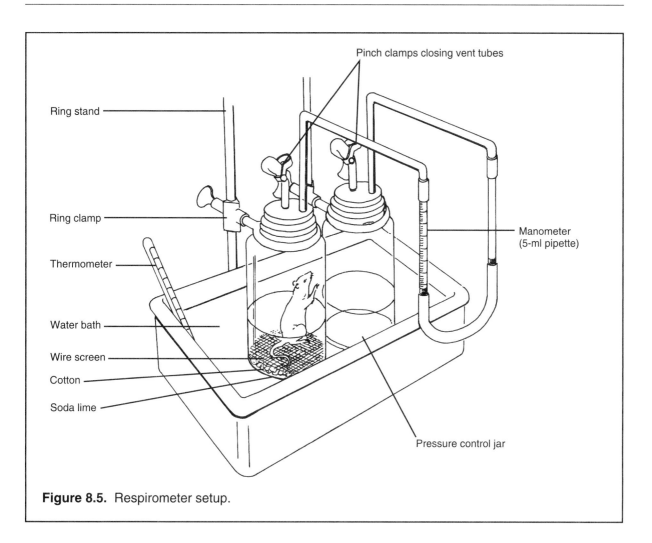

Figure 8.5. Respirometer setup.

Per lab
Cotton, absorbent
Crushed ice
Soda lime
Triple-beam balance

Assignment 5

1. Set up a respirometer as shown in Figure 8.5. Place soda lime in the bottom of each chamber to a depth of about 2 cm. Cover the soda lime with about 1 cm of cotton and add a wire screen. Place the respirometer in a water bath at the assigned temperature.
2. Measure the mass of the animal assigned to you to the nearest 0.1 g. (Mice should be picked up by their tails.)

3. Place the animal in the experimental chamber. *With the vent tubes open*, loosely replace the stopper. **Caution:** *Failure to keep the vent tubes open when inserting the stopper may injure the test animal due to a sudden increase in air pressure.*
4. After 5–10 min for temperature equilibration, and *with the vent tubes open*, insert the stopper snugly into the jar. Close the vent tubes, record the time, and take the first reading from the manometer. Exactly 3 min later, take the second reading. *Open the vent tubes, remove the stopper, and place the stopper loosely on top of the jar.*
5. After 3–5 min, insert the stopper, close the vent tubes, and take the first reading of the second replica of the experiment. Proceed as before. Repeat to make at least five replicates. **Record your data in item 5a on the laboratory report.**

6. Discard the lowest and highest values, and calculate the average oxygen consumption (milliliters per 3-min interval) and the ml O_2/hr/g of body weight that the mouse or frog consumes.

7. Return the animal to its cage, and clean the respirometer and your workstation.

8. Exchange data with groups doing the experiment at different temperatures.

9. **Plot the respiration level (ml O_2/hr/g) for each of the organisms studied in item 5d on the laboratory report.**

10. **Complete the laboratory report.**

9

Cell Division

OBJECTIVES

After completion of the laboratory session, you should be able to:
1. Name the stages of the cell cycle and describe their characteristics.
2. Name the phases of mitosis and meiosis and describe their characteristics.
3. Identify the phases of mitosis when viewed with a microscope.
4. Compare the processes and end products of mitotic and meiotic cell division.
5. Describe the significance of mitotic and meiotic cell division.
6. Define all terms in bold print.

All new cells are formed by the division of pre-existing cells. In **prokaryotic cells**, cell division is relatively simple. The process is known as **binary fission**, and it occurs by (1) replication of the circular DNA molecule and (2) the formation of additional cell membrane and cell wall material to separate the original cell into two new cells. Figure 9.1 depicts the process of binary fission. The cells are too small for you to observe this process in the laboratory, however.

In **eukaryotic cells**, two different processes of cell division produce distinctly different types of cells. Study Table 9.1. Cells formed by **mitotic cell division** contain the same number and composition of chromosomes as the parent cell. In contrast, cells formed by **meiotic cell division** have only one-half the number of chromosomes as the parent cell. Thus, these two types of cell division differ in the way the chromosomes are dispersed to the new cells that are formed. The terms **mitosis** and **meiosis** refer to the orderly process of separating and distributing the replicated chromosomes to the new cells. **Cytokinesis** (division of the cytoplasm) is the process of actually forming the **daughter cells**.

Each organism has a characteristic number of chromosomes in the nuclei of its cells. If a single set of chromosomes is present, the cell is **haploid (n)**, and it contains only one chromosome of each chromosome pair. If two sets of chromosomes are present, the cell is **diploid (2n)**, and each chromosome pair is composed of **homologous chromosomes**.

The body cells of animals and most higher plants are diploid. For example, fruit flies have 8 chromosomes (4 pairs), onions have 16 (8 pairs), and humans have 46 (23 pairs). Gametes (eggs and sperm) of these organisms are always haploid, and contain 4, 8, and 23 chromosomes,

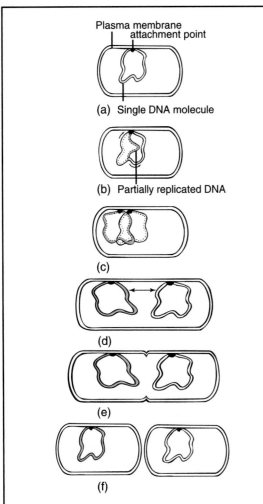

Plasma membrane
attachment point

(a) Single DNA molecule

(b) Partially replicated DNA

(c)

(d)

(e)

(f)

Figure 9.1. Cell division in a prokaryotic cell. (a) Cell before replication of DNA. Note attachment of DNA to cell membrane. (b) Replication of DNA moving in both directions from starting point. (c) DNA replication completed. (d) Growth of cell membrane and cell wall occurs between the points of DNA attachment. (e) Cleavage furrow of cell membrane and cell wall begins. (f) New cells formed.

respectively. The body cells of most simple organisms are haploid.

MITOTIC CELL DIVISION

In unicellular organisms, mitotic cell division serves as a means of reproduction. In multicellular organisms, it serves as a means of growth and repair. As worn-out or damaged cells die, they are replaced by new cells formed by mitotic division in the normal processes of maintenance and healing. Millions of new cells are formed in the human body each day in this manner.

Mitotic cell division is an orderly, controlled process, but it sometimes breaks out of control to form massive numbers of nonfunctional, rapidly dividing cells that constitute either a benign tumor or a cancer. Seeking the causes of uncontrolled mitotic cell division is one of the major efforts of current biomedical research.

The Cell Cycle

A cell passes through several recognizable stages during its life span. These stages constitute the **cell cycle**. There are two major stages in the cell cycle. **Mitosis**, the M stage, accounts for only 5–10% of the cell cycle. The **interphase** forms the remainder. See Figure 9.2.

The interphase stage of the cell cycle has three subdivisions. Immediately after mitosis is a growth period, the G_1 **stage**. Next is the **synthesis (S) stage**, when chromosome and centriole (if present) replication occurs. Each replicated chromosome consists of two **sister chromatids** joined at the **centromere**. See Figure 9.3. A second growth stage, the G_2 **stage**, follows and prepares the cell for the next mitotic division. Cells that will not divide again remain in the G_1 stage and carry out their normal functions.

Mitotic Phases in Animal Cells

The process of mitosis is arbitrarily divided into recognizable stages or phases to facilitate understanding of it, although the process is actually a continuous one. These phases are **prophase**, **metaphase**, **anaphase**, and **telophase**. The characteristics of each phase as observed in animal cells are noted here to aid your study. Interphase is also included for comparative purposes. Compare these descriptions with Figure 9.4.

Interphase

Cells in interphase have a distinct nucleus and two pairs of **centrioles**. The chromosomes are uncoiled and are visible only as **chromatin granules**.

TABLE 9.1
Significant Differences in Mitotic and Meiotic Cell Divisions

Mitotic Cell Division	*Meiotic Cell Division*
1. Occurs in both haploid (n) and diploid (2n) cells.	1. Occurs in diploid (2n) cells, but not in haploid (n) cells.
2. Completed when one cell divides to form two cells.	2. Requires two successive cell divisions to produce four cells from the single parent cell.
3. Duplicated chromosomes do not align themselves in homologous pairs during division.	3. Duplicated chromosomes arrange themselves in homologous pairs during the first cell division.
4. The two daughter cells contain (a) the same genetic composition as the parent cell and (b) the same chromosome number as the parent cell.	4. The four daughter cells contain (a) different genetic compositions and (b) one-half the chromosome number of the parent cell.

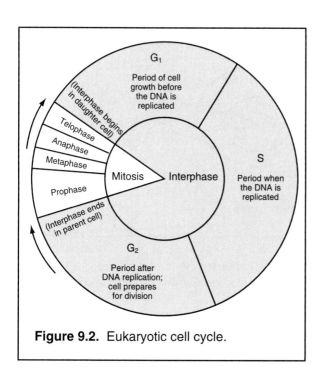

Figure 9.2. Eukaryotic cell cycle.

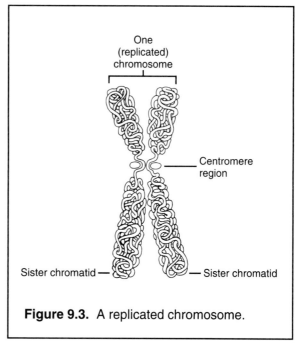

Figure 9.3. A replicated chromosome.

Prophase

During prophase, (1) the nuclear membrane and nucleolus disappear, (2) the chromosomes coil tightly to appear as rodshaped structures, (3) each pair of centrioles migrates to opposite ends of the cell, and (4) the **spindle** forms. Each pair of centrioles and its radiating **astral rays** constitute an **aster** at each end (pole) of the spindle.

Metaphase

This brief phase is characterized by the chromosomes lining up at the equator of the spindle. The sister chromatids of each replicated chromosome

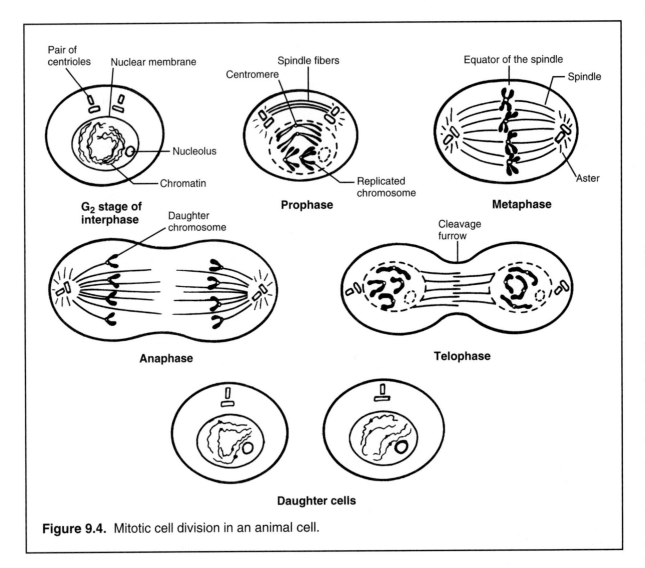

Figure 9.4. Mitotic cell division in an animal cell.

are attached to separate spindle fibers by their **centromeres**.

Anaphase

Anaphase begins with the separation of the centromeres of the sister chromatids, which migrate toward opposite poles of the spindle. Once the sister chromatids separate, they are called **daughter chromosomes**. Thus, a cell in anaphase contains two complete sets of chromosomes.

Telophase

In telophase, (1) a new nuclear membrane forms around each set of chromosomes, yielding two new nuclei, (2) a nucleolus reappears, and (3) the chromosomes start to uncoil. Also, cytokine-

sis usually occurs during telophase. A **cleavage furrow** forms to divide the parent cell into two **daughter cells**.

Mitotic Division in Plant Cells

Mitotic division in plants follows the same basic pattern that occurs in animals, with some notable exceptions. The cells of most plants do not have centrioles, although a spindle of fibers is present. Cytokinesis usually, but not always, occurs during telophase. The rigid cell wall prevents the formation of a cleavage furrow during cytokinesis; instead, a **cell plate** forms to separate the parent cell into two daughter cells, and a new cell wall forms along the cell plate. See Figure 9.5.

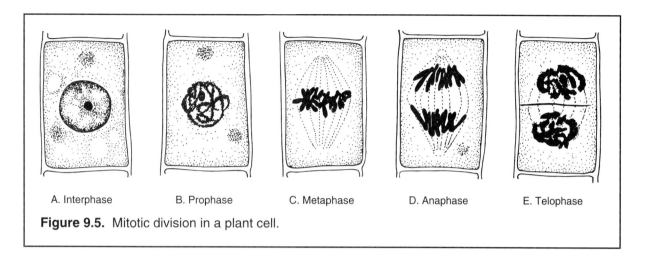

A. Interphase   B. Prophase   C. Metaphase   D. Anaphase   E. Telophase

Figure 9.5. Mitotic division in a plant cell.

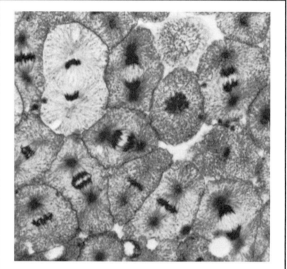

Figure 9.6. Various stages of mitotic division in whitefish blastula cells. (Courtesy of WARD'S Natural Science Establishment, Inc.)

Microscopic Study

The rapidly dividing cells of whitefish blastula, an early fish embryo, are excellent for studying mitotic division in animals. See Figure 9.6. On the prepared slide that you will use are several thin sections of a whitefish blastula, and each contains many cells in various stages of the cell cycle, including mitosis. A prepared slide of onion (*Allium*) root tip is used to study mitotic division in plants. Each slide usually contains

three longitudinal sections of onion root tip. The region of cell division is near the pointed tip.

Materials

Per student
Compound microscope

Per lab
Prepared slides of:
  whitefish blastula, x.s.
  onion root tip, l.s.

Assignment 1

Complete items 1, 2a, and 2b on Laboratory Report 9 that begins on page 339.

Assignment 2

1. Obtain a prepared slide of whitefish blastula. Locate a section for study with the 4× objective. Then switch to the 10× objective to find mitotic phases for observation with the 40× objective. **Locate cells in each phase of mitosis, and draw them in the space for item 2c on the laboratory report.** You may have to examine all the sections on your slide, or even additional slides, to observe each phase of mitosis.
2. Examine a prepared slide of onion root tip. Locate the cells in mitotic phases near the tip of the section. Observe cells in metaphase,

anaphase, and telophase. Note the cell plate. Note how mitotic division in an onion root tip differs from that observed in a whitefish blastula.

3. **Complete item 2 on the laboratory report.**

MEIOTIC CELL DIVISION

In contrast to mitotic cell division, meiotic cell division consists of *two* successive divisions but only *one* chromosome replication. This results in the formation of four cells that have only half the number of chromosomes of the diploid (2n) parent cell. Thus, the daughter cells have a haploid (n) number of chromosomes, since they each contain only *one member of each chromosome pair*. In addition to reducing the chromosome number in the daughter cells, meiosis also reshuffles the genes—hereditary units formed of small segments of DNA within a chromosome—and this greatly increases the genetic variability among the daughter cells.

In humans and most animals, cells formed by meiotic division become gametes (sperm or eggs). In plants, meiotic cell division results in the formation of meiospores that grow into haploid gametophytes. These gametophytes in turn produce gametes by mitotic division. In either case, the basic result of meiosis is the same: haploid cells with increased genetic variation.

Meiotic Phases in Animal Cells

Study Figure 9.7 as you read the following description of meiotic cell division in an animal cell. Chromosome and centriole replication occur in the S stage of interphase, prior to the start of meiosis.

Meiosis I

Prophase I exhibits the following characteristics. Each chromosome is composed of two sister chromatids joined together at the centromere. The replicated members of each chromosome pair join together in a side-by-side pairing called **synapsis**. Chromosomes in synapsis are often called **tetrads**, since they consist of four chromatids. An exchange of chromosome segments (crossover) frequently occurs between members of the tetrad, and increases the genetic variabil-

ity of the cells produced by meiotic division. The chromosomes coil tightly to appear as rodshaped structures, the nuclear membrane and nucleolus disappear, and a spindle forms.

Metaphase I is characterized by the synapsed chromosomes lining up at the equatorial plane, where they attach to spindle fibers by their centromeres.

Anaphase I begins with the separation of the members of each chromosome pair. The centromeres do *not* separate, and therefore each chromosome still consists of two chromatids joined at their centromeres. Members of each chromosome pair migrate to opposite poles of the spindle in the replicated state.

Telophase I proceeds to form a nuclear membrane around each set of chromosomes. The chromosomes untwist and the nucleolus reappears. Cytokinesis separates the parent cell into two daughter cells. Keep in mind that the nucleus of each daughter cell contains only *one member of each chromosome pair* in a replicated state. Thus, each daughter cell is haploid (n).

Meiosis II

Both cells formed by meiosis I divide again in meiosis II, but for discussion purposes we will follow only one of these cells in the second division. In interphase between meiosis I and II, the centrioles replicate but chromosomes do *not* replicate again. Recall that the chromosomes are already replicated.

Prophase II is characterized by the usual loss of the nuclear membrane and nucleolus, spindle formation, and the appearance of rodshaped chromosomes.

Metaphase II is characterized by the chromosomes lining up at the equator of the spindle. Each chromosome consists of two sister chromatids joined together at the centromere, which is attached to a spindle fiber.

Anaphase II begins with the separation of the centromeres. The sister chromatids, now called daughter chromosomes, move toward opposite poles of the spindle.

Telophase II proceeds as usual to form the new nuclei, and cytokinesis divides the cell to form two haploid (n) daughter cells.

Since each cell entering meiosis II forms two daughter cells, a total of four haploid (n) cells are produced from the original diploid (2n)

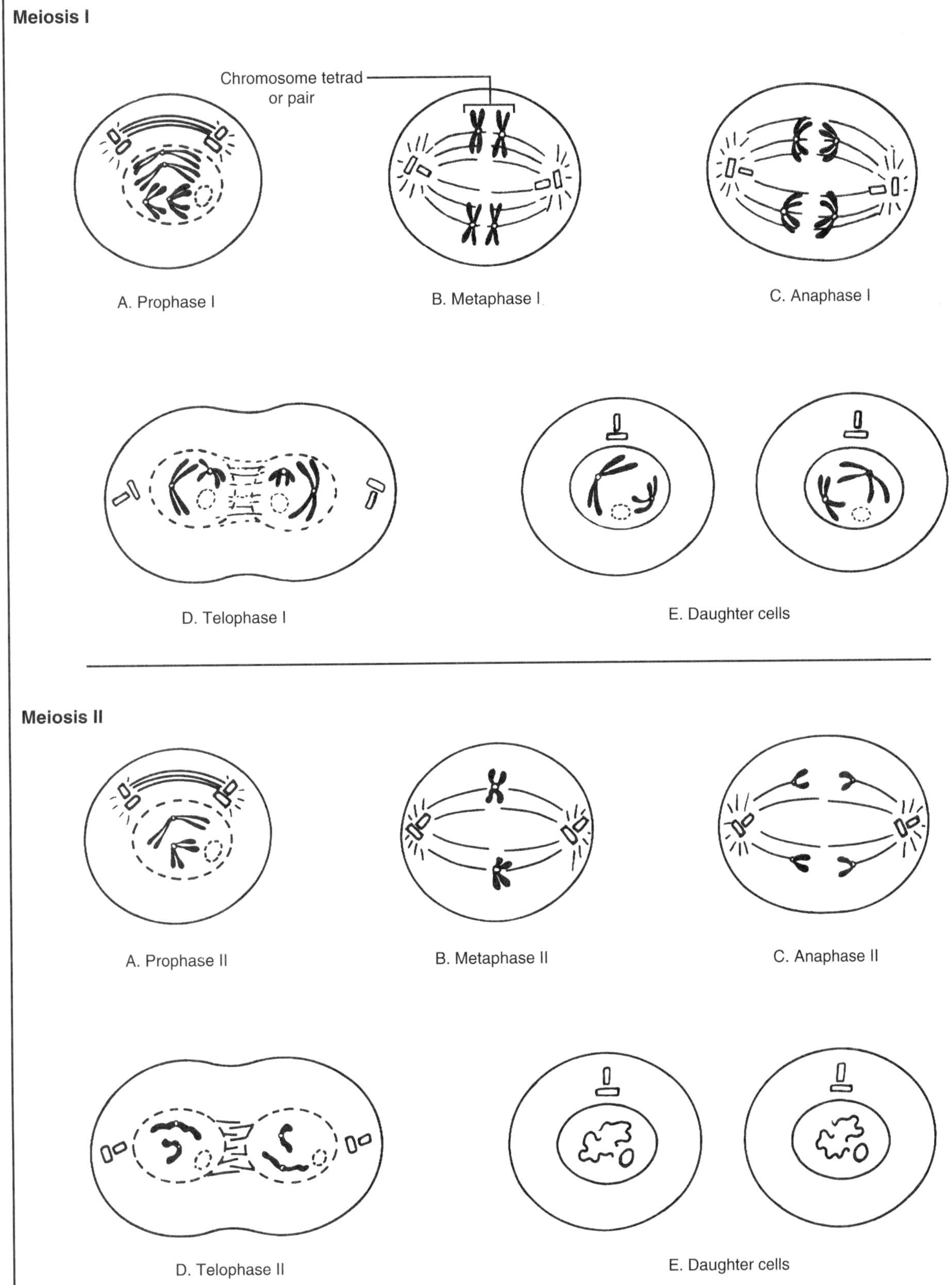

Meiosis I

Chromosome tetrad or pair

A. Prophase I

B. Metaphase I

C. Anaphase I

D. Telophase I

E. Daughter cells

Meiosis II

A. Prophase II

B. Metaphase II

C. Anaphase II

D. Telophase II

E. Daughter cells

Figure 9.7. Meiotic cell division. For simplicity, only one of the cells formed in meiosis I is shown in meiosis II.

parent cell entering meiosis I. Thus, meiotic cell division may be summarized as:

$$1 \text{ cell } (2n) \xrightarrow{\text{M I}} 2 \text{ cells } (n) \xrightarrow{\text{M II}} 4 \text{ cells } (n)$$

Materials

Per student group
Chromosome simulation kits or colored pipe cleaners

Assignment 3

1. Study Figure 9.7.
2. Using colored pipe cleaners to represent chromosomes or a chromosome simulation kit, simulate the replication and distribution of chromosomes in both mitosis and meiosis where 2n = 4.
2. **Complete item 3 on the laboratory report.**

Assignment 4

Your instructor has set up several microscopes as a mini-practicum so that you can check your understanding of mitotic cell division and cell structures. Your task is to identify the stage of mitotic division and the cell structures identified by the microscope pointers. **Complete item 4 on the laboratory report.**

Part 2

Inheritance

10

Heredity

After completion of the laboratory session, you should be able to:

1. Explain Mendel's principle of segregation and principle of independent assortment and give examples of each.
2. Solve simple genetic problems involving dominance, recessiveness, codominance, and sex linkage.
3. Determine gametes from genotypes where genes are linked or nonlinked.
4. Perform a chi-square analysis.
5. Define all terms in bold print.

A human baby begins with the fusion of a haploid (n) egg and a haploid (n) sperm to form a diploid (2n) zygote (fertilized egg). The zygote contains one haploid set of chromosomes from the mother and one haploid set from the father. At the moment of egg and sperm fusion, the baby's **inherited characteristics** (traits) are determined. All that is left is for it to grow and develop in accordance with the inherited instructions received from its mom and dad. This is also the case in all sexually reproducing organisms: Inherited traits are determined at the moment of zygote formation. Subsequent growth and development enable the expression of those traits.

The genetic information that determines hereditary traits is found in the structure of the **DNA molecules** in the **chromosomes**. A short segment of DNA that codes for a particular protein constitutes a **gene**, a hereditary unit. In diploid organisms both genes and chromosomes occur in homologous pairs. See Figure 10.1.

In the simplest situation, an inherited trait, such as flower color, is determined by a single pair of genes. The members of a **gene pair** may be identical (e.g., each member may code for purple flowers) or may code for a different variation of the trait (e.g., one member of the gene pair codes for purple flowers, and the other codes for white flowers). Again, in the simplest case, only two forms of a gene exist. Alternate forms of a gene are called **alleles**.

Biologists use symbols (usually letters like "P" or "p") to represent alleles when solving genetic problems. When both members of a gene pair consist of the same allele, such as PP or pp, the individual is **homozygous** for the expressed trait. When the members of the gene pair consist of unlike alleles, such as Pp, the individual is **heterozygous** (hybrid) for the expressed trait.

The genetic composition of the gene pair (e.g., PP, Pp, or pp) is known as the **genotype** of the

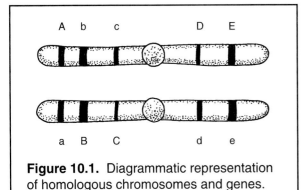

Figure 10.1. Diagrammatic representation of homologous chromosomes and genes.

different chromosome pairs (i.e., the genes are not linked).

DOMINANT-RECESSIVE TRAITS

When a gene pair consists of two alleles, and one of these alleles is expressed and the other is not, the expressed allele is said to be **dominant**. The unexpressed allele is **recessive**. Many traits are inherited in this manner. For example, the following traits in garden peas exhibit a dominant/recessive pattern of inheritance. In each case, the dominant trait is in italics, and the dominant allele is capitalized in the genotype.

Flower color: *Purple flowers* (PP, Pp) or white flowers (pp)
Plant height: *Tall plants* (TT, Tt) or dwarf plants (tt)

Table 10.1 shows the genotypes and phenotypes that are possible for purple or white flowers in peas. Note that the dominant allele is assigned an uppercase P, while the recessive allele is represented by a lowercase p. Only one dominant allele is required for the expression of purple flowers. In contrast, both recessive alleles must be present for white flowers to be expressed in the phenotype. *This relationship is true for all dominant and recessive alleles.*

Solving Genetic Problems

Consider this genetic problem: What are the expected genotype and phenotype ratios (probabilities) in the progeny of a cross between purple-flowering and white-flowering pea plants when each parent is homozygous? Steps used to

individual. The observable (expressed) form of a trait (e.g., purple flowers or white flowers) is called the **phenotype** of the individual.

An understanding of inheritance patterns enables the prediction of an **expected ratio** for the occurrence of a trait in the progeny (offspring) of parents of known genotypes.

MENDEL'S PRINCIPLES

Gregor Mendel, an Austrian monk, worked out the basic patterns of simple inheritance in 1860, long before chromosomes or genes were associated with inheritance. Mendel's work correctly identified the existence of the units of inheritance that are today known as genes.

Mendel proposed two principles concerning the activity of genes, and these principles form the basis for the study of inheritance. Look for evidence of these principles as you work through this exercise. In modern terms, these principles may be stated as follows:

1. The **principle of segregation** states that (1) genes occur in pairs and exist unchanged in the heterozygous state, and (2) members of a gene pair are segregated (separated) from each other during gamete formation by meiosis, ending up in separate gametes.
2. The **principle of independent assortment** states that genes for one trait are assorted (segregated into the gametes) independently from genes for other traits. This principle applies *only* to traits whose genes are located on

TABLE 10.1
Genotypes and Phenotypes for Flower Color in Garden Peas

Genotype	Phenotype
PP	Purple
Pp	Purple
pp	White

TABLE 10.2
Steps Used to Solve Standard Genetics Problem

1. Be sure that you understand what you are to solve. Write down what is known.
2. Write out the cross, using genotypes of the parents.
3. Determine the possible gametes that may be formed.
4. Use a Punnett square to establish the genotypes of all possible progeny.
5. Determine the genotype ratio of the progeny. Count the number of identical genotypes and express them as a ratio of the total genotypes (e.g., $\frac{1}{4}$ PP: $\frac{2}{4}$ Pp: $\frac{1}{4}$ pp).
6. Use the information obtained in step 1 to determine the phenotype ratio from the genotypes (e.g., $\frac{3}{4}$ purple flowers to $\frac{1}{4}$ white flowers).

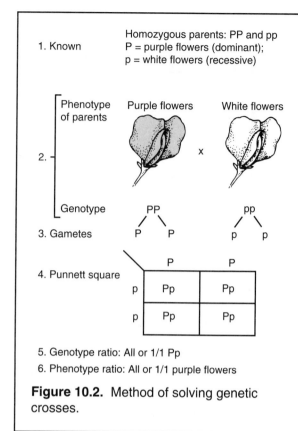

Figure 10.2. Method of solving genetic crosses.

solve genetic problems such as this are listed in Table 10.2. Note how the steps are used in Figure 10.2 to solve this problem.

The determination of the gametes is a critical step. Recall that meiosis separates homologous chromosomes (and genes) into different gametes. Thus, the members of the gene pair are separated into different gametes. In Figure 10.2, the gametes formed by each parent are identical, since each parent is homozygous.

In setting up the Punnett square shown in Figure 10.2, the gametes of one parent are placed on the vertical axis, and the gametes of the other parent are placed on the horizontal axis. The number of squares composing a Punnett square depends on the number of different classes of gametes formed by the parents. Four squares are used in Figure 10.2 to enable you to understand the setup, although only one square is actually needed since each parent forms only one class of gametes.

All possible combinations of gametes are simulated by recording the gametes on the vertical axis in each square that lies to the right of those gametes, and by recording those on the horizontal axis in each square that lies below them. Note that uppercase letters (dominant alleles) always compose the first letter in each gene pair.

When the Punnett square is complete, the individual squares contain the expected genotypes of progeny in the F_1 (first filial) generation. The genotype and phenotype ratios may then be determined. Study Figure 10.2.

Test Cross

It is usually not possible to distinguish between homozygous and heterozygous phenotypes exhibiting a dominant trait, but they may be determined by a **test cross**. In a test cross, the individual exhibiting the dominant phenotype is crossed with an individual exhibiting the recessive phenotype. Recall that an individual exhibiting a recessive phenotype is *always* homozygous for that trait. If all progeny exhibit the dominant trait, the parent with the dominant phenotype is homozygous. If half the progeny exhibit the dominant trait and half exhibit the recessive trait, the parent with the dominant phenotype is heterozygous.

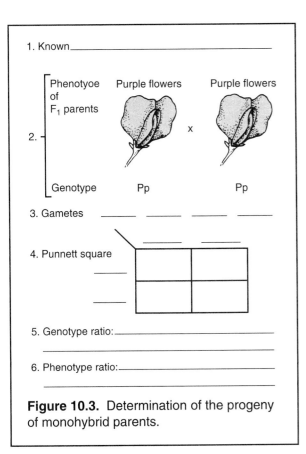

1. Known_____

Phenotyoe
of
F_1 parents   Purple flowers   Purple flowers

2.

Genotype   Pp   Pp

3. Gametes _____ _____   _____ _____

4. Punnett square

5. Genotype ratio:_____

6. Phenotype ratio:_____

Figure 10.3. Determination of the progeny of monohybrid parents.

fractions of the total number of genotypes. The different types of genotypes are then expressed as a proportion to establish the expected ratios.

$$\tfrac{1}{4}\,\text{PP} : \tfrac{2}{4}\,\text{Pp} : \tfrac{1}{4}\,\text{pp}$$
or
$$1\,\text{PP} : 2\,\text{Pp} : 1\,\text{pp}$$

The phenotype ratio may then be determined from the genotype ratio. Remember that the presence of a single dominant allele in a genotype produces a dominant phenotype.

$\tfrac{3}{4}$ purple-flowering plants : $\tfrac{1}{4}$ white-flowering plants

or

3 purple-flowering plants : 1 white-flowering plant

These ratios are always obtained in a monohybrid cross where the gene consists of only two alleles and one allele is dominant.

3. **Complete items 2a–2d on the laboratory report.**
4. Now that you know how to predict the genotype and phenotype ratios of progeny when the genotypes of the parents are known, examine the tray of corn seedlings. These are progeny of a monohybrid cross. Count the number of tall and dwarf plants and calculate the ratio of tall to dwarf plants. **Complete items 2e–2g on the laboratory report.**
5. Now use the knowledge you have gained to determine the phenotype and genotype of parents when the progeny are known. **Complete item 2h on the laboratory report.**
6. Examine Table 10.3, which shows several human traits that are inherited in a dominant/recessive manner. Using this information, **complete items 2i–2n on the laboratory report.**

INCOMPLETE DOMINANCE AND CODOMINANCE

In some cases, alternate alleles are always expressed, so dominant and recessive alleles are not involved. Incomplete dominance and codominance are examples of such inheritance patterns.

Incomplete dominance is characterized by expression of alternate alleles in the heterozy-

Assignment 1

Complete item 1 on Laboratory Report 10 that begins on page 343.

Materials

Per lab
Trays of tall-dwarf corn seedlings from monohybrid crosses

Assignment 2

1. Using Figure 10.2 as a guide, determine the expected progeny in the F_2 generation by crossing two members of the F_1 generation, both of which are monohybrids (Pp). A monohybrid is heterozygous for one trait. Complete the Punnett square in Figure 10.3.
2. Once you have completed the Punnett square, determine the genotypes by counting the identical genotypes and recording them as

TABLE 10.3
Dominant and Recessive Phenotypes for a Few Human Traits

Trait	Dominant Phenotype	Recessive Phenotype
Ear lobes	Free	Attached
Pigment distibution	Freckles	No Freckles
Hairline	Widow's peak	Straight
Little finger	Bent	Straight
Tongue roller	Yes	No

gote, and the heterozygote's phenotype appears as a blended intermediate between the phenotypes of the two homozygous parents. This type of inheritance occurs in flower color in snapdragons. A homozygous red-flowering snapdragon (RR) crossed with a homozygous white-flowering snapdragon (rr) yields all pink-flowering heterozygous snapdragons (Rr). In spite of the blended appearance of the heterozygous phenotype, the alleles are not altered and segregate unchanged in the next generation.

Codominance is characterized by the expression of both alleles in the heterozygote, but no apparent blending occurs in the heterozygote's phenotype. Sickle-cell anemia, a disorder affecting some Black Americans, is inherited in this manner. Persons heterozygous for sickle-cell ($Hb^A Hb^S$) produce both normal and abnormal hemoglobin in their red blood cells. They rarely experience illness since sufficient normal hemoglobin is present to carry oxygen. However, persons homozygous for sickle-cell ($Hb^S Hb^S$) exhibit sickle-cell anemia and die prematurely without medical intervention.

Assignment 3

Complete item 3 on the laboratory report.

TABLE 10.4
Phenotypes and Genotypes of the ABO Blood Types

Blood Type	Genotype
O	ii
A	$I^A I^A$ or $I^A i$
B	$I^B I^B$ or $I^B i$
AB	$I^A I^B$

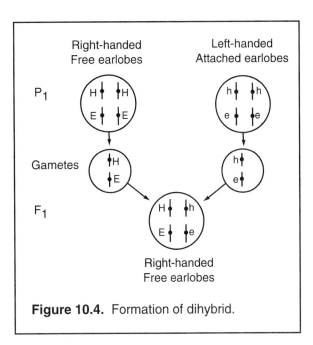

Figure 10.4. Formation of dihybrid.

MULTIPLE ALLELES

Some traits are controlled by genes with more than two alleles. The inheritance of ABO blood groups in humans is an example of this. Three alleles are involved: I^A codes for type A blood, I^B codes for type B blood, and i codes for type O blood. Table 10.4 shows the relationship between blood genotypes and phenotypes. Note that alleles I^A and I^B are both dominant over i, but that they are codominant with each other.

Assignment 4

Complete item 4 on the laboratory report.

DIHYBRID CROSS

Let's see how to predict progeny ratios when considering two traits at the same time. In humans, handedness and earlobe type are controlled by genes located on separate (nonhomologous) chromosomes. Right-handedness (H) is dominant over left-handedness (h), and free earlobes (E) are dominant over attached earlobes (e). Therefore, the genotypes of persons homozygous for these dominant and recessive traits are:

Phenotype	Genotype
Right-handed-Free earlobes	HHEE
Left-handed-Attached earlobes	hhee

What is the predicted progeny ratio for these traits in children of parents with these genotypes?

Mom	Dad
HHEE	× hhee

The alleles for handedness are located on one chromosome pair, and the alleles for earlobes are located on a different chromosome pair. Recall that the members of each gene pair are separated into different gametes when the chromosomes are segregated by meiotic division, and that therefore the only possible genotypes of gametes produced by the parents are:

Parents' genotypes:	HHEE	hhee
Gametes' genotypes:	HE	he

Figure 10.4 shows the relationships between homologous chromosomes and genes in the parents, their gametes, and the offspring. A Punnett square may be used to predict the ratio in the offspring, as shown below. Note that when combining genotypes of gametes in the square, the alleles of homologous genes are grouped together in the resulting genotype. The alleles for handedness have arbitrarily been placed first.

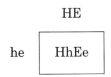

Therefore, all children of these parents will have a genotype of HhEe for these traits and a phenotype of right-handed and free earlobes. Each child is heterozygous for each trait, and

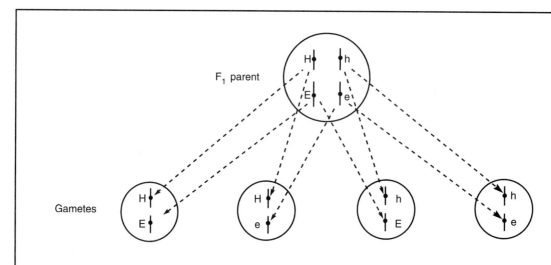

Figure 10.5. Gamete formation in a dihybrid with nonlinked genes.

TABLE 10.5

Gamete Determination in a Dihybrid with Non-linked Genes

Parent Genotype	First Gene Pair	Second Gene Pair		Possible Gametes
HhEe		E	=	HE
	H			
		e	=	He
		E	=	hE
	h			
		e	=	he

since we are considering two traits, each child is a **dihybrid**. (Actually, all people are hybrid for many traits.)

Now consider a dihybrid cross. What is the predicted ratio for these traits in children of parents who are both dihybrid (HhEe)? This can be solved with a Punnett square. The most difficult part of this problem is determining the genotypes of the gametes, but it is rather simple if you keep in mind Mendel's Principle of Independent Assortment.

Figure 10.5 shows the distribution of homologous genes and chromosomes in gamete formation. Table 10.5 shows an easier way to determine the genotypes of the gametes in such crosses. Once you set up the table with the first three columns filled in, multiply algebraically each of the first pair of alleles by both of the second pair of alleles as shown, to yield the four types of genotypes found in the gametes.

Assignment 5

1. **Work out this cross in item 5a on the laboratory report.** In writing the genotypes of the offspring in the squares, group the homologous genes together and place the alleles for handedness first.
2. The predicted phenotype ratio of progeny from dihybrid crosses is always 9:3:3:1 when each trait is determined by two alleles and a dominant/recessive mode of inheritance, and when genes for the two traits are located on different chromosome pairs. **Complete item 5 on the laboratory report.**

LINKED GENES

Each chromosome contains many genes that are linked together in a definite sequence. When members of a chromosome pair are separated in gamete formation, the genes of each chromosome tend to remain linked together as a unit. See Figure 10.6.

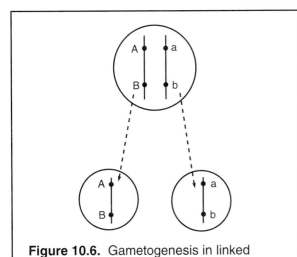

Figure 10.6. Gametogenesis in linked genes.

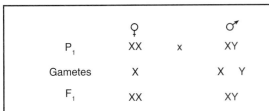

Figure 10.7. Sex inheritance in humans.

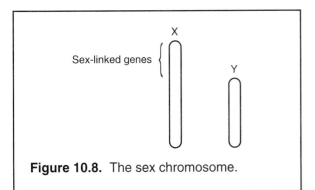

Figure 10.8. The sex chromosome.

Sex Linked Traits

In humans, sex is determined by a single pair of sex chromosomes. Females possess two X chromosomes (XX), and males possess an X and a Y (XY). Sex in humans is inherited as shown in Figure 10.7.

The Y chromosome is shorter than the X chromosome and lacks some of the genes present on the X chromosome. Those genes that are present on the X chromosome but absent on the Y chromosome are the **sex-linked** or **x-linked genes** that control inheritance of sex-linked traits. See Figure 10.8.

For the sex-linked genes, a female is diploid and a male is haploid. Therefore, a recessive allele on the X chromosome of a male will be expressed, whereas this same recessive allele must be present on both X chromosomes of a female to be expressed. Common recessive sex-linked traits in humans are red-green color blindness and hemophilia.

Assignment 6

Complete item 6 on the laboratory report.

Pedigree Analysis

Now that you understand the fundamentals of simple inheritance patterns, it is possible to trace a trait in a pedigree (family tree) to determine whether it is inherited in a simple dominant/recessive pattern, a sex-linked pattern of inheritance, or some other pattern.

Assignment 7

Complete item 7 on the laboratory report.

POLYGENIC INHERITANCE

Traits inherited as dominant or recessive traits are known as **qualitative traits**. For example, people are either left-handed or right-handed, and they either have freckles or not. General observations suggest that some traits are not inherited in this manner, that is, some are **quantitative** in nature. For example, people are not either short or tall, but show a gradation of heights typical of a normal (bell-shaped) curve. Such traits exhibit **polygenic inheritance**, in which (1) several genes control the same trait and (2) codominance is evident among the alleles of these genes.

TABLE 10.6
Chi-Square Determination

Phenotype	Actual Results	Expected Results	Deviation (d)	(d^2)	(d^2/e)
Purple flowers	78	75	3	9	9/75 = 0.12
White flowers	22	25	3	9	9/25 = 0.36
					$\Sigma(d^2/e)$ = 0.48
					χ^2 = 0.48

Assignment 8

1. Refer to item 3f on Laboratory Report 1, where you graphed the frequency of little-finger lengths for your class.
2. **Complete item 8 on the laboratory report.**

CHI-SQUARE ANALYSIS

To this point in the exercise, you have learned how to predict the expected genotype and phenotype ratios of progeny. However, biologists must verify the expected ratio of a cross to establish the pattern of inheritance. This is done by using the **chi-square** (χ^2) test. This statistical test indicates the probability (p) that differences between the expected ratio and the actual ratio are due to chance alone, or whether a different hypothesis (expected ratio) is more appropriate to explain the results (observed ratio). The formula for the chi-square test is $\chi^2 = \Sigma(d^2/e)$, where

χ^2 = chi square
Σ = sum of
d = deviation (difference) between expected and observed results
e = expected results

Consider a monohybrid cross involving flower color in garden peas. The predicted phenotype ratio is 3 purple-flowering plants to 1 white-flowering plant. Thus, if 100 plants were produced from the cross, 75 should have purple flowers and 25 should have white flowers. Table 10.6 shows the results of such a cross and the calculation of the chi square value.

Comparing the calculated value of chi-square (χ^2) with the values in Table 10.7 is necessary to

TABLE 10.7
Chi-Square Values

	Probability (p)						
	Deviation Insignificant: Hypothesis Supported					Deviation Significant: Hypothesis Not Supported	
$C - 1$	.99	.80	.50	.20	.10	.05	.01
1	.0016	.064	.455	1.642	2.706	3.841	6.635
2	.0201	.446	1.386	3.219	4.605	5.991	9.210
3	.115	1.005	2.366	4.642	6.251	7.815	11.341
4	.297	1.649	3.357	5.989	7.779	9.488	13.277

determine the probability (p) that the deviation from the expected ratio is either (1) by chance, and verifies the predicted ratio, or (2) greater than would occur by chance, and does not support the predicted ratio. Note that the chi-square values are arranged in columns headed by probability values and in horizontal rows by phenotype classes minus one ($C - 1$).

The two classes of progeny in the example are purple flowers and white flowers. Since 2 classes $- 1 = 1$, you must look for the calculated chi-square value in the first horizontal row of values. A χ^2 value of 0.48 falls between the columns of 0.50 and 0.20 probability. This means that by random chance, the deviation between the expected and actual results will occur between 20% and 50% of the time. Thus, the predicted ratio for the progeny is supported. Probabilities greater than 5% ($p \geq 0.05$) are generally accepted as supporting the hypothesis (expected ratio), while those of 5% or less indicate that the results could not be due to chance.

Materials

Per lab
Corn ears with purple and white kernels from a monohybrid cross

Assignment 9

1. **Complete item 9a on the laboratory report.**
2. Examine a corn ear with both purple and white kernels that have resulted from a monohybrid cross. The predicted ratio of kernels is 3 purple to 1 white. Count the purple and white kernels to determine the actual ratio. Mark the row of kernels where you start counting with a pin stuck into the cob under the first kernel. Then **do a chi-square analysis of the results in item 8b on the laboratory report.**
3. **Complete the laboratory report.**

11

Molecular and Chromosomal Genetics

OBJECTIVES

After completion of the laboratory session, you should be able to:

1. Describe the basic structure of DNA and RNA.
2. Describe the process of information transfer in (a) DNA replication, (b) RNA synthesis, and (c) protein synthesis.
3. Explain how mutations involving base substitution, addition, or deletion affect protein synthesis.
4. Prepare a karyotype from a metaphase smear of human chromosomes.
5. Describe the basis of the chromosomal abnormalities studied.
6. Define all terms in bold print.

Chromosomes are responsible for transmitting the hereditary material from cell to cell in cell division and from an organism to its progeny in reproduction. This is why the distribution of replicated chromosomes in mitotic and meiotic cell divisions is so important in eukaryotic cells. The genetic information carried within a cell is contained in the structure of **deoxyribonucleic acid (DNA)**, which forms the hereditary portion of the chromosomes.

DNA AND THE GENETIC CODE

DNA is a long, thin molecule consisting of two strands twisted in a spiral arrangement to form a double helix, somewhat like a twisted ladder. See Figure 11.1. The sides of the ladder are formed of sugar and phosphate molecules, and the rungs are formed by nitrogeneous bases joined to one another by hydrogen bonds.

Each strand of DNA consists of a series of **nucleotides** that are joined together to form a polymer of nucleotides. Each nucleotide of DNA is formed of three parts: (1) a deoxyribose (C_5) sugar, (2) a phosphate group, and (3) a nitrogenous base. Four kinds of nitrogenous bases are present in DNA. The purine bases which have a (double-ring structure) are **adenine** (A) and **guanine** (G). The pyrimidine bases (single-ring structure) are **thymine** (T) and **cytosine** (C). Note the **complementary pairing** of the bases in Figure 11.1. Can you discover a pattern to their pairing? It is the sequence of nucleotides, with their respective purine or pyrimidine bases, that contains the genetic information within the DNA molecule.

DNA Replication

Each DNA molecule can replicate itself during interphase of the cell cycle and can thereby

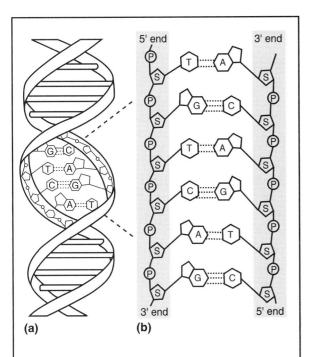

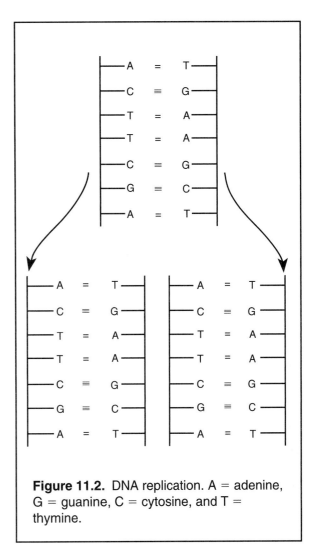

Figure 11.1. DNA structure. (a) The double helix of a DNA molecule. Complementary pairing of the nitrogenous bases joins the sides like rungs of a twisted ladder. A = adenine, G = guanine, C = cytosine, and T = thymine. (b) If a DNA molecule is untwisted, it would resemble a ladder in shape. The sides of the ladder are formed of deoxyribose sugar and phosphate, and the rungs consist of nitrogenous bases.

Figure 11.2. DNA replication. A = adenine, G = guanine, C = cytosine, and T = thymine.

maintain the constancy of the genetic information of the parent cell in new cells that are formed. **Replication** begins with the breaking of the weak hydrogen bonds that join the nitrogen bases of the nucleotides. This results in the separation of the DNA molecule into two strands of nucleotides. See Figure 11.2. Each strand then serves as a **template** for the synthesis of a complementary strand of nucleotides that is formed from nucleotides available within the cell nucleus. The complementary pairing of nitrogen bases determines the sequence of nucleotides in the new strands, and results in the formation of two DNA molecules that are identical. Since each new DNA molecule contains one "old" strand and one "new" strand, replication is said to be **semiconservative**. Occasionally, errors are made during replication, and such er-

rors are a type of **mutation**. Replication is controlled by a series of enzymes that catalyze the process.

Materials

Per student group
Colored pencils
DNA, RNA, and protein synthesis kit

Assignment 1

1. Color-code the nitrogenous bases in Figure 11.1 and circle one nucleotide.
2. **Complete items 1a–1c on Laboratory Report 11 that begins on page 349.**

3. Use a DNA kit to construct a segment of a DNA molecule that matches the base sequence of the DNA segment in item 1c on the laboratory report.
4. **Complete item 1d on the laboratory report.**
5. Use a DNA kit to construct a segment of a DNA molecule that matches the "old" non-replicated DNA segment in item 1d on the laboratory report. Then separate the strands and construct the replicated strands as shown in item 1d.

RNA Synthesis

DNA serves as the template for the synthesis of **ribonucleic acid (RNA)**. RNA differs from DNA in three important ways: (1) it consists of only a single strand of nucleotides, (2) its nucleotides contain ribose sugar instead of deoxyribose sugar, and (3) **uracil** (U) is substituted for thymine as one of the four nitrogenous bases.

To synthesize RNA, a segment of a DNA molecule untwists and the hydrogen bonds between the nucleotides in the two strands of the DNA are broken. The nucleotides of one strand pair up with complementary RNA nucleotides in the nucleus. When the RNA nucleotides are joined by sugar-phosphate bonds, the RNA strand is complete, and it separates from the DNA strand. Few or many RNA molecules may be formed before the DNA strands reunite.

Assignment 2

1. **Complete item 2 on the laboratory report.**
2. Use a DNA-RNA kit to synthesize an RNA molecule with a base sequence identical to the hypothetical RNA molecule in item 2b on the laboratory report.

Protein Synthesis

The genetic information of DNA functions by determining the kinds of protein molecules that are synthesized in the cell. A sequence of three bases—a base triplet—in a DNA molecule has been shown to code indirectly for an amino acid. By controlling the sequence of amino acids, DNA determines the kind of protein produced. Recall that enzymes are proteins and that the chemical reactions in a cell are controlled by its enzymes. Thus, DNA indirectly controls cellular functions by controlling enzyme production.

Each of the three types of RNA molecules plays an important role in protein synthesis. **Messenger RNA (mRNA)** is a complement of the genetic information of DNA. A **transcription** of the genetic information in DNA is made when mRNA is synthesized. A base triplet of mRNA is called a **codon**, and the codons for the 20 amino acids composing proteins have been determined. The genetic information is carried by mRNA as it passes from the nucleus to the **ribosomes** which are sites of protein synthesis in the cytoplasm.

Transfer RNA (tRNA) carries amino acids to the ribosomes. At one end of a tRNA molecule are three nitrogenous bases, an **anticodon**, which is complementary to a codon of bases contained in the mRNA. At the other end of the tRNA molecule is an attachment site for 1 of the 20 types of amino acids.

The codon of mRNA and anticodon of tRNA briefly join, to place a specific amino acid in its position in a polypeptide chain. This interaction takes place on the surface of a ribosome containing the necessary enzymes for the reaction. **Ribosomal RNA (rRNA)**, the third type of RNA, is an integral component of ribosomes, and plays an important role in decoding the mRNA message to enable protein synthesis. The formation of an amino acid chain represents the **translation** of the genetic information contained in the DNA of a cell.

Figure 11.3 depicts the interaction of mRNA, tRNA, and rRNA in the formation of a polypeptide. Note how the sequence of the amino acids is controlled by the pairing of the codons and anticodons. The process may be simplified as follows:

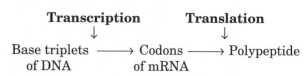

Transcription		**Translation**
↓		↓
Base triplets ⟶	Codons ⟶	Polypeptide
of DNA	of mRNA	

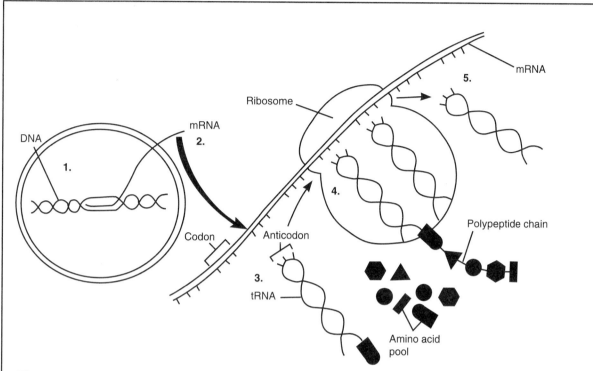

Figure 11.3. A summary of protein synthesis. 1. Chromosomal DNA is the template for the synthesis of mRNA. 2. mRNA exits the nucleus and moves to a ribosome carrying the genetic code in its codons. 3. Each tRNA transports a specific amino acid. 4. Temporary pairing of mRNA codons and tRNA anticodons at a ribosome determines the position of the trasported amino acid in the forming polypeptide chain. 5. After adding its amino acid and transfering the polypeptide chain to the next tRNA, the tRNA molecule departs the ribosome.

The Genetic Code

In protein synthesis, the genetic information inherent in the sequence of base triplets in DNA is transcribed into the sequence of codons in mRNA, which in turn are translated into the sequence of amino acids in a polypeptide chain. In this way, DNA determines both the kinds of amino acids and their sequences in the different proteins that are synthesized by a cell.

There are 64 possible combinations of nucleotide bases in mRNA codons. Their translation is shown in Table 11.1. Note that AUG specifies the amino acid methionine and also is the start signal for protein synthesis. Three codons, UAA, UAG, and UGA, do not specify an amino acid, but instead signal the ribosomes to stop assembling the polypeptide chain. Most amino acids are specified by more than one codon, but no codon specifies more than one amino acid. Thus, the code is **redundant** but not **ambiguous**.

Mutations

The substitution of one base pair for another, the deletion of a base pair, or the addition of a base pair constitutes a mutation in a gene. The effect of such mutations is variable, depending on how the mutation is translated via the genetic code.

For example, if a base substitution mutation resulted in a codon change from GCU to GCC, there would be no effect, since both of these triplet codons specify the amino acid alanine. But if the change was from GCU to GUU, valine would be substituted for alanine in the polypeptide chain, which might have a marked effect on

TABLE 11.1
The Codons of mRNA and the Amino Acids That They Specify

AAU AAC	Asparagine	CAU CAC	Histidine	GAU GAC	Aspartic acid	UAU UAC	Tyrosine
AAA AAG	Lysine	CAA CAG	Glutamine	GAA GAG	Glutamic acid	UAA UAG	(Stop)*
ACU ACC ACA ACG	Threonine	CCU CCC CCA CCG	Proline	GCU GCC GCA GCG	Alanine	UCU UCC UCA UCG	Serine
AGU AGC	Serine	CGU CGC CGA CGG	Arginine	GGU GGC GGA GGG	Glycine	UGU UGC	Cysteine
AGA AGG	Arginine					UGA UGG	(Stop)* Tryptophan
AUU AUC AUA	Isoleucine	CUU CUC CUA CUG	Leucine	GUU GUC GUA GUG	Valine	UUU UUC	Phenylalanine
AUG	Methionine and start					UUA UUG	Leucine

*Signals the termination of the polypeptide chain.

the protein. Similarly, if UAU mutated to UAA, it would terminate the polypeptide chain at that point instead of adding tyrosine to the chain. This likely would form a nonfunctional protein.

The mutation involving the addition or deletion of one or two base pairs will cause a **frameshift** in the reading of the codons that may either terminate the polypeptide chain or insert different amino acids into the chain. Usually, addition or deletion mutations have a more disastrous effect than substitution mutations.

Assignment 3

1. Add the bases of the DNA template and the anticodons of tRNA in Figure 11.4.
2. **Complete items 3a–3c on the laboratory report.**

3. Use a DNA-RNA-protein synthesis kit to synthesize an amino acid sequence as shown in item 3c on the laboratory report, starting with the DNA template.
4. **Complete item 3 on the laboratory report.**
5. Use a DNA–RNA-protein synthesis kit to synthetize the amino acid sequences determined in items 3e and 3f on the laboratory report.

HUMAN CHROMOSOMAL DISORDERS

In the study of cell division, you learned that (1) chromosomes occur in pairs in diploid cells, (2) chromosomes are faithfully replicated and equally distributed in mitotic cell division, and (3) cells formed by meiotic cell division receive only one member of each chromosome pair. In

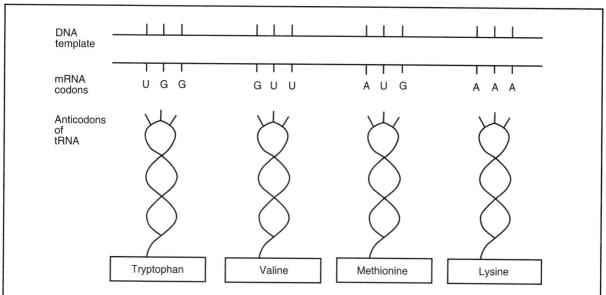

Figure 11.4. Interaction of DNA, mRNA, and tRNA in protein synthesis. Complete the figure by adding the bases in the anticodons of tRNA and the base triplets of DNA.

both types of cell division, the distribution of the chromosomes is systematically controlled, but errors sometimes occur. In this section of the exercise, you will consider human chromosomal abnormalities, in which whole chromosomes or large parts thereof are missing or added. Since you now understand the role of DNA, you can appreciate the effect that the deletion or addition of large amounts of DNA would have on normal cellular function.

The 46 chromosomes of human body cells are classified as 22 pairs of **autosomes**, nonsex chromosomes, and 1 pair of **sex chromosomes**, XX in females and XY in males. Normally, the separation of chromosomes in the meiosis of gametogenesis places 22 autosomes and 1 sex chromosome in each gamete, but occasionally errors place both members of a chromosome pair in the same gamete. As a result, another gamete lacks this chromosome entirely. If either of these gametes participates in fertilization, the resulting zygote will have an abnormal chromosome number and will be minimally or severely affected, depending on the chromosome involved. Severe defects or death usually occur. For example, if

the zygote contains an extra copy of chromosome 21, the presence of three 21 chromosomes (trisomy 21) results in Down's syndrome. See Table 11.2. In contrast, the loss of a chromosome 21, like the loss of any autosome, is lethal.

Chromosomal abnormalities also may stem from the **translocation** of a portion of one chromosome to a member of a different chromosome pair, resulting in reduced or extra chromosomal material in a gamete.

Cytogeneticists can identify some of the abnormalities among chromosomes by examining them at the metaphase stage of mitosis. A photograph of the spread chromosomes is then taken and enlarged. Then, the chromosomes are cut out one at a time from the photo and sorted on an analysis sheet to form a **karyotype**, an arrangement of chromosome pairs by size that allows determination of the chromosome number and any abnormalities that can be visually identified.

Examine the normal male karyotype in Figure 11.5. Note that the chromosomes are sorted into seven groups, A through G, on the basis of their size and the location of the centromere. This

TABLE 11.2
Examples of Chromosomal Abnormalities

Chromosomal Abnormality	Effect
Trisomy 18	E syndrome; usually fatal within 3 mo due to multiple congenital defects
Trisomy 21	Down's syndrome: mental retardation, short and incurved fifth finger, marked creases in palm, characteristic facial appearance
Deletion from short arm of chromosome 5	Cri-du-chat syndrome: mental and physical retardation, round face, plaintive cat-like cry, death by early childhood
XXY	Klinefelter's syndrome (male): underdeveloped testes, breasts enlarged, usually sterile, mentally retarded
XO	Turner's syndrome (female): underdeveloped ovaries, no ovulation or menstruation

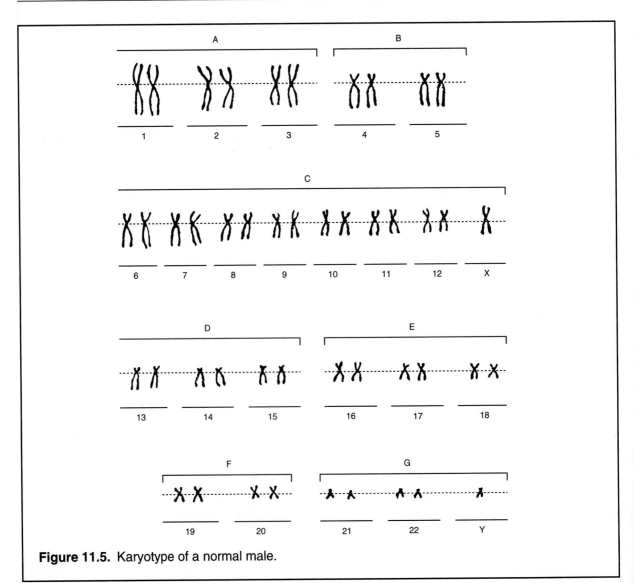

Figure 11.5. Karyotype of a normal male.

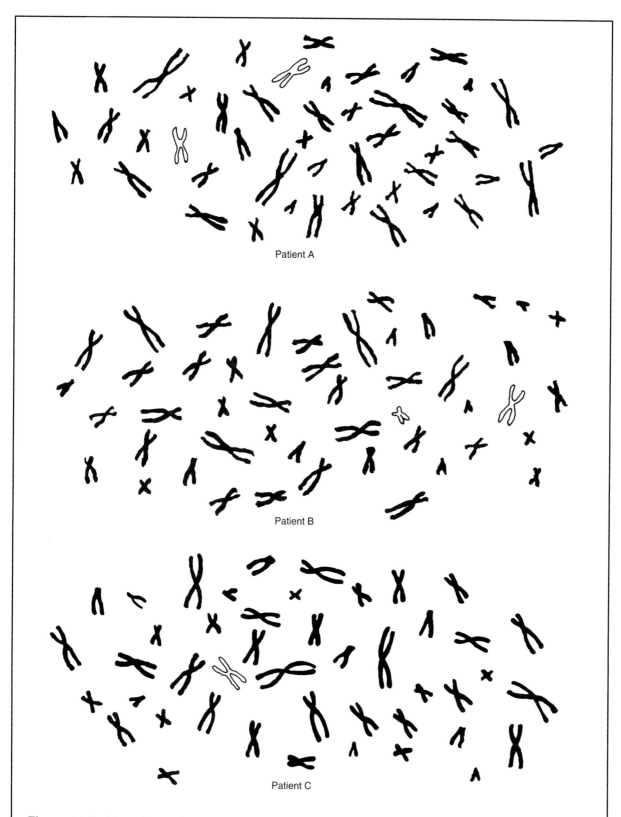

Figure 11.6. Metaphase chromosomes of three patients. Sex chromosomes are shown in outline.

process separates the sex chromosomes, X and Y, from each other in the karyotype.

Materials

Per student
Compound microscope
Dissecting instruments
Glue stick

Per lab
Prepared slides of human chromosomes

Assignment 4

1. Examine a prepared slide of human chromosomes. Note their small size and replicated state.

2. Study Figure 11.5 to understand how chromosomes are sorted in preparing a karyotype.

3. Prepare a karyotype from the chromosome spread for each patient in Figure 11.6, *one patient at a time.* Cut out the chromosomes one by one and place them on the karyotype analysis forms on the laboratory report, with the centromere on the dashed line. Use the size of the chromosome and the position of the centromere to determine the correct position of each chromosome. Do not glue the chromosomes firmly in place until you are sure of their correct positions. Refer to Figure 11.5 as needed.

4. If an abnormality exists, refer to Table 11.2 to determine the specific defect.

5. **Complete item 4 on the laboratory report.**

12 DNA Fingerprinting

After completing the laboratory session, you should be able to:
1. Describe the use of restriction endonucleases in DNA technology.
2. Describe how electrophoresis separates DNA restriction fragments.
3. Describe the procedures used in electrophoresis of DNA restriction fragments.
4. Describe the basis of DNA fingerprinting.
5. Recognize simple identical DNA fingerprints.
6. Define all terms in bold print.

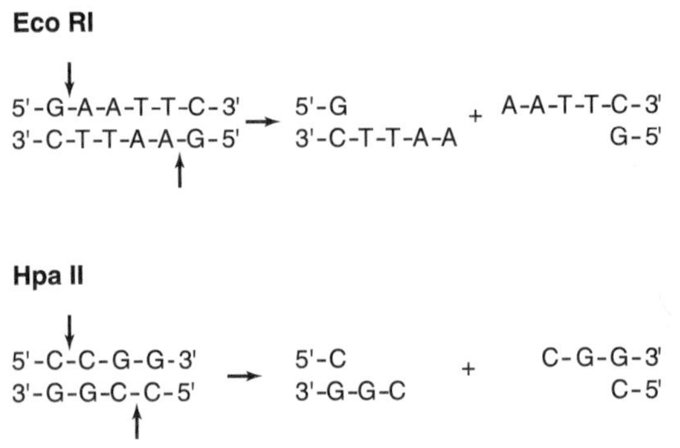

Eco RI

```
      ↓
5'-G-A-A-T-T-C-3'        5'-G              A-A-T-T-C-3'
3'-C-T-T-A-A-G-5'   →    3'-C-T-T-A-A   +      G-5'
      ↑
```

Hpa II

```
    ↓
5'-C-C-G-G-3'        5'-C                C-G-G-3'
3'-G-G-C-C-5'   →   3'-G-G-C     +         C-5'
    ↑
```

Figure 12.1. Recognition palindromes and cutting sites of two restriction endonucleases. Nucleotide sequences are read from the 5' to the 3' direction. Both of these endonucleases produce restriction fragments with "sticky ends."

DNA technology is based on the ability of **restriction endonucleases**, or **restriction enzymes**, to cut DNA molecules at specific points. The resulting DNA fragments are called **restriction fragments**. Restriction endonucleases are naturally produced by bacteria as protection against foreign DNA, such as bacteriophage DNA. Over 2,500 restriction endonucleases have been identified, and many are mass produced for commercial and scientific use.

Each restriction endonuclease recognizes a specific sequence of nucleotides, usually a palindrome—a sequence of four to eight nucleotides that reads the same when reading forward in one strand and in the opposite direction in the complementary strand. For example, the commonly used restriction endonuclease Eco R1 recognizes the sequence GAATTC. The cut is made at specific points within the palindrome. See Figure 12.1.

DNA cleavage by most restriction endonucleases leaves single-stranded, complementary ends ("sticky ends") on the restriction fragments (Figure 12.1). If two different DNA molecules are cut by the same restriction endonuclease, the "sticky ends" enable DNA from one source to combine with DNA from the other source. This is how a DNA fragment (a gene) from one organism is inserted into DNA of a different organism to form **recombinant DNA**. For example, human genes for the production of insulin and growth hormone have been inserted into certain bacteria, and these bacteria are used to mass produce these human hormones for medical use.

The size (length) of the restriction fragments produced by a restriction endonuclease depends upon the frequency of the recognition sites and the distances between them. Thus, the longer a DNA molecule is, the more recognition sites are likely to be present. And, the fewer the nucleotides composing a recognition site, the more frequently the site is likely to occur. Thus, a four-nucleotide recognition site will occur more frequently than a six-nucleotide recognition site.

A large portion of mammalian DNA consists of **tandemly arranged repeats**, repetitious nucleotide sequences that occur between genes. Tandemly arranged repeats have no known function but they are genetically determined and vary in number from individual to individual. Therefore, when mammalian DNA is cut with a restriction endonuclease, the resulting restriction fragments are of different lengths as determined by the distances between recognition sites due to the number of tandemly arranged repeats.

DNA from the same individual, cut with the same restriction endonuclease, always yields the same pattern of restriction fragment lengths. Similarly, DNA from the same individual, cut with two or more restriction endonucleases, will yield a distinctive pattern of restriction fragment lengths for each endonuclease. Since these patterns depend on the sequence of nucleotides, which is genetically determined, they constitute the unique **DNA fingerprint** for the individual.

If the same endonucleases are used to cut DNA from two different individuals, a different pattern of restriction fragment lengths is produced from each individual's DNA. The distinctive patterns are known as **RFLPs** (restriction fragment length polymorphisms). RFLP differences reflect the genetic differences in the sequence of nucleotides in DNA of different individuals. Since each individual's genetic composition is unique (except for identical twins), a person's DNA fingerprint is different from the DNA fingerprint of all other persons, and it can be used to identify an individual with utmost precision.

After DNA is cut by a restriction endonuclease, the restriction fragments can be separated by **agarose gel electrophoresis**. An agarose gel provides a thin meshwork or sieve through which DNA fragments migrate when exposed to an electrical field. Since DNA has a negative charge, the restriction fragments migrate toward the positive pole, and their rate of movement varies with their size. Smaller fragments migrate faster than larger fragments. Thus, restriction fragments separate according to their molecular weights. The separated restriction fragments may be used for DNA fingerprinting, producing recombinant DNA, or mapping DNA.

OVERVIEW OF THE EXERCISE

In this exercise, a DNA fingerprinting simulation will be used to solve a hypothetical crime. DNA fingerprinting has become important evidence in criminal cases because it can positively identify a suspect as being present at a crime scene. You will carry out agarose gel electrophoresis to separate restriction fragments of DNA simulating samples collected at a crime scene and from two suspects. The DNA samples have been cut with two different endonucleases.

The electrophoresis apparatus (Figure 12.2) consists of a gel bed, an electrophoresis chamber, a safety cover, and a direct current power

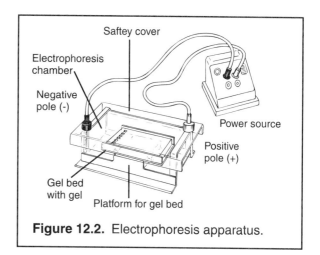

Figure 12.2. Electrophoresis apparatus.

source. The sequence of steps that you will perform is as follows:

1. Prepare an agarose gel on the gel bed.
2. Place the gel bed in the electrophoresis chamber.
3. Add buffer solution to the electrophoresis chamber to cover the gel.
4. Transfer samples of restriction fragments to the gel using an automatic micropipetter.
5. Run the gel (expose it to an electric current) to separate the restriction fragments.
6. Destain the gel.
7. Read and interpret the results.

Materials

(Materials based on Edvotek kit 109)
Beakers, tip-collection, 250 ml
Destaining trays
Flasks of sterile:
   distilled or deionized water
   agarose gel solution (0.8%), containing methylene blue, melted and cooled to 55°C
   buffer solution, containing methylene blue
Gels for practice loading
Gel viewing box, white-light
Gloves, vinyl
Horizontal electrophoresis setups with DC power source
Microcentrifuge tubes of restriction fragments labeled A to F
Microcentrifuge tube racks
Micropipetters, 5–50 μl size
Micropipetter tips, sterile, 50, μl
Spatula, plastic gel-removing
Tape, labeling or masking
Tubes of practice loading dye
Water bath, 55°C

PREPARING THE GEL

Your instructor has prepared the agarose gel solution and the electrophoresis buffer solution. Your task is to form a gel on the gel bed and set up the electrophoresis chamber for loading the gel. See Figure 12.3. Your instructor will describe and demonstrate how to prepare the gel. Follow the steps listed below unless directed otherwise by your instructor. **Wear the safety gloves provided throughout the exercise.**

1. Rinse the gel bed with distilled water and dry with paper towels.
2. If your gel bed has rubber dams, install a rubber dam at each end, making sure that they are firmly attached to the bottom and sides of the gel bed. If your gel bed does not have rubber dams, use masking tape or labeling tape to close off the open ends. Fold the ends (about 1 in.) of the tape back on the sides of the gel bed and be sure that they are firmly secured. Make sure that the tape is firmly adhered to the edges of the bottom and sides by running your fingertip over the tape on these surfaces.
3. Place a six-tooth comb in the set of notches near one end of the gel bed. Note that there is a small space between the bottom of the teeth and the gel bed. Be sure that the comb is positioned evenly across the gel bed. The comb is used to form wells (depressions) in the gel.
4. Obtain one of the small flasks of 0.8% agarose gel solution containing methylene blue stain from the water bath at 55°C. Place the gel bed on a level surface and pour the gel solution into the gel bed until it is filled. The gel bed must remain motionless while the gel is solidifying, which takes 15–20 min. The gel will become translucent when solidified. *While the gel is solidifying, skip down to the next section and practice loading wells in a practice gel. Then, return here to complete the gel preparation.*
5. After the gel has solidified, gently remove the rubber dams or tape, being careful not to damage the gel. Running a plastic knife between the gel and the rubber dams or tape helps to prevent the gel from tearing.
6. Gently remove the comb by lifting it straight up while keeping it level. This will prevent damage to the wells.
7. Place the gel bed in the electrophoresis chamber, centered on the platform, with the wells near the negative pole (black).
8. Obtain a flask of buffer plus methylene blue stain. The flask contains the amount of buffer needed for the electrophoresis chamber. Pour the buffer into the electrophoresis chamber until the gel is covered by about 2 mm of buffer.

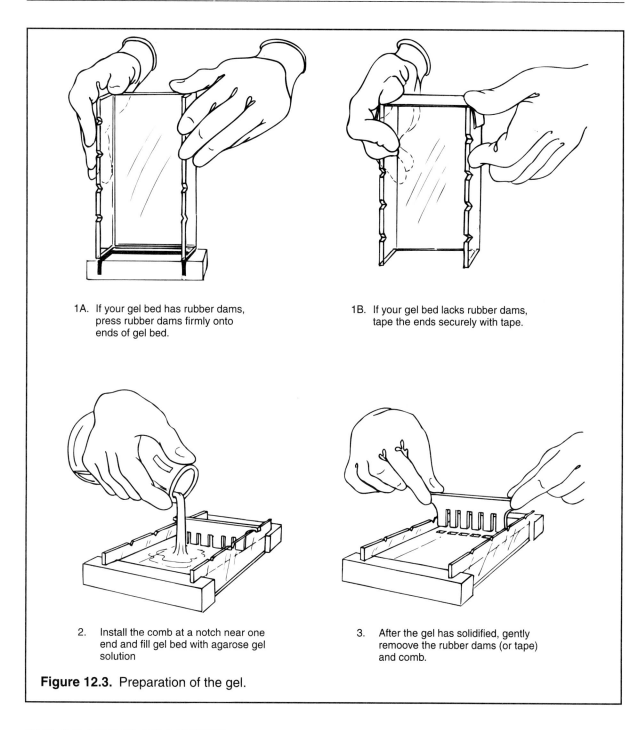

1A. If your gel bed has rubber dams, press rubber dams firmly onto ends of gel bed.

1B. If your gel bed lacks rubber dams, tape the ends securely with tape.

2. Install the comb at a notch near one end and fill gel bed with agarose gel solution

3. After the gel has solidified, gently remoove the rubber dams (or tape) and comb.

Figure 12.3. Preparation of the gel.

GEL LOADING PRACTICE

Loading the wells of a gel can be a bit tricky, so a little practice is helpful. Figure 12.4 shows how to hold a micropipetter with your thumb on the plunger at the top. A removable micropipette tip is attached at the bottom of the micropipetter. When filling and dispensing fluid, fluid moves into and out of the tip, only. Micropipetters differ in the volume of fluid that they can handle. Your micropipetter may be adjusted to pipette automatically small volumes from 5μl to

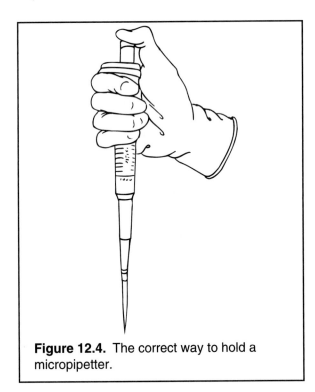

Figure 12.4. The correct way to hold a micropipetter.

50µl, but it is set at 40µl since that is the volume needed to fill the wells of the agarose gel. If you need to change the volume setting, see your instructor.

Here's how the micropipetter works. The plunger at the top is used to withdraw and expel fluid.

1. To withdraw fluid, the plunger is depressed to the first stop, the removable tip is inserted into the fluid, and then the plunger is slowly released to draw the fluid into the tip.
2. Dispensing fluid is just the reverse. The tip is inserted into the receiving chamber, and the plunger is slowly depressed to the first stop to expel the fluid. When pipetting fluid into microcentrifuge tubes, the plunger is depressed to the second stop, which ejects air to blow out any remaining fluid. ***Do not do this filling gel wells.***
3. Pressing the plunger to the third stop ejects the tip from the micropipetter. Some micropipetters have a separate plunger to eject the tip.

Your instructor will demonstrate the correct use of a micropipetter to get you started. Here are some general rules to keep in mind.

1. When not in use, always keep the micropipetter on its stand. Never lay it down on a table or countertop.
2. Always hold the micropipetter with the tip pointing down. This will prevent fluid from running into the micropipetter and contaminating or damaging it.
3. Always use a new sterile tip for each solution that you transfer.

Procedure

Practice filling the wells of the practice gel with the practice loading dye using these steps unless directed otherwise by your instructor. See Figure 12.5.

1. Obtain a practice gel to practice loading and a tube of practice loading dye. Add water to cover the gel to a depth of 2–4 mm. Note the wells for practice loading. Placing the gel on a piece of black construction paper will make the wells more visible.
2. Depress the plunger of the micropipetter to the first stop, insert the tip into the practice loading dye and *slowly* release the plunger allowing it to move up to its highest position. This will draw 40µl of dye into the micropipette tip.
3. Filling the gel wells requires a steady hand, precision, and care. Holding the micropipette with the tip down, and holding the base of the tip between thumb and forefinger of your other hand to guide it with precision, place the pipette tip in the water over a gel well.
4. Insert the tip *just barely below* the top of the well, being careful not to touch the sides or bottom of the well. *Slowly* press the plunger to the first stop and note how the dense dye sinks into the well. Keeping the plunger at the first stop position, remove the tip from the well and water.
5. Repeat steps 2–4 several times until you feel comfortable that you can load a gel well correctly without damaging the gel or losing the dye.
6. When finished practicing, hold the micropipetter over the tip collection beaker and depress the plunger to the third position ejecting the tip into the beaker.
7. Return the micropipetter to its rack.

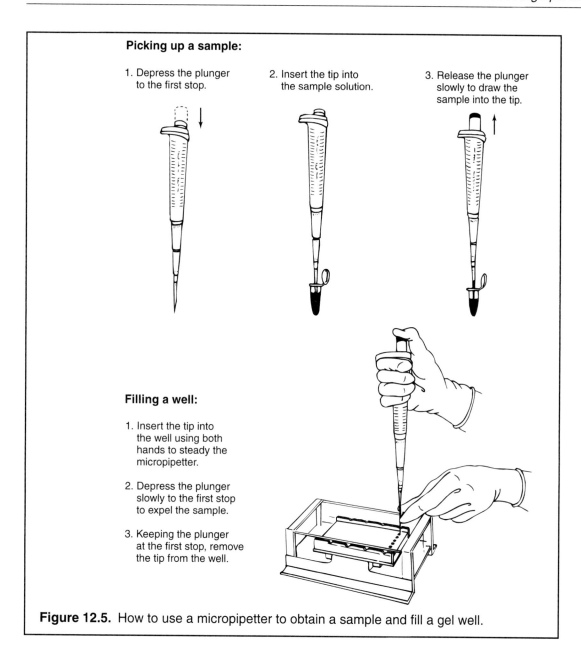

Picking up a sample:

1. Depress the plunger to the first stop.

2. Insert the tip into the sample solution.

3. Release the plunger slowly to draw the sample into the tip.

Filling a well:

1. Insert the tip into the well using both hands to steady the micropipetter.

2. Depress the plunger slowly to the first stop to expel the sample.

3. Keeping the plunger at the first stop, remove the tip from the well.

Figure 12.5. How to use a micropipetter to obtain a sample and fill a gel well.

Now, return to your gel to see if its solidified. Complete the preparation of the gel by following the directions in the section above.

LOADING THE GEL

There are six microcentrifuge tubes (A through F) containing restriction fragments to be loaded into the gel. The hypothetical sources of these tubes are as follows.

Tube A: DNA from the crime scene cut with endonuclease 1.

Tube B: DNA from the crime scene cut with endonuclease 2.

Tube C: DNA from suspect 1 cut with endonuclease 1.

Tube D: DNA from suspect 1 cut with endonuclease 2.

Tube E: DNA from suspect 2 cut with endonuclease 1.

Tube F: DNA from suspect 2 cut with endonuclease 2.

Separating the restriction fragments by agarose gel electrophoresis will enable you to determine if DNA from either suspect matches the DNA from the crime scene.

1. Be sure that the wells of the gel are nearer the negative pole (black) of the electrophoresis chamber and that the leads will reach from the chamber to the power source. *Once the wells are filled, the chamber must not be moved until electrophoresis is completed.*

2. When viewing a gel from above with the wells farthest from you, the wells are read left to right, either 1 through 6 or, in this case, A through F.

3. Place a sterile tip on the micropipetter.

4. Depress the plunger of the micropipetter to the first stop, insert the tip into the fluid in tube A, and *slowly* release the plunger drawing a sample into the pipette tip.

5. Insert the tip just below the top of well A, being careful not to touch the sides or bottom of the well. *Slowly* press the plunger to the first stop to fill the well with the sample. Keeping the plunger at the first stop position, remove the tip from the well and buffer.

6. Hold the micropipetter over the tip collection beaker and eject the tip into the beaker.

7. Place a new sterile tip on the micropipetter, and use the same procedure to place a sample from tube B into well B of the gel. Then, eject the tip into the tip collection beaker.

8. Use the same procedure to place samples from tubes C through F into wells C through F, respectively, *being certain to use a new sterile pipette tip for each sample.*

9. *When you have loaded the last well, eject the pipette tip into the tip collection beaker and return the micropipetter to its rack.*

RUNNING THE GEL

1. Place the safety cover on the electrophoresis chamber, lining up the electrode terminals correctly so that the (+) and (–) terminal indicators on the cover and chamber match.

2. Insert the plug of the black (–) wire into the black (–) jack of the power source.

3. Insert the plug of the red (+) wire into the red (+) jack of the power source.

4. Set the voltage indicator of the power source on 50 V unless directed otherwise by your instructor.

5. Turn on the switch of the power source and run the gel for 1.5–2.0 hr as directed by your

instructor. While the gel is running, you can complete Assignment 1 on the laboratory report.

Assignment 1

Complete item 1 on Laboratory Report 12 which begins on page 357.

READING THE GEL

1. Since both the gel and buffer contain methylene blue stain, you will be able to see the faint bands of the restriction fragments as they move down the gel.

2. After running the gel, turn off the power source and unplug the wires from the jacks of the power source.

3. Remove the safety cover from the electrophoresis chamber.

4. Grasp each end of the gel bed to prevent the gel from sliding off, and gently lift the gel bed and gel from the electrophoresis chamber.

5. Place the gel bed and gel on a white-light gel viewing box and read the positions of the bands of restriction fragments in the lanes below each well. If the bands are not easily visible, destaining will be necessary as described in step 6.

6. Using a plastic spatula, gently slide the gel into a destaining tray containing distilled water. Use the plastic spatula to nip off the upper left corner of the gel to allow easy recognition of the orientation of the gel wells. Destain for 20 min with frequent stirring of the water, and change the distilled water every 5 min or so.

7. Use a plastic spatula to remove the gel to a white-light gel viewing box and read the positions of the bands of restriction fragments in the lanes below the wells.

8. After reading the gel and completing Assignment 2, clean your apparatus as directed by your instructor.

Assignment 2

Complete item 2 on the laboratory Report.

Part 3 Human Biology

13 Organization of the Human Body

OBJECTIVES

After completion of the laboratory session, you should be able to:
1. List the organ systems of the body, their general functions, and the major organs that compose them.
2. Identify the organs in the body cavities.
3. Describe the characteristics and functions of the four types of tissues in the body.
4. Recognize the types of epithelial, connective, muscle, and nerve tissue when viewed microscopically, and give examples of their locations and functions.
5. Define all terms in bold print.

The human body exhibits the levels of structural organization that are found in most animals. The major organizational levels from simplest to most complex are:

Chemical   Organic and inorganic compounds composing the body (e.g., protein).

Cellular   Cells are structural and functional units of life that are specialized in different ways to perform specific functions (e.g., nerve cells).

Tissue   Aggregations of similar cells performing similar functions (e.g., muscle tissue).

Organ   Defined structures formed of two or more tissues and performing specific functions (e.g., stomach).

Organ system   Groups of organs functioning in a coordinated manner to perform specific functions (e.g., digestive system).

ORGANS AND ORGAN SYSTEMS

Body functions are carried out by organs of the organ systems working in a coordinated fashion. In this section, your task is to learn the names of the organ systems of the body, their general functions, and the organs that compose them. In later exercises, you will investigate in more detail the structure and function of organ systems.

Integumentary System

The skin, including hair, nails, and sweat glands, composes the integumentary system. It consists of a **dermis** (inner layer) and an **epidermis** (outer layer) that protect the body from ultraviolet radiation, mild abrasions, excessive evaporative water loss, and invasion by microorganisms. The evaporation of perspiration secreted by sweat glands helps to cool the body surface.

Skeletal System

The skeletal system consists of **bones, cartilages**, and **ligaments**. It provides support for the body and protection for vital organs like the heart, lungs, and brain.

Muscle System

Muscles of the body form the muscle system. **Skeletal muscles** are attached to bones by **tendons** to form levers that enable movement when the muscles contract. About half of the body weight consists of skeletal muscles.

Nervous System

The nervous system consists of the **brain, spinal cord, cranial nerves, spinal nerves**, and **sensory receptors**. The formation and transmission of neural impulses enable the nervous system to provide rapid perception of environmental changes and coordination of body functions. Self-awareness, intelligence, will, and emotions are functions of the human brain.

Cardiovascular System

The cardiovascular system consists of the **heart, blood vessels**, and **blood**. As it is pumped by the heart through blood vessels blood is the transporting medium that carries materials throughout the body. Blood also provides a primary defense against invasion by disease-causing microorganisms.

Lymphatic System

Lymphatic vessels, spleen, tonsils, and **lymph nodes** compose the lymphatic system. Extracellular fluid is collected by lymphatic vessels; filtered through lymph nodes, where microorganisms and cellular debris are removed; and then returned to the blood. The lymphatic system plays an important role in immunity.

Respiratory System

The respiratory system consists of the lungs and the air passages. The **lungs** are gas-exchange organs, where atmospheric air is separated from blood by thin membranes, enabling oxygen in the air to diffuse into the blood and carbon dioxide in the blood to diffuse into the air in the lungs. The air passages through which air enters and leaves the lungs are the **nasal cavity, pharynx, larynx, trachea**, and **bronchi**.

Digestive System

The digestive system consists of the alimentary canal, through which food passes, and the accessory digestive glands. The parts of the alimentary canal are the **mouth, pharynx, esophagus, stomach, small intestine, large intestine** or **colon, rectum**, and **anus**. The digestive glands are the **salivary glands, pancreas**, and **liver**. The function of the digestive system is to convert large, nonabsorbable food molecules into small, absorbable nutrient molecules.

Urinary System

The urinary system removes metabolic wastes and excessive minerals from the body as urine. Wastes and excessive minerals are removed from the blood by the paired **kidneys**, and are then carried through **ureters** to the **urinary bladder** for temporary storage, and voided from the body through the **urethra**.

Endocrine System

The endocrine system consists of hormone-producing endocrine glands: **pituitary, thyroid, parathyroid, thymus, adrenal, pancreas, pineal, ovaries**, and **testes**. Hormones are chemical messengers that provide a relatively slow-acting, but long-lasting, coordination of body functions.

Reproductive System

The function of the male and female reproductive systems is the production of babies. The male reproductive system consists of sperm-producing **testes** located in the **scrotum**, the exterior pouch; two **vas deferens**, two tubes carrying sperm to the **urethra**; **accessory glands** providing fluid for sperm transport; and the **penis**, the male copulatory organ. The female reproductive system consists of egg-producing **ovaries**; two **oviducts** that carry eggs and, after egg–sperm fusion, early embryos to the **uterus**, where prenatal development occurs; **accessory glands** that secrete lubricating fluids; and the **vagina**, the female copulatory organ and birth canal.

Assignment 1

Complete item 1 on Laboratory Report 13 that begins on page 359.

BODY CAVITIES

Many organs are located in body cavities. There are two body cavities that are named for their location. See Figure 13.1. The **ventral body cavity** is the larger cavity, and is located near the anterior (front) of the body. It is divided by a dome-shaped sheet of muscle, the **diaphragm**, into the **thoracic cavity** and the **abdominopelvic cavity**. The thoracic cavity is further subdivided into left and right **pleural cavities** containing the lungs and the **pericardial cavity** containing the heart. The abdominopelvic cavity is further subdivided into an **abdominal cavity** and a **pelvic cavity**.

The **dorsal body cavity** is the smaller cavity, and it consists of the **cranial cavity** containing the brain and the **spinal cavity** containing the spinal cord. Note that the dorsal body cavity is surrounded by bones.

Materials

Per lab
Colored pencils
Human torso models with removable parts

Assignment 2

1. Label Figure 13.1, color-code parts of the dorsal and ventral body cavities, and **complete item 2a on the laboratory report**.
2. Study and color-code the organs in Figures

13.2 and 13.3. Note the location of the organs in the ventral cavity and identify the organ system to which each belongs.
3. Examine the human torso model. Identify the organs in the ventral cavity and the organ system to which each belongs. Remove the anterior organs as necessary to locate the more posterior organs. Note how the organs fit together in the available space. Replace the organs carefully when finished.
4. **Complete item 2 on the laboratory report.**

TISSUES

Organs of the body are formed of two or more tissues, and each tissue is formed of similar cells that are specialized to perform their particular functions. There are four basic types of tissues in the body.

Epithelial tissue	Covers surfaces and lines cavities; forms secretory portions of glands; functions include protection, secretion, and absorption.
Connective tissue	Binds tissues together; provides protection and support for organs and body.
Muscle tissue	Specialized for contraction enabling body movements.
Nerve tissue	Transmission of neural impulses enabling rapid coordination of body functions.

In this section your task is to learn the recognition characteristics of body tissues and their general functions, and to correlate their microscopic structure with their functions.

Epithelial Tissues

The cells of epithelial tissues are tightly packed together, forming single or multiple sheetlike layers. The innermost layer is attached to underlying connective tissue by a very thin, noncellular **basement membrane**. The tissue sur-

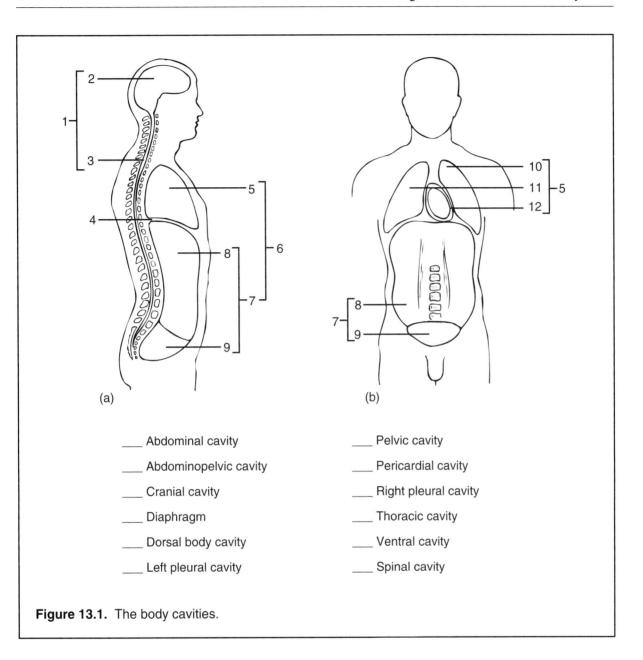

Figure 13.1. The body cavities.

___ Abdominal cavity

___ Abdominopelvic cavity

___ Cranial cavity

___ Diaphragm

___ Dorsal body cavity

___ Left pleural cavity

___ Pelvic cavity

___ Pericardial cavity

___ Right pleural cavity

___ Thoracic cavity

___ Ventral cavity

___ Spinal cavity

face opposite the basement membrane is always exposed, i.e., it is not attached to other tissues.

Epithelial tissues are named according to (1) whether they have only one layer (simple epithelium) or more than one layer (stratified epithelium) of cells, (2) the shape of the cells in the outermost cell layer, and (3) the presence or absence of cilia.

Simple Epithelium

Simple epithelium (Figure 13.4) consists of a single layer of cells that form a thin layer cover-

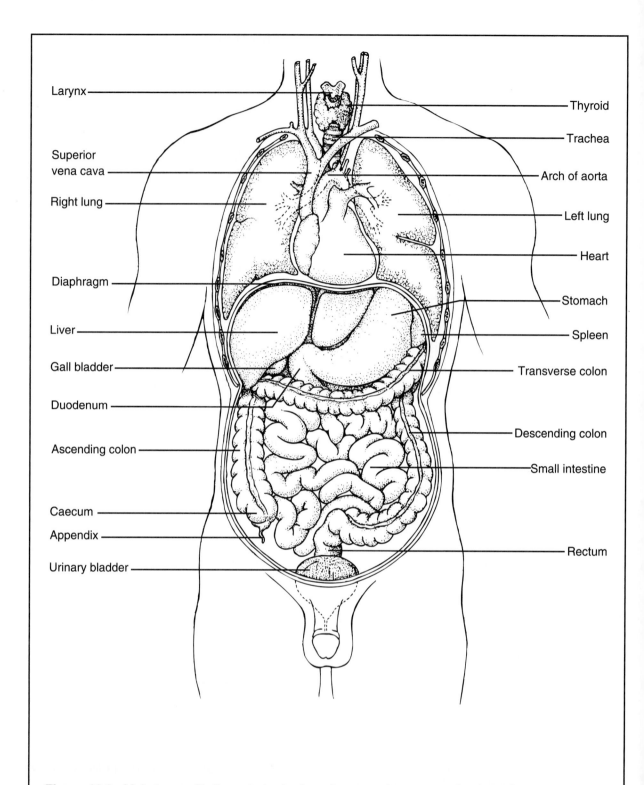

Figure 13.2. Male torso with the anterior body wall removed to expose the thoracic and abdominopelvic organs.

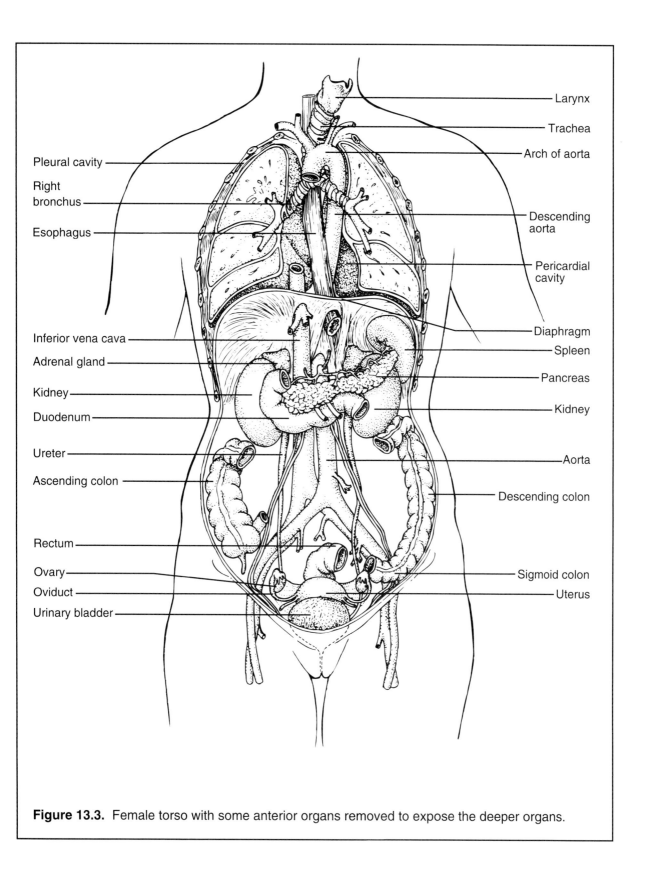

Larynx

Trachea

Arch of aorta

Pleural cavity

Right
bronchus

Esophagus

Descending
aorta

Pericardial
cavity

Inferior vena cava

Adrenal gland

Kidney

Duodenum

Ureter

Ascending colon

Diaphragm

Spleen

Pancreas

Kidney

Aorta

Descending colon

Rectum

Ovary

Oviduct

Urinary bladder

Sigmoid colon

Uterus

Figure 13.3. Female torso with some anterior organs removed to expose the deeper organs.

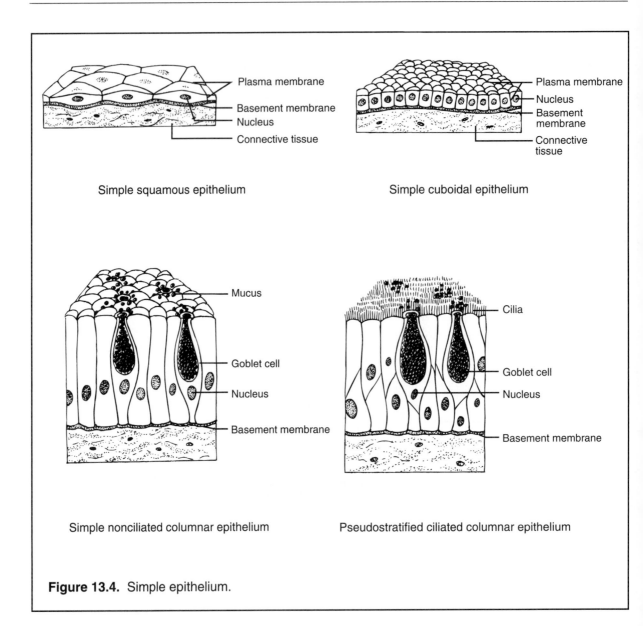

Figure 13.4. Simple epithelium.

ing underlying tissues. It is involved in the diffusion, secretion, and absorption of materials.

Simple squamous epithelium consists of a single layer of thin, flat, tilelike cells arranged in a mosaic pattern. It forms the tiny air sacs in the lungs and the capillaries (smallest blood vessels), where its thin, flat cells aid the diffusion of substances. Simple squamous epithelium also provides a smooth, friction-reducing interior lining of the heart, blood vessels, lymphatic vessels, and body cavities.

Simple cuboidal epithelium occurs as a single layer of cubelike cells. It forms the secretory portions of glands, and it composes kidney tubules, where it is involved in both secretion and absorption.

Simple ciliated columnar epithelium lines the interior of the oviducts, and its beating cilia move eggs released from the ovaries toward the uterus. Scattered goblet cells secrete a protective layer of mucus.

Simple nonciliated columnar epithelium lines the interior of the digestive tract. It is involved in the secretion of digestive juices and the absorption of nutrients. Scattered goblet cells secrete a protective layer of mucus.

Pseudostratified ciliated columnar epithelium looks as if it consists of more than one cell layer, but it does not. All of the cells extend from the basement membrane to the surface of the tissue. Scattered goblet cells secrete a protective layer of mucus. This epithelium lines the interior of the upper respiratory passages, where its beating cilia remove airborne particles trapped in mucus released from goblet cells.

Stratified Epithelium

Stratified epithelium consists of multiple layers of cells, which makes this epithelium resistant to abrasion. The innermost layer of cells continuously forms new cells by mitotic division, so as old cells are worn and lost from the tissue surface, new cells are pushed upward to replace them. As you might expect, protection of underlying tissues is an important function of stratified epithelium. We will consider only one example.

Stratified squamous epithelium (Figure 13.5) occurs in two forms. The keratinized form composes the outer layer (epidermis) of the skin. As cells migrate to the surface they are filled with **keratin**, a waterproofing substance that prevents evaporative water loss directly through the skin. The nonkeratinized form of stratified squamous epithelium lines the mouth, esopha-

gus, and vagina. The cells from your mouth that you observed microscopically in Exercise 3 were cells from surface layers of stratified squamous epithelium.

Materials

Per student
Compound microscope

Per lab
Prepared slides of epithelium:
   pseudostratified ciliated columnar
   simple columnar
   simple cuboidal
   stratified squamous

Assignment 3

1. Examine microscopically the prepared slides of simple columnar, simple cuboidal, pseudostratified ciliated columnar, and stratified squamous epithelia. These slides have been stained to make it easier to observe cellular structure. Start at 100× to locate the tissue, and then use 400× for observing cellular details. Compare your observations with the corresponding figures. **Draw each of the tissues as they appear on your slide at 100× in item 3a on the laboratory report.**
2. **Complete item 3 on the laboratory report.**

Connective Tissues

Connective tissues are characterized by the presence of relatively few cells embedded in a large amount of intercellular, nonliving material called **matrix**. The matrix may be jellylike, fibrous, solid, or combinations of these. Your task is to learn the characteristics of the connective tissues studied and correlate their functions with their structure.

Loose fibrous connective tissue (Figure 13.6) is the most widespread connective tissue in the body. It provides flexible support within and around internal organs, muscles, and nerves, and it attaches the skin to underlying muscles. It is characterized by having a jelly-like matrix in which relatively few elastic and nonelastic (collagen) fibers are embedded, along with a few cells (fibroblasts) that produce the fibers.

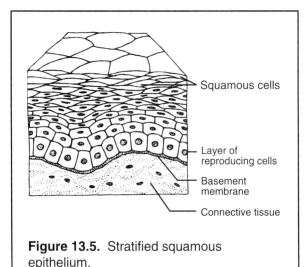

Figure 13.5. Stratified squamous epithelium.

Squamous cells

Layer of reproducing cells

Basement membrane

Connective tissue

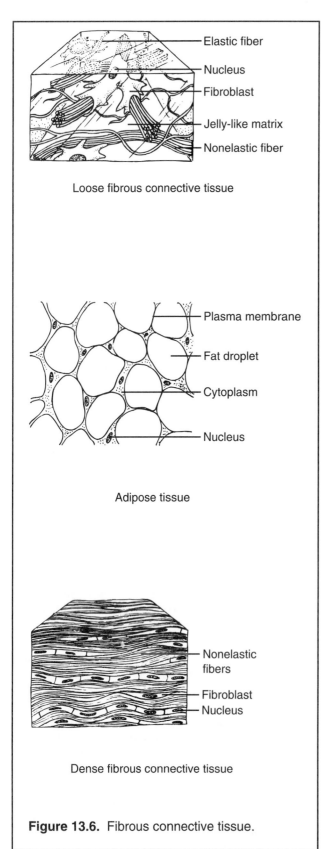

Loose fibrous connective tissue

Adipose tissue

Dense fibrous connective tissue

Figure 13.6. Fibrous connective tissue.

Adipose tissue (Figure 13.6) is a special type of loose fibrous connective tissue that contains a large number of fat cells, in which surplus energy is stored as fat. Vacuoles of fat cells are filled with fat droplets that push the cytoplasm and nucleus of these cells against the cell membrane. Adipose tissue is abundant under the skin, where it provides insulation, and around internal organs, where it provides a protective cushion.

Dense fibrous connective tissue (Figure 13.6) has a matrix filled with nonelastic (collagen) fibers that give it great tensile strength. Cells (fibroblasts) occur in rows among the fibers. Dense fibrous connective tissue forms the ligaments that attach bones to each other at joints and tendons that join muscles to bones. One form, with interwoven fibers, composes the dermis of the skin.

Hyaline cartilage (Figure 13.7) has a white, glassy, flexible matrix, and the scattered cartilage cells are located in tiny spaces in the matrix called **lacunae**. Most of the skeleton of a developing baby is first formed of hyaline cartilage and later replaced by bone. Hyaline cartilage supports the nose, larynx, and trachea, and it covers the ends of long bones at joints.

Elastic cartilage (Figure 13.7) is more flexible than hyaline cartilage because of the many elastic fibers in the matrix. It forms the supporting framework of the external ear.

Fibrocartilage (Figure 13.7) contains many nonelastic fibers that make it tough and strong. Cartilage cells in lacunae are arranged in short rows. Fibrocartilage forms the intervertebral discs, where it serves as a cushioning shock absorber.

Bone (Figure 13.8) is the hardest and most rigid connective tissue, because its matrix is formed of calcium salts. The matrix is deposited in concentric rings (lamellae) around an **osteonic canal** that contains small blood vessels and nerves. **Osteocytes** (bone cells) are located within **lacunae**, which are tiny spaces in the matrix. Lacunae are arranged in concentric rings between concentric rings of matrix around an osteonic canal. **Canaliculi** (tiny canals) extending from each lacuna enable the diffusion of materials to and from the cells. The concentric rings of matrix and osteocytes around an osteonic canal form an **os-**

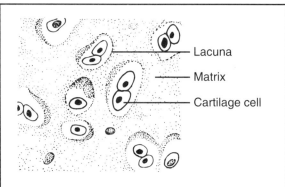

Hyaline cartilage

— Lacuna
— Matrix
— Cartilage cell

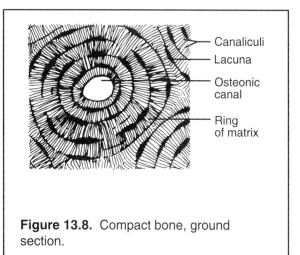

— Canaliculi
— Lacuna
— Osteonic canal
— Ring of matrix

Figure 13.8. Compact bone, ground section.

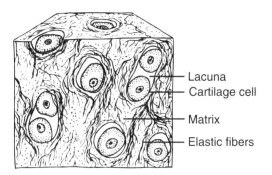

Elastic cartilage

— Lacuna
— Cartilage cell
— Matrix
— Elastic fibers

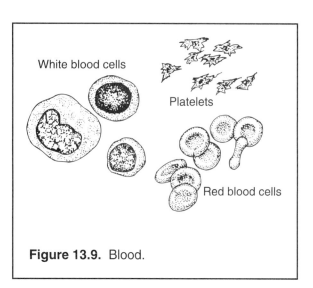

White blood cells

Platelets

Red blood cells

Figure 13.9. Blood.

teonic system, and many osteonic systems are packed together to form solid bone.

Blood (Figure 13.9) is a special form of connective tissue, since it consists of several types of blood cells—**red blood cells**, **white blood cells**, and **platelets**—in a liquid matrix, the **plasma**. Blood transports materials throughout the body and is involved in fighting disease organisms.

Materials

Per lab
Prepared slides of:
    loose fibrous connective tissue
    adipose tissue
    dense fibrous connective tissue

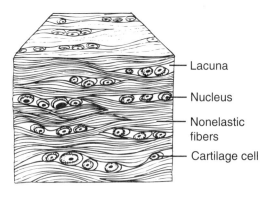

Fibrocartilage

— Lacuna
— Nucleus
— Nonelastic fibers
— Cartilage cell

Figure 13.7. Cartilage.

hyaline cartilage
fibrocartilage
bone, ground, x.s.

Assignment 4

Examine the prepared slides of connective tissues, correlating their structure with their functions. **Draw the tissues as they appear on your slide at 100× in item 4a on the laboratory report and complete item 4.**

Muscle Tissue

Muscle tissue is composed of muscle cells joined by fibrous connective tissue. Muscle cells are adapted for contraction (shortening), which results from an interaction between protein fibrils (small fibers) within muscle cells. There are three types of muscle tissue that are distinguished from one another by their (1) location, (2) structure, and (3) contraction characteristics. Your task is to learn these distinguishing characteristics. See Figure 13.10.

Skeletal muscle tissue occurs in organs known as skeletal muscles that are attached to bones. Individual skeletal muscle cells are elongate and cylindrical in shape. Each cell (fiber) contains several peripherally located nuclei and exhibits **striations**—alternating light and dark bands extending across the width of the cells. Functionally, skeletal muscle is said to be **voluntary** because its contraction is under conscious control.

Cardiac muscle tissue is the muscle tissue that forms the walls of the heart. It consists of branching and interwoven muscle cells that are joined end-to-end to one another by **intercalated discs**. Striations are present, and each cell contains a single, centrally located nucleus. Functionally, cardiac muscle is **involuntary** because its contractions are not consciously controlled.

Smooth muscle tissue is the muscle tissue in the walls of hollow organs other than the heart, e.g., in intestines and blood vessels. Its spindle-shaped cells lack striations and have a single, centrally located nucleus. Functionally, smooth muscle tissue is involuntary because its contractions are not under conscious control.

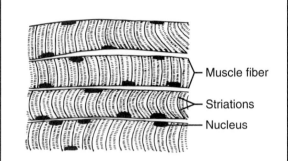

Skeletal muscle tissue

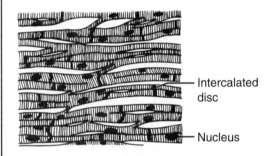

Cardiac muscle tissue

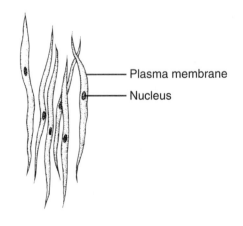

Smooth muscle cells

Figure 13.10. Muscle tissue.

Materials

Per lab
Prepared slides of:
  cardiac muscle tissue
  skeletal muscle tissue
  smooth muscle tissue, teased

Assignment 5

Examine the prepared slides of muscle tissues, noting their structural characteristics. **Draw the muscle tissues as they appear on your slides in item 5a, and complete item 5 on the laboratory report.**

Nerve Tissue

Nerve tissue composes the brain, spinal cord, and nerves. It consists of **neurons** (nerve cells) (Figure 13.11), which form and transmit neural impulses, and **neuroglial cells**, which support the neurons. A neuron is the structural and functional unit of nerve tissue. A neuron consists of a cell body, which is where the cell nucleus of the neuron is located, and two types of neuron processes: (1) an axon that carries impulses away from the cell body and (2) one or more dendrites that carry impulses toward the cell body. Neuron processes may be very short (less than 1 mm) or very long (over 1 m). Axons and dendrites of different neurons form complex interconnections that are essential for the nervous system to perform its functions.

Materials

Per lab
Prepared slides of:
  neurons, giant multipolar

Assignment 6

Examine the prepared slides of multipolar neurons, noting their structural characteristics.

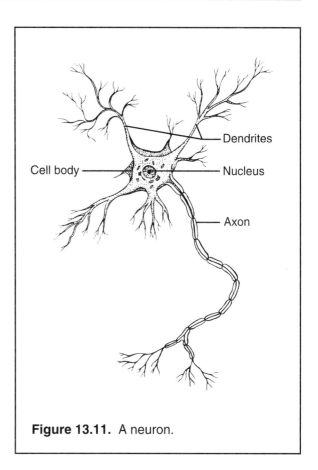

Figure 13.11. A neuron.

Draw a neuron as it appears on your slide in item 6a, and complete the laboratory report.

Assignment 7

Your instructor has set up several microscopes as a mini-practicum so that you can check your recognition of tissues. Your task is to identify the tissues indicated by the microscope pointers. **Complete item 7 on the laboratory report.**

14

Dissection
of the Fetal Pig

<table>
<tr><td colspan="2" align="center">OBJECTIVES</td></tr>
</table>

OBJECTIVES

After completion of the laboratory session, you should be able to:
1. Identify the major internal organs of the fetal pig and indicate to which organ system each belongs.
2. Describe the relative positions of the major organs.
3. Name the body cavities and the organs each one contains.
4. Define all terms in bold print.

TABLE 14.1
Anatomical Terms of Direction

Anterior	Toward the head
Posterior	Toward the hind end
Cranial	Toward the head
Caudal	Toward the hind end
Dorsal	Toward the back
Ventral	Toward the belly
Superior	Toward the back
Inferior	Toward the belly
Lateral	Toward the sides
Medial	Toward the midline

In this exercise you will study the basic body organization of mammals through the dissection of a fetal (unborn) pig. The dissection will focus on the internal organs located in the ventral body cavity. Your dissection will proceed more easily if you understand the common directional terms used to describe the location of organs. These terms are noted in Table 14.1. Figure 14.1 shows the relationship of most of the directional terms, as well as the three common planes associated with a bilaterally symmetrical animal. The integumentary, skeletal, muscular, nervous, and endocrine systems will not be studied. Study the organ systems outlined in Table 14.2 so that

you know their functions and the major organs of each.

GENERAL DISSECTION GUIDELINES

The purpose of the dissection is to expose the various organs of the fetal pig for study, and this should be done in a manner that causes minimal damage to the specimen. *Follow the directions carefully and in sequence.* Read the entire description of each incision and be sure that you understand it *before* you attempt it. Work in

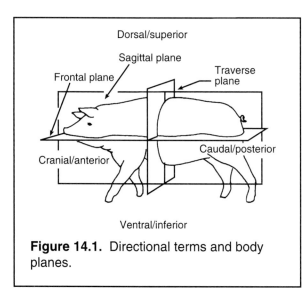

Figure 14.1. Directional terms and body planes.

pairs, alternating the roles of reading the directions and performing the dissection. Use scissors for most of the incisions. *Use a scalpel only when absolutely necessary*. Be careful not to damage organs that must be observed later.

Your instructor may have you complete the dissection over several sessions. If so, wrap the specimen in wet paper towels at the end of each session, and seal it in a plastic bag to prevent it from drying out.

Protect your hands from the preservative by applying a lanolin-based hand cream before starting the dissection or by wearing disposable plastic gloves. Wash your hands thoroughly at the end of each dissection session and reapply the hand cream.

TABLE 14.2
Major Organs and Functions of the Organ Systems

System	Major Organs	Major Functions
Intergumentary	Skin, hair, nails	Protection, cooling
Skeletal	Bones	Support, protection
Muscle	Muscles	Movement
Nervous	Brain, spinal cord, sense organs, nerves	Rapid coordination via impulses
Endocrine	Pituitary, thyroid, parathyroid, testes, ovaries, pancreas, adrenal, pineal	Slower coordination via hormones
Digestive	Mouth, pharynx, esophagus, stomach, intestines, liver, pancreas	Digestion of food, absorption of nutrients
Respiratory	Nasal cavity, pharynx, larynx, trachea, bronchi, lungs	Exchange of oxygen and carbon dioxide
Cardiovascular	Heart, arteries, capillaries, veins, blood	Circulation of blood, transport of materials
Lymphatic	Lymphatic vessels, lymph, spleen, thymus, lymph nodes	Cleansing and return of extracellular fluid to bloodstream
Urinary	Kidneys, ureters, urinary bladder, urethra	Formation and removal of urine
Reproductive		
Male	Testes, epididymis, vas deferens, seminal vesicles, bulbourethral gland, prostate gland, urethra, penis	Formation and transport of sperm and semen
Female	Ovaries, oviducts, uterus, vagina, vulva	Formation of eggs, sperm reception, intrauterine development of offspring

Materials

Per student group
Dissecting instruments and pins
Dissecting microscope
Dissecting pan, wax-bottomed
Fetal pig (*Sus scrofa*), double-injected

Per lab
Hand cream, lanolin-based
String

EXTERNAL ANATOMY

Examine the fetal pig and locate the external features shown in Figure 14.2. Determine the sex of your specimen. The **urogenital opening** in the female is immediately ventral to the anus and has a small **genital papilla** marking its location. A male is identified by the **scrotal sac** ventral to the anus and a **urogenital opening** just posterior to the **umbilical cord**. Two rows of nipples of mammary glands are present on the ventral abdominal surface of both males and females, but the mammary glands later develop only in maturing females. Mammary glands and hair are two distinctive characteristics of mammals.

Make a transverse cut through the umbilical cord and examine the cut end. Locate the two **umbilical arteries** that carry blood from the fetal pig to the placenta, and the single **umbili-cal vein** that returns blood from the placenta to the fetal pig.

POSITIONING THE PIG FOR DISSECTION

Position your specimen in the dissection pan as follows:

1. Tie a piece of heavy string about 20 in. long around each of the left feet.
2. Place the pig on its back in the pan. Run the strings under the pan, and tie them to the corresponding right feet to spread the legs and expose the ventral surface of the body. This also holds the pig firmly in position.

HEAD AND NECK

This portion of the dissection will focus on structures of the mouth and pharynx. If your instructor wishes you to observe the salivary glands, a demonstration dissection will be prepared for you to observe.

Dissection Procedure

To expose the organs of the mouth and pharynx, start by inserting a pair of scissors in the angle of the lips on one side of the head and cut posteriorly through the cheek. Open the mouth as you

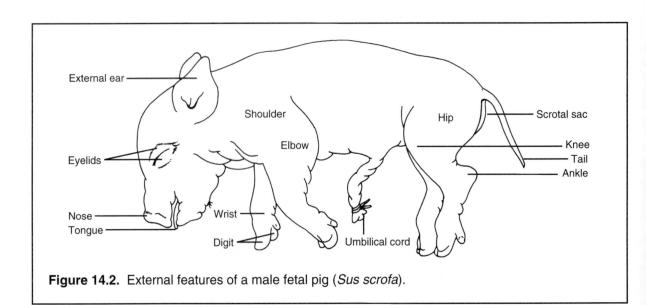

Figure 14.2. External features of a male fetal pig (*Sus scrofa*).

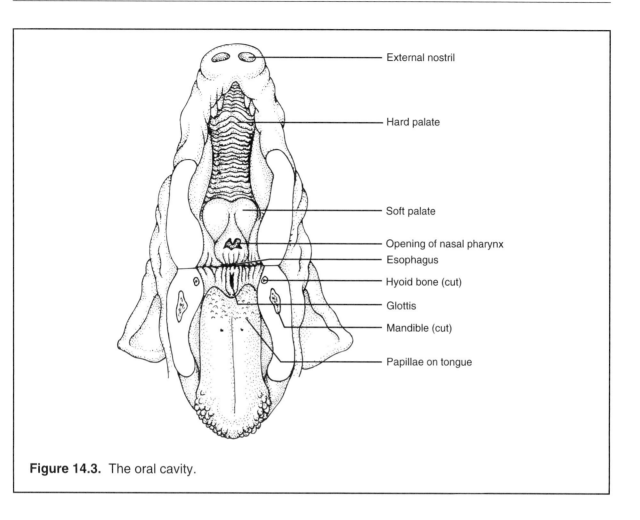

Figure 14.3. The oral cavity.

make your cut, and follow the curvature of the tongue to avoid cutting into the roof of the mouth. Continue through the angle of the mandible (lower jaw). Just posterior to the base of the tongue, you will see a small whitish projection, the **epiglottis**, extending toward the roof of the mouth. Hold down the epiglottis and surrounding tissue and continue your incision dorsal to it and on into the opening of the **esophagus**. Now, repeat the procedure on the other side so that the lower jaw can be pulled down to expose the structures of the mouth and pharynx, as shown in Figure 14.3.

Observations

Only a few deciduous teeth will have erupted, usually the third pair of incisors and the canines. Other teeth are still being formed and may cause bulging of the gums. Make an inci-

sion in one of these bulges to observe the developing tooth.

Observe the tongue. Note that it is attached posteriorly and is free anteriorly, as in all mammals. Locate the numerous **papillae** on its surface, especially near the base of the tongue and along its anterior margins. Papillae contain numerous microscopic taste buds.

Observe that the roof of the mouth is formed by the anteriorly located **hard palate**, supported by bone and cartilage, and the posteriorly located **soft palate**. Paired nasal cavities lie dorsal to the roof of the mouth. The space posterior to the nasal cavities is the **nasal pharynx**, and it is contiguous with the **oral pharynx**, or throat, located posterior to the mouth. Locate the opening of the nasal pharynx posterior to the soft palate. The oral pharynx may be difficult to visualize because your incision has cut through it on each side, but it is the region where the glottis, esophageal opening, and opening of the nasal pharynx are located.

Make a midline incision through the soft palate to expose the nasal pharynx. Try to locate the openings of the **eustachian tubes** in the dorsolateral walls of the nasal pharynx. Eustachian tubes allow air to move into or out of the middle ear to equalize the pressure on the eardrums.

When you have completed your observations, close the lower jaw by tying a string around the snout.

GENERAL INTERNAL ANATOMY

In this section, you will open the **ventral cavity** to expose the internal organs of the fetal pig. The ventral cavity consists of the **thoracic cavity**, which is located anterior to the diaphragm, and the **abdominopelvic cavity**, which is located posterior to the diaphragm.

Dissection Procedure

Using a scalpel, carefully make an incision, through the skin only, from the base of the throat to the umbilical cord along the ventral midline. This is incision 1 in Figure 14.4. Separate the skin from the body wall along the incision just enough to expose the body wall. Do this by lifting the cut edge of the skin with forceps while separating the loose fibrous connective tissue that attaches the skin to the body wall with the handle of a scalpel or a blunt probe.

While lifting the body wall at the ventral midline with forceps, use scissors to make a small incision through the body wall about halfway between the **sternum** and the **umbilical cord**. (Always cut away from your body when using scissors. This gives you better control. Rotate the dissecting pan as necessary.) Once the incision is large enough, insert the forefinger and middle finger of your left hand and lift the body wall from below with your fingertips as you make the incision. This position gives you good control of the incision and prevents unnecessary damage.

Extend the incision along the ventral midline anteriorly to the tip of the sternum and posteriorly to the umbilical cord. Insert the blunt blade of the scissors under the body wall as you make

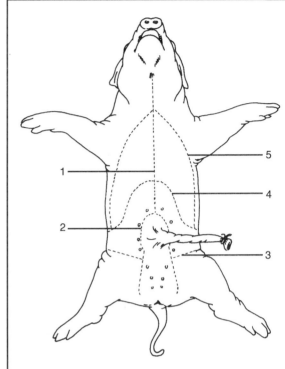

Figure 14.4. Sequence of the incisions (ventral view).

your cuts. Be careful not to cut any underlying organs.

In this manner, continue cutting through both skin and body wall around the anterior margin of the umbilical cord and posteriorly along each side, as shown for incision 2. Then, make the lateral incisions just in front of the hind legs (incision 3). Make incision 4 from the midline laterally, following the lower margin of the ribs as shown. This will allow you to fold out the resulting lateral flaps.

The **umbilical vein**, which runs from the umbilical cord to the liver, must be cut in order to lay out the umbilical cord and attached structures posteriorly. Tie a string on each end of the umbilical vein so that you will recognize the cut ends later. You will have an unobstructed view of the abdominal organs when this is done.

Extend incision 1 anteriorly through the sternum, using heavy scissors. Locate the **diaphragm** and cut it free from the ventral body wall. Lift the left half of the ventral thoracic wall

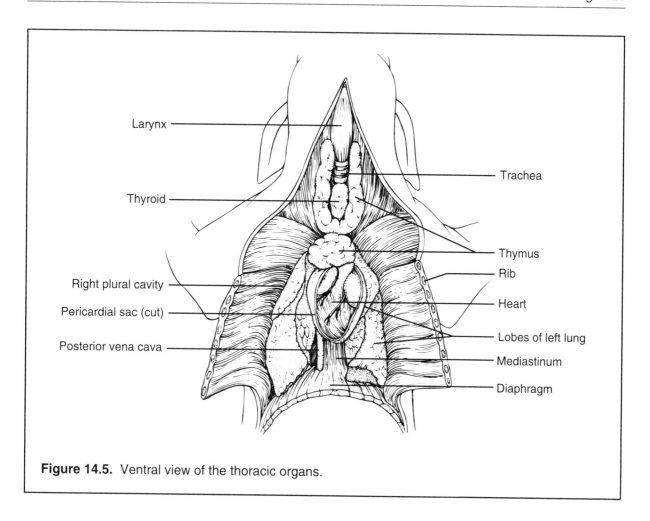

Figure 14.5. Ventral view of the thoracic organs.

as you cut the connective tissue forming the mediastinal septum that extends from the ventral wall to and around the heart. While lifting the left ventral wall, cut through the thoracic wall (ribs and all) at its lateral margin (incision 5), and cut through the muscle tissue at the anterior end of the sternum so that the left thoracic ventral wall can be removed. Repeat the process for the right thoracic ventral wall. The heart and lungs are now exposed.

Continue incision 1 anteriorly to the chin, and separate the neck muscles to expose the thymus gland, thyroid gland, larynx, and trachea. Remove muscle tissue as necessary to expose these structures.

Wash out the cavities of the pig in a sink while being careful to keep the organs in place.

THE THORACIC ORGANS

The thoracic cavity is divided into left and right pleural cavities containing the **lungs**. See Figure 14.5. The left lung has three lobes and the right lung has four lobes. The inner thoracic wall is lined by a membrane, the **parietal pleura**, and each lung is covered by a **visceral pleura**. The potential space between these membranes is the **pleural cavity**. In life, the pleural membranes secrete a small amount of fluid into the pleural cavity, and this fluid reduces friction between the pleural membranes as the lungs expand and contract during breathing.

At the midline, the parietal pleurae join with fibrous membranes to form a connective tissue partition between the pleural cavities. This partition is called the **mediastinum**. The **heart** is located within the mediastinum. The heart is enclosed in a pericardial sac formed of the **parietal pericardium** and an outer fibrous membrane that is attached to the diaphragm. Remove this sac to expose the heart, if you have not already done so. The **visceral pericardium** tightly adheres to the surface of the heart.

In the neck region, locate the **larynx** (voice box), which is composed of cartilage and contains the vocal folds (cords). The **trachea** (windpipe) extends posteriorly from the larynx and divides dorsal to the heart to form the **bronchi** that enter the lungs. You will see these structures more clearly later, after the heart has been removed.

Locate the whitish **thymus gland** that lies near the anterior margin of the heart and extends into the neck on each side of the trachea.

The thymus is an endocrine gland, and it is the maturation site for blood cells known as T-lymphocytes which are important components of the immune system.

Locate the small, dark-colored **thyroid gland** located on the anterior surface of the trachea at the base of the neck. The thyroid is an endocrine gland that controls the rate of metabolism in the body.

Remove the thymus and thyroid glands as necessary to get a better view of the trachea and larynx, but don't cut major blood vessels. Carefully remove the connective tissue supporting the trachea so that you can move the trachea to one side to expose the **esophagus** located dorsal to it. The esophagus is a tube that carries food from the pharynx to the stomach. In the thorax, it descends through the mediastinum dorsal to the trachea and heart, and then penetrates the diaphragm to open into the stomach. You will get a better view of the esophagus later on.

DIGESTIVE ORGANS IN THE ABDOMEN

The inner wall of the abdominal cavity is lined by the **parietal peritoneum**, while the internal organs are covered by the **visceral peritoneum**. The internal organs are supported by thin membranes, the **mesenteries**. The mesenteries consist of two layers of peritoneum, between which are located blood vessels and nerves that serve the internal organs.

The digestive organs are the most obvious organs within the abdominal cavity, especially the large liver that fits under the dome-shaped diaphragm. See Figure 14.6.

Stomach

Lift the liver anteriorly on the left side to expose the **stomach**. The stomach receives food from the esophagus and releases partially digested food into the small intestine. Food is temporarily stored in the stomach, and the digestion of proteins begins there. The sac-like stomach is roughly J-shaped. The longer, curved margin on the left side is known as the **greater curvature**. The shorter margin between the openings of the esophagus and small intestine on the right side is the **lesser curvature**.

A sac-like fold of mesentery, the **greater omentum**, extends from the greater curvature of the stomach to the dorsal body wall. The elongate, dark organ supported by the greater omentum is the **spleen**, a lymphatic organ that stores and filters the blood in an adult. Locate the **lesser omentum**, a smaller fold of mesentery extending from the lesser curvature of the stomach and small intestine to the liver. The peritoneum also attaches the liver to the diaphragm.

Cut open the stomach along the greater curvature, from the esophageal opening to the junction with the small intestine. The greenish material within the stomach and throughout the digestive tract is the **meconium**, which is composed mostly of epithelial cells sloughed off from the lining of the digestive tract, mucus, and bile from the gallbladder. The meconium is passed out of the body in the first bowel movement of a newborn pig.

Wash out the stomach and observe the interior of the stomach. Note the numerous folds in the stomach lining that allow the stomach to expand when filled with food. Observe that the stomach wall is thicker just anterior to the junction with the small intestine. This thickening is the **pyloric sphincter** muscle, which is usually closed, but it opens to release material from the stomach into the small intestine. Locate a similar, but less obvious, thickening at the esophagus–stomach junction. This is the **cardiac sphincter** muscle, which opens to allow food to enter the stomach, but usually remains closed to prevent regurgitation.

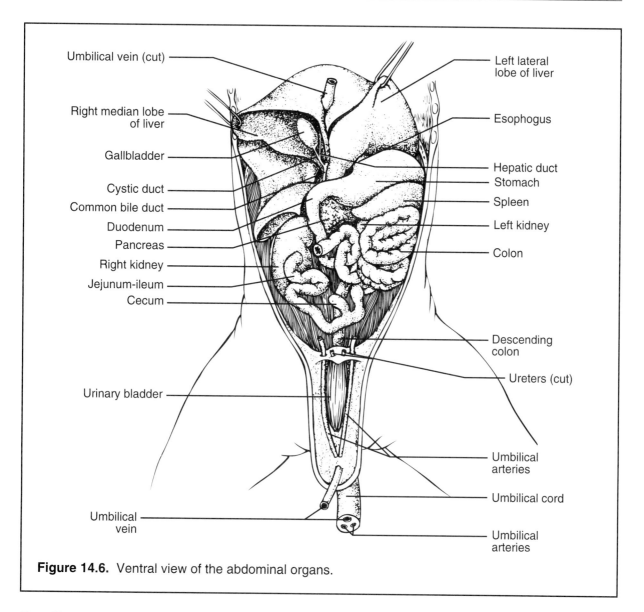

Figure 14.6. Ventral view of the abdominal organs.

Small Intestine

The small intestine consists of two parts: the duodenum and the jejunum-ileum. The first part of the small intestine is the short **duodenum**. It extends from the stomach, curves posteriorly, then turns anteriorly toward the stomach. The duodenum ends and the **jejunum-ileum** begins where the small intestine again turns posteriorly. In life, most of the digestion of food and absorption of nutrients occurs within the small intestine.

Lift out the small intestine and observe how it is supported by mesenteries that contain blood vessels and nerves. Without damaging the blood vessels, remove a 1-in. segment of the jejunum-ileum, cut through its wall lengthwise, open it up, and wash it out. The inner lining has enormous numbers of microscopic projections, known as **villi**, that give it a velvety appearance. Place a small, flattened section of the jejunum-ileum under water in a small petri dish and examine the inner lining under a dissecting microscope for a better look. The absorption of nutrients occurs through the villi.

Pancreas and Liver

Locate the pennant-shaped **pancreas**, which is situated in the mesentery within the curve of the duodenum. It secretes the hormone insulin as well as digestive enzymes that are emptied into the duodenum via a tiny **pancreatic duct**. Try to find this small duct by carefully dissecting away the peritoneum from a little "finger" of pancreatic tissue that follows along the descending duodenum.

The **liver** is divided into five lobes. It performs numerous vital metabolic functions and stores nutrients. Nutrients absorbed from the digestive tract are processed by the liver before being released into the general circulation. In addition, the liver produces **bile**. Bile consists of bile pigments, by-products of hemoglobin breakdown, and bile salts. Bile salts help emulsify fats, which facilitates their digestion in the small intestine.

Lift the right median lobe of the liver to locate the **gallbladder** on its undersurface, just to the right of the site at which the umbilical vein enters the liver. Careful dissection of the peritoneum will reveal the **hepatic duct**, which carries bile from the liver, and the **cystic duct**, which carries bile to and from the gallbladder. These two ducts merge to form the **common bile duct** that carries bile into the anterior part of the duodenum. A sphincter muscle at the end of the common bile duct prevents bile from entering the duodenum except when food enters the duodenum from the stomach. When this sphincter is closed, bile backs up and enters the gallbladder via the cystic duct, where it is stored until needed. Food entering the duodenum triggers a hormonal control mechanism that causes contraction of the gallbladder, forcing bile into the duodenum.

Large Intestine

Follow the small intestine to where it joins with the **large intestine** or **colon**. At this juncture, locate the small side pouch, the **cecum**, which is small and nonfunctional in pigs and humans. The **ileocecal valve** is a sphincter muscle that prevents food material in the colon from reentering the small intestine. Make a longitudinal incision through this region to observe the ileocecal opening and valve.

In life, the large intestine contains large quantities of intestinal bacteria that decompose the nondigested food material that enters it. Water is reabsorbed from the large intestine into the blood, leaving the feces that are expelled in defecation.

The pig's large intestine forms a unique, tightly coiled spiral mass. From the spiral mass, it extends anteriorly to loop over the duodenum before descending posteriorly against the dorsal wall into the pelvic region, where its terminal portion, the **rectum**, is located. The external opening of the rectum is the **anus**. The rectum will be observed in a later portion of the dissection.

CARDIOVASCULAR SYSTEM

The cardiovascular system consists of the heart, arteries, capillaries, veins, and blood. **Blood** is the carrier of materials that are transported by the circulatory system. It is pumped by the **heart** through **arteries** that carry it away from the heart and into the small vessels known as **capillaries**, which carry blood to the body tissues. Blood is collected from the capillaries by **veins** that return the blood to the heart.

In this section, you are to (1) locate the major blood vessels, noting their location and function, and (2) identify the external features of the heart. In a double-injected fetal pig, the arteries are injected with red latex and the veins are injected with blue latex. However, a few veins in your pig may not be injected because valves in veins tend to restrict the flow of latex. Also, some blood vessels may not be in the exact locations shown in the figures. Sometimes it is necessary to trace a vessel to the organ it serves in order to positively identify it.

As you study the heart, major veins, and major arteries, carefully remove tissue as necessary to expose the vessels. This is best done by separating tissues with a blunt probe and by picking away connective tissue from the blood vessels with forceps. Make your observations in accordance with the descriptions and sequence that follow.

Before starting your dissection of the circulatory system, it is important to understand the basic pattern of circulation in an adult mammal, as shown in Figure 14.7, and to contrast that pattern with the circulation in a fetal mammal, as shown in Figure 14.8.

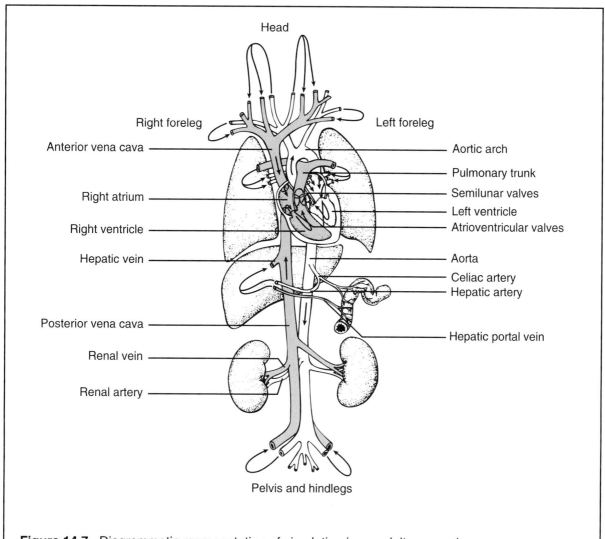

Figure 14.7. Diagrammatic representation of circulation in an adult mammal.

Adult Circulation

In the adult mammal, deoxygenated blood is returned from the body to the **right atrium** of the heart by two large veins. The **anterior vena cava** returns blood from regions anterior to the heart. The **posterior vena cava** returns blood from areas posterior to the heart. Blood from the digestive tract is carried by the **hepatic portal vein** to the liver, whose metabolic processes modify the nutrient content of the blood before it is carried by the **hepatic vein** into the posterior vena cava.

At the same time that deoxygenated blood enters the right atrium, oxygenated blood from the lungs is carried by the **pulmonary veins** into the **left atrium** of the heart.

When the atria contract, blood from each atrium is forced into its corresponding ventricle. Immediately thereafter, the ventricles contract. During ventricular contraction, blood pressure in the ventricles closes the **atrioventricular valves**, preventing a backflow of blood into the atria, and opens the **semilunar valves** at the bases of the two large arteries that exit the heart. Deoxygenated blood in the **right ventricle** is pumped through the **pulmonary trunk**, which branches into the **pulmonary arteries** carrying blood to the lungs. Oxygenated blood in the **left ventricle** is pumped into the **aorta**,

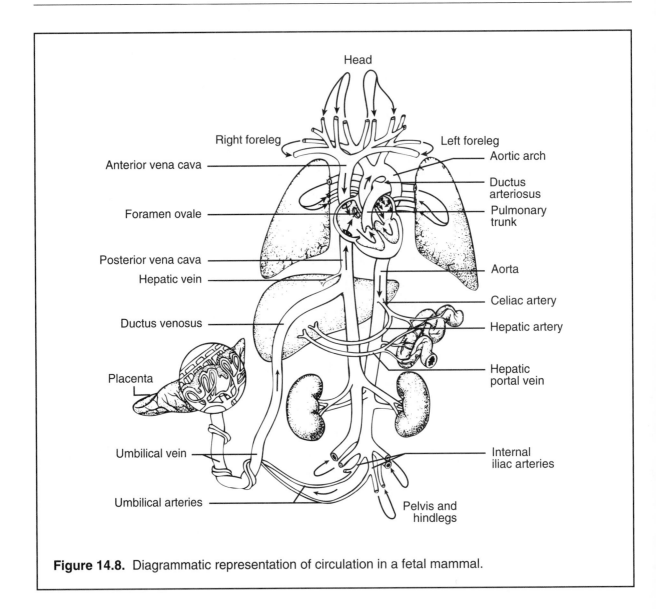

Figure 14.8. Diagrammatic representation of circulation in a fetal mammal.

which divides into numerous branches to carry blood to all parts of the body except the lungs.

At the end of ventricular contraction, the semilunar valves close, preventing a backflow of blood into the ventricles, and the atrioventricular valves open, allowing blood to flow from the atria into the ventricles in preparation for another heart contraction.

Fetal Circulation

In a fetus, the placenta is the source of oxygen and nutrients, and also removes metabolic wastes from the blood. The lungs, digestive tract, and kidneys are nonfunctional. Circulatory adapta-

tions to this condition make the circulation of blood in the fetus quite different from that in an adult.

Deoxygenated blood is returned from the body via the anterior and posterior vena cavae, as in the adult. However, blood rich in oxygen and nutrients is carried from the placenta by the **umbilical vein** directly through the liver by a segment of the umbilical vein called the **ductus venosus**, and on into the posterior vena cava. Therefore, the posterior vena cava returns blood to the right atrium, which has a high oxygen concentration, but less than that in the umbilical vein.

The fetal heart contains an opening between

the right and left atria called the **foramen ovale**, which allows much of the blood entering the right atrium from the posterior vena cava to pass directly into the left atrium. This enables blood with a high oxygen content to enter the left ventricle, which pumps it through the aorta to the body.

Much of the blood pumped from the right ventricle through the pulmonary trunk passes through a fetal vessel called the **ductus arteriosus**, which leads into the aorta, bypassing the lungs and augmenting the blood supply to the body. Very little blood is carried by the pulmonary arteries to the nonfunctional lungs, or returned to the heart by the pulmonary veins.

In the pelvic region, the **umbilical arteries** arise from the internal iliac arteries and carry blood to the placenta, where wastes are removed and oxygen and nutrients are picked up from the maternal blood.

At birth, the lungs are inflated by breathing, and the umbilical blood vessels, ductus venosus, and ductus arteriosus constrict, causing blood pressure changes that close the foramen ovale. These changes produce the adult pattern of circulation. Growth of fibrous connective tissue subsequently seals the foramen ovale and converts the constricted vessels into ligamentous cords.

The Heart and Its Great Vessels

Study the structure of the heart and the great vessels shown in Figures 14.9, 14.10, and 14.11. Be sure that you know the relative positions of these vessels before trying to locate them in your specimen.

If you haven't done so, carefully cut away the pericardial sac from the heart, and the attachment of this sac to the great vessels. Locate the **left** and **right atria**, blood-receiving chambers, and the **left** and **right ventricles**, blood-pumping chambers. Note the **coronary arteries** and **cardiac veins** that run diagonally across the heart at the location of the ventricular septum, a muscular partition separating the two ventricles. Coronary arteries supply blood to the heart muscle, and blockage of these arteries results in a heart attack.

While viewing the ventral surface of the heart, locate the large **anterior vena cava** that returns blood from the head, neck, and forelegs to the right atrium. By lifting up the apex of the heart you will see the large **posterior vena cava** that returns blood to the right atrium from regions posterior to the heart. Later, when you remove the heart, you will see where these veins enter the right atrium.

Now, locate the **pulmonary trunk** that exits the right ventricle and, just dorsal to it, the **aorta**, which exits from the left ventricle. These large arteries appear whitish because of their thick walls. Move the apex of the heart to your left and locate the whitish **ductus arteriosus** that carries blood from the pulmonary trunk into the aorta, bypassing the nonfunctional fetal lungs. You will see the **pulmonary arteries** and **pulmonary veins** later, when you remove the heart.

Major Vessels of the Head, Neck, and Thorax

As you read the description of each major vessel, locate it first in Figures 14.10 and 14.11; then locate it in your dissection specimen.

As noted earlier, the large **anterior vena cava** returns blood from the head, neck, and forelegs into the right atrium. It is formed a short distance anterior to the heart by the union of the left and right brachiocephalic veins. The **brachiocephalic veins** are large but very short. Each one is formed by the union of veins draining the head, neck, and forelegs.

There are two jugular veins on each side of the neck. The **external jugular vein** is more laterally located and drains superficial tissues. The **internal jugular vein** is more medially located and drains deep tissues including the brain. The external and internal jugular veins may join just before their union with the subclavian vein. A **cephalic vein** drains part of the shoulder and joins with the external jugular vein near the union of the jugular veins.

The short, paired **subclavian veins** drain the shoulders and forelegs. Each is formed by the merging of a **subscapular vein**, which drains the posterior shoulder muscles, and an **axillary vein**, which receives blood from a **brachial vein** (non shown) of the foreleg.

Inferior to the brachiocephalic veins, the anterior vena cava receives blood from (1) a pair of **internal thoracic veins**, which return blood from the internal thoracic wall; (2) a pair of

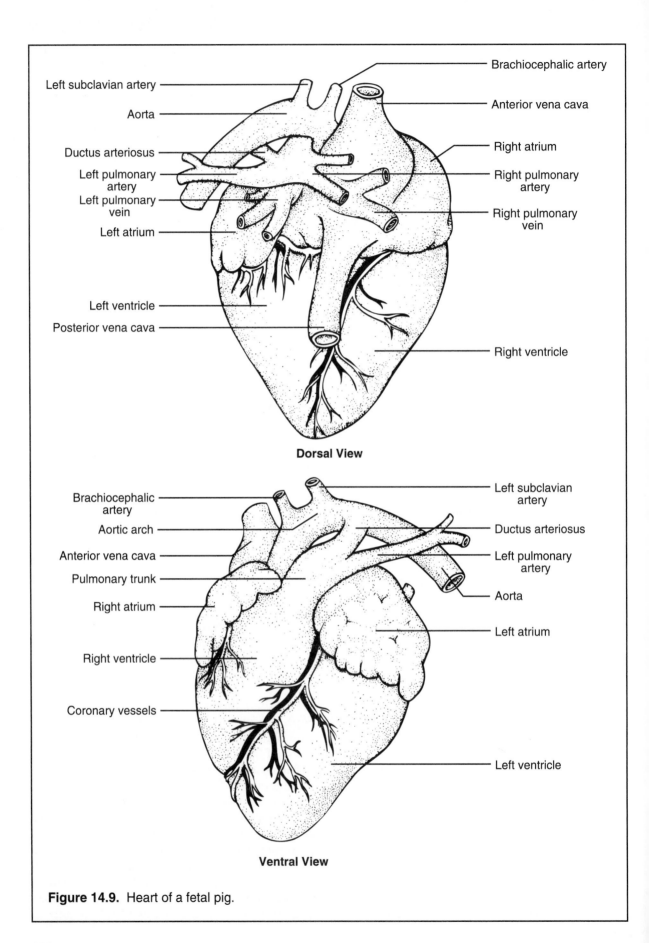

Left subclavian artery

Aorta

Ductus arteriosus

Left pulmonary artery

Left pulmonary vein

Left atrium

Left ventricle

Posterior vena cava

Brachiocephalic artery

Anterior vena cava

Right atrium

Right pulmonary artery

Right pulmonary vein

Right ventricle

Dorsal View

Brachiocephalic artery

Aortic arch

Anterior vena cava

Pulmonary trunk

Right atrium

Right ventricle

Coronary vessels

Left subclavian artery

Ductus arteriosus

Left pulmonary artery

Aorta

Left atrium

Left ventricle

Ventral View

Figure 14.9. Heart of a fetal pig.

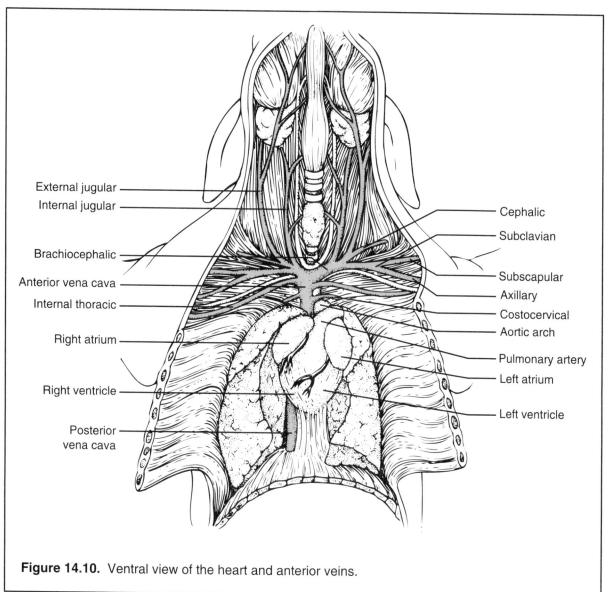

Figure 14.10. Ventral view of the heart and anterior veins.

Labels (left side, top to bottom): External jugular, Internal jugular, Brachiocephalic, Anterior vena cava, Internal thoracic, Right atrium, Right ventricle, Posterior vena cava

Labels (right side, top to bottom): Cephalic, Subclavian, Subscapular, Axillary, Costocervical, Aortic arch, Pulmonary artery, Left atrium, Left ventricle

external thoracic veins (not shown), which return blood from the external thoracic wall; and (3) a pair of **costocervical veins**, which return blood from back and neck muscles.

Once you have located these veins, cut the superior vena cava, leaving a stub at the heart, and lift it and the attached veins anteriorly to expose the underlying arteries. Locate the **aorta** and note that it forms the aortic arch as it curves posteriorly, dorsal to the heart. Two major arteries branch from the aortic arch. The first and larger branch is the **brachiocephalic artery**.

The second and smaller branch is the **left subclavian artery**.

Follow the brachiocephalic artery anteriorly, where it branches to form the **right subclavian artery** and a **carotid trunk** that divides to form a pair of **common carotid arteries**. The common carotid arteries parallel the internal jugular veins to carry blood to the head and neck. At the base of the head, each common carotid artery branches into **internal** and **external carotid arteries**.

Each subclavian artery gives off (1) a **costo-**

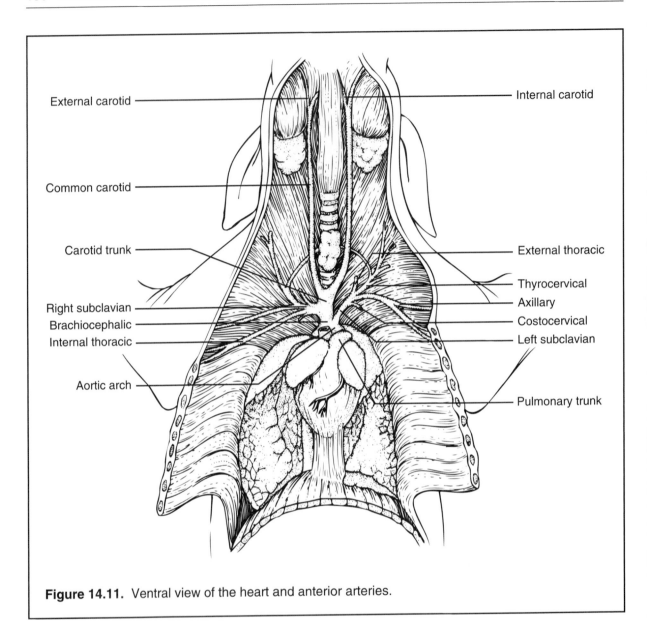

Figure 14.11. Ventral view of the heart and anterior arteries.

cervical artery, which supplies the back and neck; (2) an **external thoracic artery**, which supplies the external thoracic wall; (3) a **thyrocervical artery**, which supplies the thyroid gland and neck; and (4) an **internal thoracic artery**, which supplies the internal thoracic wall. Each subclavian artery becomes an **axillary artery** in the axillary or armpit region.

After you have located the major anterior arteries and veins, carefully cut through the major vessels to remove the heart. Leave stubs of the vessels on the heart and identify them by comparing your preparation with Figure 14.9.

Especially locate the pulmonary arteries and veins.

Major Vessels Posterior to the Diaphragm

Follow the aorta and posterior vena cava posteriorly through the diaphragm. Compare your observations with Figures 14.12 and 14.13.

Just posterior to the diaphragm, locate the **hepatic vein**, which emerges from the liver and enters the posterior vena cava. Immediately posterior to this union, the **ductus venosus** enters the posterior vena cava. Recall that the ductus

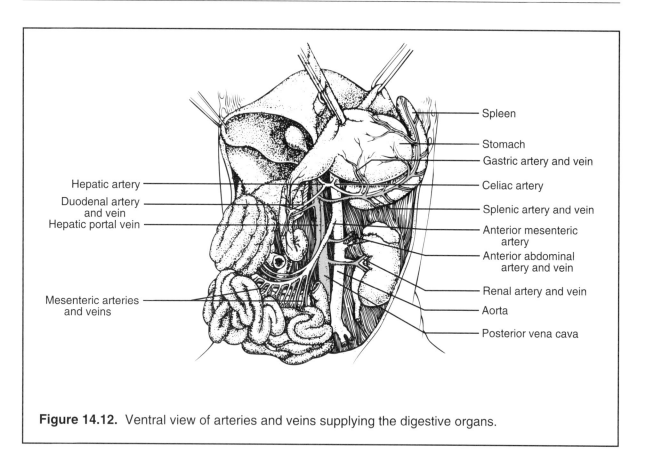

Figure 14.12. Ventral view of arteries and veins supplying the digestive organs.

venosus is an extension of the **umbilical vein** in the fetal pig. Dissect away liver tissue as necessary to expose the unions of the hepatic vein and ductus venosus with the posterior vena cava. Save the liver tissue adjacent to the hepatic vein and ductus venosus, but remove and discard the rest. Cut through the left side of the diaphragm, and push the abdominal organs to your left (specimen's right) to expose the posterior vena cava and aorta.

Lift the stomach anteriorly to expose the first branch from the aorta posterior to the diaphragm. This is the **celiac artery**. It gives off branches to the stomach, spleen, and liver (**hepatic artery**). Posterior to the celiac artery is a large branch, the **anterior mesenteric artery**, that in turn gives off numerous branches to the intestine. Note that many small veins from the intestine join to form the **hepatic portal vein** (often not injected), which runs anteriorly to enter the liver. This vein carries nutrient-rich blood from the intestine to the liver for processing, prior to its being released into the general circulation.

Now remove most of the intestines, but save a small section associated with the hepatic portal vein. This will expose the lower portion of the aorta and posterior vena cava.

Locate the **renal arteries** and **renal veins**, which serve the kidneys. Since the kidneys are located dorsal to the parietal peritoneum, you will have to strip away some of this tissue to observe the vessels and kidneys clearly. The **adrenal glands** (endocrine glands), narrow strips of tissue, should be visible on the anterior margin of the kidneys.

Carefully dissect away the connective tissue to expose the posterior portions of the aorta, posterior vena cava, and the attached vessels. Locate the ureters, which carry urine from the kidneys to the urinary bladder. The bladder is located in the reflected tissue containing the umbilical arteries and umbilical cord. Free the ureters from surrounding tissue so that they are movable, but don't cut through them.

The small arteries and veins posterior to the renal vessels serve the gonads—the ovaries or

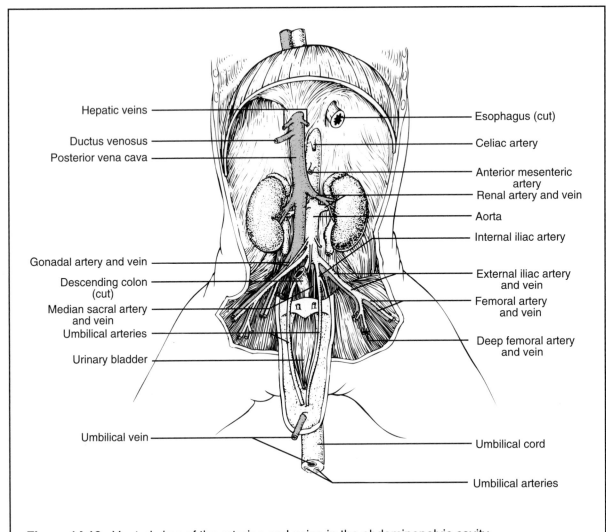

Hepatic veins

Ductus venosus

Posterior vena cava

Gonadal artery and vein

Descending colon
(cut)

Median sacral artery
and vein

Umbilical arteries

Urinary bladder

Umbilical vein

Esophagus (cut)

Celiac artery

Anterior mesenteric
artery

Renal artery and vein

Aorta

Internal iliac artery

External iliac artery
and vein

Femoral artery
and vein

Deep femoral artery
and vein

Umbilical cord

Umbilical arteries

Figure 14.13.   Ventral view of the arteries and veins in the abdominopelvic cavity.

testes, as the case may be. Just below the gonadal arteries, the **posterior mesenteric artery** arises from the ventral surface of the aorta and supplies most of the large intestine.

Locate the pair of **external iliac arteries** that branch laterally from the posterior end of the aorta. After giving off small branch arteries to the body wall, each external iliac artery extends into the thigh, where it divides into the medially located **deep femoral artery** and the laterally located **femoral artery**. Each of these arteries is associated with a corresponding vein, the **deep femoral vein** and the **femoral vein**, which drain the leg and merge to form the **external iliac vein**.

Now locate the medial terminal branches of the aorta, the **internal iliac arteries**. The major extensions of these arteries form the **umbilical arteries**, which pass over the urinary bladder to enter the umbilical cord. The other branches of the internal iliac arteries supply the pelvic area and lie alongside the **internal iliac veins**, which drain the pelvic area.

The external and internal iliac veins join to form a **common iliac vein** on each side of the body. The union of the common iliac veins forms the posterior beginning of the posterior vena cava.

You will see, located between the commen iliac veins, the **median sacral artery** and **vein**, which serve the dorsal wall of the pelvic area.

THE RESPIRATORY SYSTEM

The respiratory system functions to bring air into the lungs, where blood releases its load of carbon dioxide and receives oxygen for transport to body cells. Air enters the lungs via a series of air passageways. Air enters the **nostrils** and flows through the **nasal cavity**, where it is warmed and moistened. It then passes into the **nasal pharynx** and on into the **oral pharynx**, where it enters the **larynx** via a slit-like opening called the **glottis**. You observed these components when you dissected the oral cavity and pharynx. From the larynx, air passes down the **trachea**, which branches to form the **primary bronchi**, which carry air into the **lungs**.

Return now to the neck and thoracic cavity to observe the respiratory organs. Locate the larynx, trachea, and primary bronchi. Since the heart has been removed, the trachea and primary bronchi may be readily seen by dissecting away some of the surrounding connective tissue and blood vessels. See Figure 14.11.

Make a mid-ventral incision in the larynx, open it, and identify the **vocal folds** located on each side. Note how the trachea branches to form the primary bronchi. Each primary bronchus divides within the lung to form smaller and smaller air passages until tiny microscopic **bronchioles** terminate in vast numbers of saclike **alveoli**, the sites of gas exchange. Dissect along a primary bronchus to locate the **secondary bronchi** that enter the lobes of a lung. Remove a section of trachea and observe the shape of the **cartilaginous rings** that hold it open. Make a section through a lung to observe its spongy nature and the air passages within it.

UROGENITAL SYSTEM

The urinary and reproductive organs compose the urogenital system. These organs are considered together because they are closely interrelated. Refer to Figures 14.14 or 14.15, depending on the sex of your specimen. You are responsible for knowing the urogenital system of each sex, so prepare your dissection carefully and exchange it when finished for a specimen of the opposite sex.

Urinary System

If you haven't removed the intestines, do so now to expose the urinary organs. Leave a stub of the large intestine because you will want to locate the rectum later. Dissect away the parietal peritoneum to expose the **kidneys**, located dorsal to the parietal peritoneum and against the dorsal body wall. The kidneys are held in place by connective tissue. An **adrenal gland** is located on the anterior surface of each kidney.

Locate the origin of a **ureter** on the medial surface of a kidney, near where the renal vessels enter it. As urine is formed by the kidney, it passes into the ureter for transport by peristalsis (wavelike contractions) to the urinary bladder. Trace the ureter posteriorly to its dorsally located entrance into the **urinary bladder**. The urinary bladder lies between the umbilical arteries on the reflected portion of the body wall containing the umbilical cord. Note that the posterior portion of the urinary bladder narrows to form the **urethra** that enters the pelvic cavity. The urethra will be observed momentarily.

If your instructor wants you to dissect a kidney, make a coronal section to expose its interior. Compare the internal structure with Figure 22.2 in Exercise 22 as you read the accompanying description.

Female Reproductive System

The **uterus** is located dorsal to the urinary bladder. It consists of two **uterine horns** that join posteriorly at the midline to form **the body of the uterus**. Locate the uterine horns and the body of the uterus.

Follow the uterine horns anterolaterally to the small, almost nodule-like **ovaries**, located just posterior to the kidneys, where they are supported by mesenteries. The uterine horns end at the posterior margin of the ovaries, where they are continuous with the tiny, convoluted **uterine tubes** that continue around the ovaries to their anterior margins. The anterior end of a uterine tube forms an expanded, funnel-shaped opening, the **infundibulum**, that receives the immature ovum when it is released from an ovary. Fertilization occurs in the uterine tubes, and fetal pigs develop in the uterine horns.

Now remove or reflect the skin from the ventral surface of the pelvis. Use your scalpel to cut carefully at the ventral midline through the

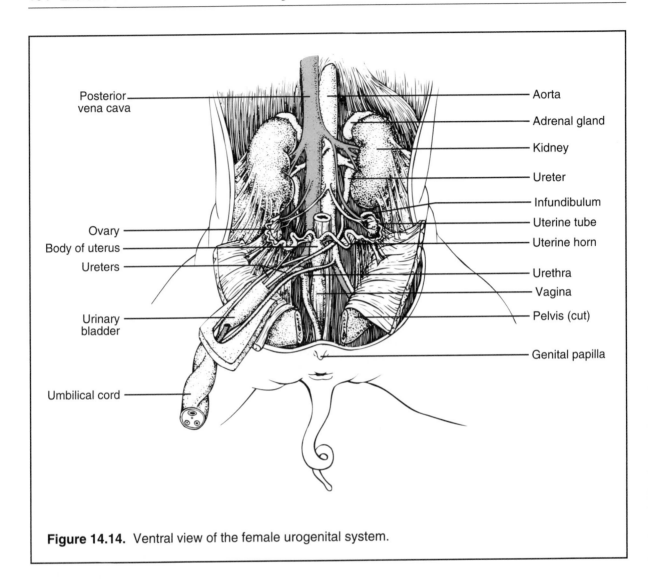

Figure 14.14. Ventral view of the female urogenital system.

muscles and bones of the pelvic girdle. Spread the legs and open the pelvic cavity to expose the **urethra** and, just dorsal to it, the **vagina**, extending posteriorly from the body of the uterus. You can now see the **rectum**, the terminal portion of the large intestine, located dorsally to the vagina. Note that the vagina and urethra unite to form the **vaginal vestibule** a short distance from the external urogenital opening, which is identified externally by the genital papilla.

Male Reproductive Organs

The **testes** develop within the body cavity just posterior to the kidneys. Later in fetal development, the testes descend through the **inguinal canals** into the **scrotum**, an external pouch. In

being located externally, the scrotum provides a temperature for the testes that is slightly less than body temperature. The lower temperature is necessary for the production of viable sperm. Follow a testicular artery and vein to locate the inguinal canal.

On one side, cut open the scrotum to expose a testis. Locate the **epididymis**, a tortuous mass of tiny tubules that begins on the anterior margin of the testis and extends along the lateral margin to join posteriorly with the **vas deferens** or sperm duct. Trace the vas deferens anteriorly from the scrotum and through the inguinal canal, to where the vas deferens loops over a ureter to enter the urethra.

Locate the **urogenital opening** just posterior to the umbilical cord in the reflected flap of

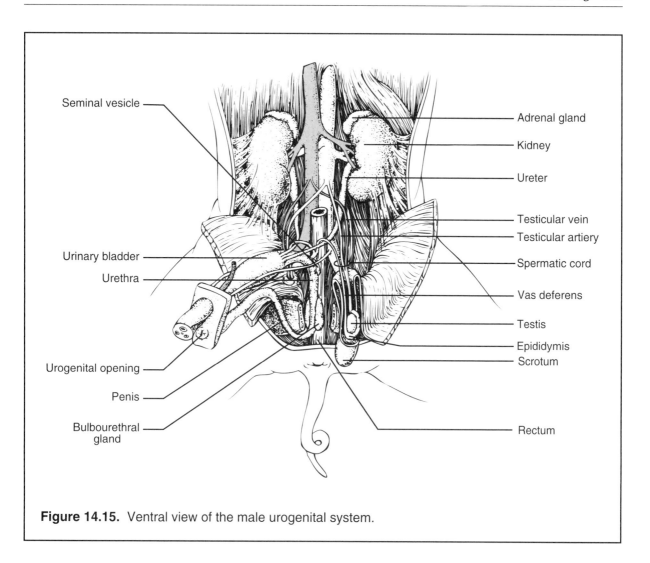

Figure 14.15. Ventral view of the male urogenital system.

body wall containing the urinary bladder. The **penis** extends posteriorly from this point. Make an incision alongside the penis and free it from the body wall. Push it to one side, and use your scalpel to make a midline incision through the pelvic muscles and bones. Spread the legs and open the pelvic cavity so that you can dissect out the pelvic organs. Locate the **urethra** at the base of the urinary bladder and carefully remove connective tissue around it, separating it from the **rectum** as it continues posteriorly into the pelvic cavity.

Where the vasa deferentia enter the urethra, locate the small glands on each side of the urethra. These are the **seminal vesicles**. Between the seminal vesicles on the dorsal surface of the

urethra is the small **prostate gland**. Follow the urethra posteriorly to locate the pair of **bulbourethral glands** on each side of the urethra where it enters the penis. At ejaculation, these accessory glands secrete the fluids that transport sperm.

CONCLUSION

This completes the dissection. If you have done it thoughtfully, you have gained a good deal of knowledge about the body organization of mammals, including humans. Dispose of your specimen as directed by your instructor. There is no laboratory report for this exercise.

15

Circulation of Blood

OBJECTIVES

After completion of the laboratory session, you should be able to:
1. Describe the structure and function of the heart.
2. Identify the parts of a dissected sheep heart and describe their functions.
3. Trace the flow of blood through the heart and the blood vessels carrying blood to and from the heart.
4. Contrast the structure and function of arteries, capillaries, and veins.
5. Determine the pulse rate and blood pressure of your lab partner.
6. Define all terms in bold print.

The **cardiovascular system** transports substances throughout the body by circulating the blood, which carries these substances, through a system of closed vessels. In this way, nutrients and oxygen are supplied to cells and organic wastes and carbon dioxide are removed from them. The components of the cardiovascular system and their functions are:

Heart   Pumps the blood through the blood vessels.

Arteries   Carry blood away from the heart.

Capillaries Microscopic vessels connecting the smallest arteries and smallest veins; sites of the exchange of materials between the blood and body cells.

Veins   Return blood to the heart.

Blood   Transporting medium.

In this exercise you will investigate the heart and blood vessels. Blood will be considered in the next exercise.

THE HEART

As you read this section, refer to Figure 15.1. The human heart, like those of all mammals and birds, consists of four chambers: two atria and two ventricles. The two **atria** receive blood returning to the heart in veins, and the two **ventricles** pump blood away from the heart into arteries.

The right atrium receives blood from two large veins. The **superior vena cava** returns blood from the head, neck, shoulders, and arms, and the **inferior vena cava** returns blood from the rest of the body and the legs. The left atrium receives blood from the left and right **pulmonary veins**. The right ventricle pumps blood into the **pulmonary trunk**, which branches into the pulmonary arteries carrying blood to the lungs.

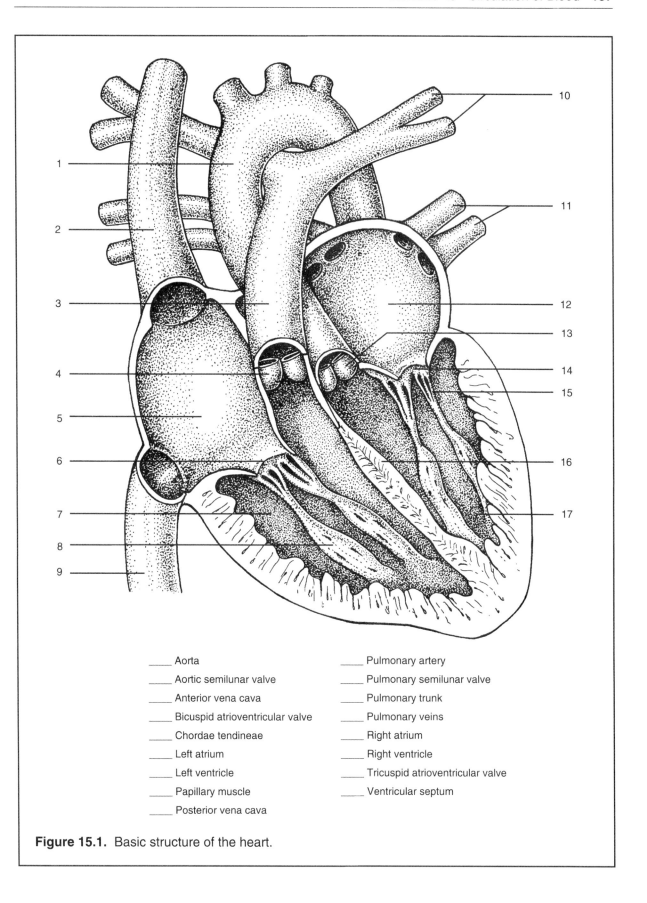

Figure 15.1. Basic structure of the heart.

_____ Aorta

_____ Aortic semilunar valve

_____ Anterior vena cava

_____ Bicuspid atrioventricular valve

_____ Chordae tendineae

_____ Left atrium

_____ Left ventricle

_____ Papillary muscle

_____ Posterior vena cava

_____ Pulmonary artery

_____ Pulmonary semilunar valve

_____ Pulmonary trunk

_____ Pulmonary veins

_____ Right atrium

_____ Right ventricle

_____ Tricuspid atrioventricular valve

_____ Ventricular septum

The left ventricle pumps blood into the **aorta**, which divides to carry blood to all parts of the body except the lungs.

The heart actually is composed of two pumps. The right atrium and right ventricle form one pump, and the left atrium and left ventricle form the other pump. There are no connections between the two pumps. That is, there are no connections between the two atria or between the two ventricles. The **ventricular septum**, a wall of cardiac muscle, separates the two ventricles.

The atrium and ventricle on each side of the heart are connected by an opening that is guarded by an atrioventricular (AV) valve. The two AV valves are named for the number of cusps or valve flaps that compose them. The **bicuspid atrioventricular valve** is located between the left atrium and left ventricle. The **tricuspid atrioventricular valve** lies between the right atrium and right ventricle. Heart valves keep the blood flowing in a single direction. Each AV valve permits blood to flow from atrium to ventricle but prevents reverse flow from ventricle to atrium.

The cusps of the AV valves are anchored by **chordae tendineae**—thin, nonelastic cords of fibrous connective tissue—to the **papillary muscles**, small mounds of heart muscle on the inner surface of the ventricles. This arrangement holds the valve flaps so that they seal off each atrium during ventricular contraction, and prevents the valve cusps from being forced into the atria. When the ventricles contract, the valve cusps restrained by the chordae tendineae resemble parachutes and their lines.

Another set of valves, the semilunar valves, are located at the base of the two large arteries carrying blood from the heart. The **pulmonary semilunar valve** and the **aortic semilunar valve** allow blood to enter their respective arteries during ventricular contraction, but prevent the backflow of blood from the arteries into the ventricles.

The wall of the heart, best observed in the ventricles, consists mostly of **cardiac muscle tissue** sandwiched between two membranes. The interior of the heart is lined by the **endocardium**, a membrane of simple squamous epithelium, and the exterior surface is covered by the **epicardium**, which is also simple squamous epithelium. In the body, the heart is enveloped by a somewhat loose-fitting, double-layered membrane, the **pericardial sac**. Fluid secreted between the pericardial sac and the epicardium reduces friction, enabling the heart to move freely within the pericardial sac as it beats.

Flow of Blood Through the Heart

The flow of blood through the heart is related to the heart cycle of relaxation and contraction. **Diastole** is the relaxation phase, and **systole** is the contraction phase.

During atrial and ventricular diastole, blood flows into the right atrium from the superior and inferior venae cavae, and into the left atrium from the pulmonary veins. The contraction of the atria in atrial systole forces blood into the relaxed ventricles until they are filled with blood. In ventricular systole, the ventricles contract and the atria relax. The sudden increase in blood pressure within the ventricles closes the AV valves and pumps blood through the semilunar valves into the arteries. The right ventricle pumps blood into the pulmonary trunk and on to the lungs, and the left ventricle pumps blood into the aorta and on to all other parts of the body. At the start of ventricular diastole, the semilunar valves close. Note that changes in blood pressure in the heart chambers cause the opening and closing of the heart valves.

Heart Sounds

The heart sounds are usually described as a *lub-dup* (pause) *lub-dup*, and so forth. These sounds are produced by the closing of the heart valves. The closure of the AV valves produces the first sound at the start of ventricular systole, and the closure of the semilunar valves produces the second sound at the start of ventricular diastole.

Materials

Per lab
Alcohol pads
Colored pencils
Heart model
Stethoscopes

Assignment 1

1. Label Figure 15.1 and color-code the major parts of the heart.
2. Identify the parts of the heart on a heart model.
3. Use a stethoscope to listen to your own heart sounds. A stethoscope amplifies the sounds so that you can hear them more clearly. Clean the ear pieces of the stethoscope with an alcohol pad. Fit them into your ears by directing them inward and forward. Place the stethoscope diaphragm over your heart just to the left of your sternum (breast bone) to hear the heart sounds. You don't need to place the stethoscope diaphragm against your skin—over a shirt or blouse usually is OK.
4. **Complete item 1 on Laboratory Report 15 that begins on page 363.**

DISSECTION OF A SHEEP HEART

A study of external and internal structure of a sheep heart will extend your understanding of heart structure and function. Refer to Figures 15.1 and 15.2 as you proceed.

Materials

Per student group
Dissecting instruments
Dissecting pan
Sheep heart, fresh or preserved

Assignment 2

1. Study the sheep heart as described below. Because a sheep walks on four legs and humans walk on two legs, descriptive directional terms pertaining to the heart are different in the two organisms. A few comparative terms for each are:

Sheep	Human
Ventral	Anterior
Dorsal	Posterior
Anterior	Superior
Posterior	Inferior

External Features

a. Usually the pericardial sac has been removed from the sheep heart, but look for remnants of it around the bases of the venae cavae, aorta, and pulmonary vessels.
b. Note that the epicardium is so tightly attached to the heart that to the naked eye, it does not appear as a distinct membrane.
c. The coronary arteries and veins that supply blood to the heart itself are obscured by fat deposits. Locate the fat deposits along the interventricular sulcus (groove) on the ventral surface of the heart. The right and left ventricles lie on either side of this sulcus. Pick away some of the fat to locate a coronary artery and cardiac vein.
d. Looking at the ventral surface of the heart and referring to Figure 15.2, locate the left and right atria, aorta, pulmonary trunk, and anterior vena cava. Compare the thickness of the vessel wall in the aorta with that of the anterior vena cava.
e. Looking at the dorsal view of the heart and referring to Figure 15.2, locate the aorta, left and right pulmonary arteries, left and right pulmonary veins (which are often embedded in fat), anterior vena cava, and posterior vena cava.

Internal Features

a. Hold the heart in your left hand, dorsal side up, with the anterior vena cava toward you. Insert a scissors blade into the anterior vena cava and cut through its wall into the right atrium. Locate the tricuspid AV valve. Holding the heart upright, pour water through the tricuspid valve into the right ventricle. Observe the action of the tricuspid valve as you gently squeeze the right ventricle.
b. Pour out the water from the right ventricle and continue your cut through the tricuspid valve and ventricle wall to the tip of the right ventricle. Spread the cut ventricular wall and locate the papillary muscles and chordae tendineae. Are the chordae tendineae fragile or tough, strong or weak, elastic or nonelastic?
c. Insert a probe into the cut end of the pulmonary trunk and into the right ventricle.

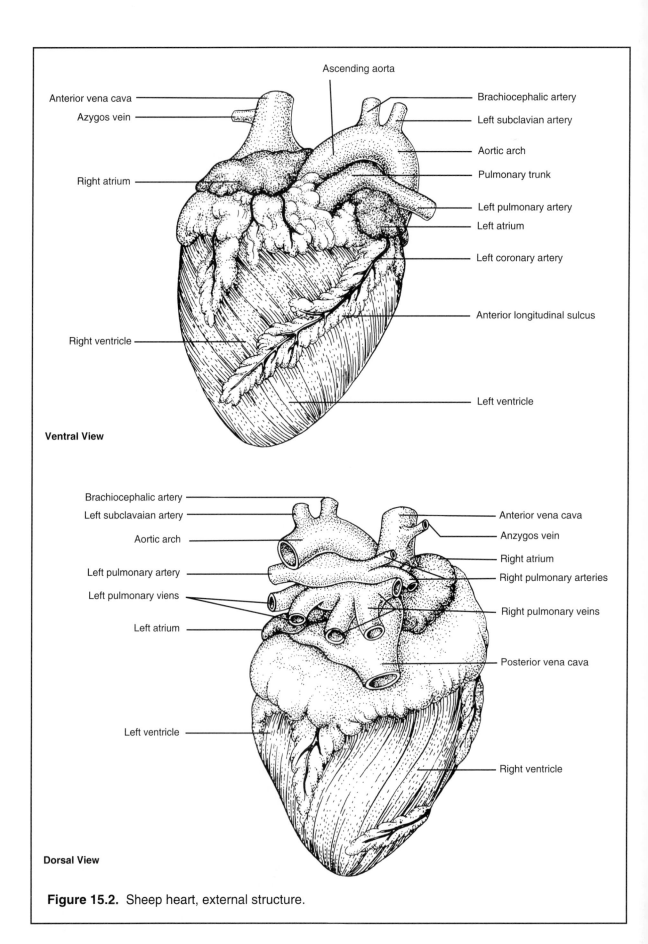

Ventral View

Ascending aorta

Anterior vena cava

Azygos vein

Right atrium

Right ventricle

Brachiocephalic artery

Left subclavian artery

Aortic arch

Pulmonary trunk

Left pulmonary artery

Left atrium

Left coronary artery

Anterior longitudinal sulcus

Left ventricle

Brachiocephalic artery

Left subclavaian artery

Aortic arch

Left pulmonary artery

Left pulmonary viens

Left atrium

Left ventricle

Anterior vena cava

Anzygos vein

Right atrium

Right pulmonary arteries

Right pulmonary veins

Posterior vena cava

Right ventricle

Dorsal View

Figure 15.2. Sheep heart, external structure.

Use scissors to cut from the right ventricle through the anterior ventricular wall along the probe and through the pulmonary semilunar valve. Note the arrangement and thickness of the cusps.

d. Insert a scissors blade into the top of the left atrium and cut through walls of the left atrium and ventricle to the tip of the left ventricle. Spread the ventricular walls and locate the bicuspid valve, chordae tendineae, and papillary muscles. Compare the thickness of the walls in the left and right ventricles.

e. Insert a probe into the cut end of the aorta and into the left ventricle. Cut from the left ventricle along the probe and through the aortic semilunar valve. Locate the two openings to coronary arteries just above (anterior to) the semilunar valve. Note the valve structure.

f. Dispose of the heart as directed by your instructor. Wash and dry your dissecting pan and instruments.

2. **Complete item 2 on the laboratory report.**

THE PATTERN OF CIRCULATION

In vertebrates, blood flows through a system of closed vessels to transport materials to and from body cells. **Arteries** carry blood from the heart, and they divide into smaller and smaller arteries ultimately leading to **arterioles**, which are nearly microscopic in size. Arterioles lead to the **capillaries**, the smallest blood vessels. The walls of the capillaries are composed of a single

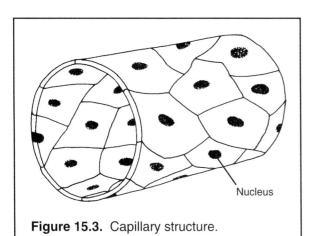

Figure 15.3. Capillary structure.

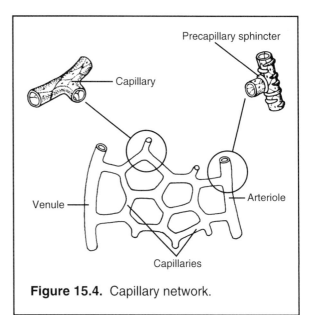

Figure 15.4. Capillary network.

layer of squamous epithelial cells. See Figure 15.3. Exchange of materials occurs between blood in capillaries and the body cells. From capillaries, blood flows into tiny veins called **venules**, which combine to form larger veins that carry blood back to the heart.

Figure 15.4 shows the relationship between an arteriole, capillaries, and a venule. The diameter of the arterioles affects the flow of blood and blood pressure, and is controlled by circular smooth muscles that respond to impulses from the autonomic nervous system. The flow of blood into capillaries is controlled by a **precapillary sphincter muscle** that is governed by impulses from the autonomic nervous system and local chemical stimuli. An increase in the local CO_2 concentration opens the sphincter and increases capillary blood flow. Similarly, a decrease in CO_2 concentration reduces capillary blood flow.

The exchange of materials between the body cells and capillary blood involves **tissue fluid** (interstitial fluid) as an intermediary. Tissue fluid is the thin layer of extracellular fluid that covers all cells and tissues. This is the pattern of the exchange:

$$\text{Blood in capillaries} \rightleftharpoons \text{Tissue fluid} \rightleftharpoons \text{Body cells}$$

Since the heart is a double pump, there are two separate circuits (pathways) of circulation:

the pulmonary circuit and the systemic circuit. Examine the diagrammatic scheme of circulation in Figure 15.5 as you read this section.

The **pulmonary circuit** carries deoxygenated (low oxygen content) blood to the lungs and returns oxygenated (high oxygen content) blood to the heart. The right ventricle pumps deoxygenated blood through the pulmonary trunk, which branches to form the pulmonary arteries carrying blood to the lungs. In the lungs, oxygen is picked up by the blood as it releases carbon dioxide into air in the lungs. The oxygenated blood is then returned through the pulmonary veins to the left atrium.

The **systemic circuit** carries oxygenated blood to all parts of the body except the lungs, and returns deoxygenated blood to the heart. The left ventricle pumps oxygenated blood through the aorta, which in turn gives off smaller arteries that carry oxygenated blood throughout the body (except to the lungs). After oxygen and carbon dioxide are exchanged with body cells, deoxygenated blood is carried in veins that merge to form the superior vena cava, which drains areas above the heart, and the inferior vena cava, draining areas below the heart. The venae cavae return deoxygenated blood into the right atrium.

Note that blood from the intestines is carried by the hepatic portal vein to the liver before it is returned to the inferior vena cava by the hepatic vein. This allows the liver to process nutrients absorbed from the intestine before they enter the systemic circulation.

Materials

Per lab

Demonstration setup of capillary flow in a frog's foot or a goldfish's caudal fin
Prepared slides of:
    artery and vein, x.s.
    atherosclerotic artery, x.s.

Assignment 3

1. **Complete items 3a–3c on the laboratory report.**
2. Examine the capillary blood in the webbing of a frog's foot that has been set up under a demonstration microscope. Note the pulsating flow in the feeding arteriole and the smooth flow in the collecting venule. Observe the blood cells moving in single file through a capillary. How would you describe the flow of blood in the capillary? Keep the foot wet with water. Turn off the microscope light when you have finished observing. **Complete items 3d–3f on the laboratory report.**
3. Examine prepared slides of artery and vein, x.s. Compare the difference in thickness of the walls of these vessels due to differences in the amount of smooth muscle and connective tissue in their walls. Locate the single layer of squamous endothelial cells forming the interior lining. **Draw these vessels in item 3g on the laboratory report.**
4. Examine a prepared slide of an atherosclerotic artery, x.s., and note the fatty deposit that partially plugs the vessel. Cholesterol is primarily responsible for this deposit. This type of obstruction in coronary arteries often causes heart attacks, and may require coronary bypass surgery. **Complete items 3g and 3h on the laboratory report.**
5. Label Figure 15.5. Add arrows to indicate the direction of blood flow. Color vessels carrying deoxygenated blood blue and those carrying oxygenated blood red.
6. **Complete item 3 on the laboratory report.**

BLOOD PRESSURE

Usually, the term "blood pressure" refers to the blood pressure within arteries of the systemic circuit. There are two types of blood pressure: systolic and diastolic. **Systolic blood pressure** occurs during ventricular contraction and normally averages 120 ± 10 mm Hg (mercury) when measured in the brachial artery of the upper arm. **Diastolic blood pressure** occurs during ventricular relaxation and normally averages 80 ± 10 mm Hg.

Pulse Pressure

The difference between the systolic and diastolic blood pressures produces the **pulse pressure**. The alternating increase and decrease of arterial blood pressures causes a corresponding expansion and contraction of the elastic arterial walls. The pulsating expansion of arterial walls

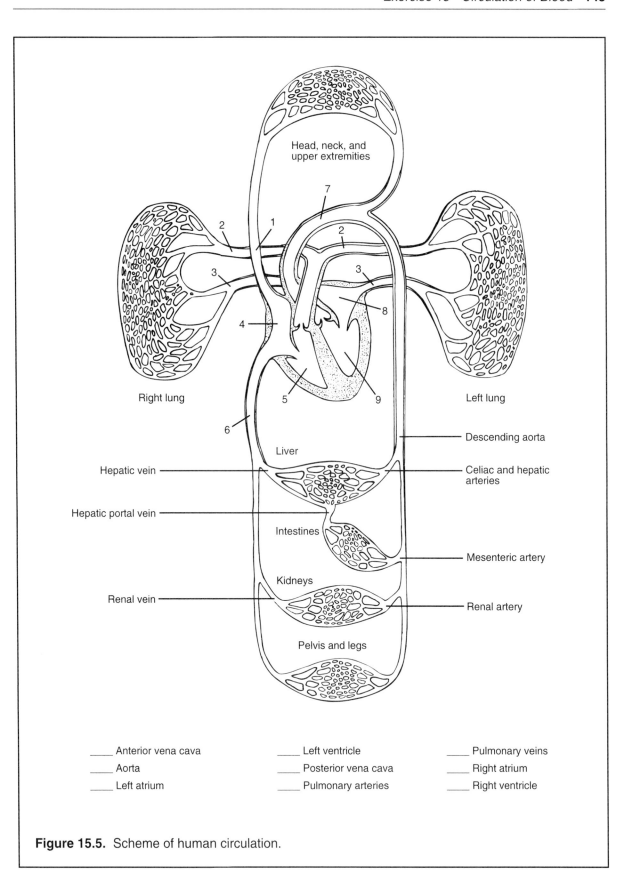

Figure 15.5. Scheme of human circulation.

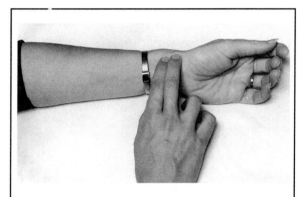

Figure 15.6. Measuring the pulse rate. Place the fingers over the radial artery at the wrist.

may be detected as the **pulse** by placing the fingers on the skin over a surface artery. See Figure 15.6.

The pulse rate indicates the number of heart contractions per minute. Normal pulse rates usually range between 65 and 80 beats per minute, but well-conditioned athletes may have rates as low as 40 per minute.

Measurement of Blood Pressure

Blood pressure is most commonly measured in the brachial artery with a **sphygmomanometer** and a **stethoscope**. The sphygmomanometer consists of an inflatable cuff that is wrapped around the upper arm, and a pressure gauge to measure the air pressure within the cuff. The stethoscope is used to hear sounds produced by blood rushing through a partially closed brachial artery. Read through the following procedures completely before you begin the process of measuring the blood pressure.

1. The arm of the "patient" should be resting palm up on the table top. Wrap the sphygmomanometer cuff around the upper arm, with the bottom edge of the cuff about an inch above the elbow joint. Secure the cuff with the Velcro® fastener and attach the pressure gauge so the dial may be easily read, as shown in Figure 15.7.
2. Insert the ear pieces of the stethoscope inward and forward into your ears. Hold the diaphragm of the stethoscope over the brachial artery with your left thumb, as shown in Fig-

ure 15.7. With your right hand, gently close the screw valve above the bulb of the sphygmomanometer and squeeze the bulb to inflate the cuff to about 150 mm Hg. This pressure closes the brachial artery.

3. Open the screw valve slightly to slowly release air from the cuff, decreasing the pressure of the cuff against the artery while listening with the stethoscope for a pulsating sound of blood squirting through the partially closed artery. As soon as you hear the first pulsating sound, read the pressure on the pressure gauge. This is the systolic blood pressure. (The pressure in the cuff equals the systolic pressure forcing blood through the partially closed artery.)
4. Continue to slowly release air from the cuff while listening to the pulsating sound as it deepens and then ceases. As soon as the sound ceases, read the pressure on the pressure gauge. This is the diastolic blood pressure.
5. Release the remaining air from the cuff and remove it from the "patient's" arm.

Materials

Per student pair
Sphygmomanometer
Stethoscope

Assignment 4

1. **Complete items 4a and 4b on the laboratory report.**
2. Measure your pulse rate (beats/min) at rest by placing your fingers over the radial artery, as shown in Figure 15.6. Count the number of beats for 15 seconds and multiply by 4 to determine the beats per minute. Add your results to the class data being recorded on the board (by gender) by your instructor. **Record your resting pulse rate in item 4c on the laboratory report.**
3. If your have no physical disability, run in place for 3 min and measure your pulse rate again. Measure your pulse rate at 1-min intervals until your pulse rate returns to its resting rate. The shorter the time required for your pulse rate to return to its resting rate,

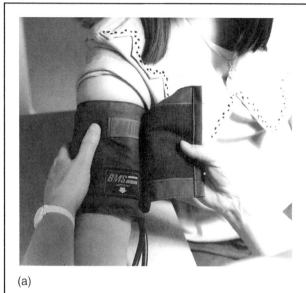

(a)

(b)

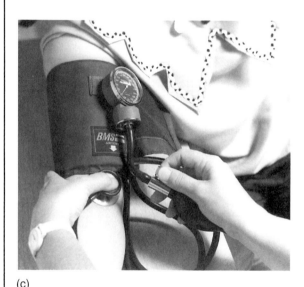

(c)

Figure 15.7. Measuring blood pressure.
(a) The sphygmomanometer cuff is placed around the upper arm with its lower edge about an inch above the elbow. Attach it snugly with the Velcro® fastener and fold back the leftover flap. (b) Attach the pressure gauge to the holding strap so that it is easily readable. (c) Close the valve with the set screw and pump air into the cuff to 150 mm Hg. Place the diaphragm of the stethoscope over the brachial artery on the medial half of the anterior elbow joint. Slightly open the valve to slowly release air. When the first pulse sound is heard, read the systolic pressure. Continue releasing air, and when the pulse sound suddenly disappears, read the diastolic pressure.

the better is your physical condition. **Complete items 4c and 4d on the laboratory report.**

4. Working with your partner, take each other's blood pressure at rest, following the procedures described above and shown in Figure 15.7. **Caution:** *Do not leave the brachial artery compressed for more than 30 seconds.* **Record your results in item 4e on the laboratory report.**

5. Measure the blood pressure after running in place for 3 min, and a second time 3 min after

the exercise. **Complete the laboratory report.**

Assignment 5

Your instructor has set up a sheep heart with numbered pins indicating certain structures as a mini-practicum. Your task is to identify the structures indicated by the pins. **Complete item 5 on the laboratory report.**

16

Blood

and **thrombocytes** or platelets, which initiate blood-clot formation.

BLOOD CELLS

In this section, you will study characteristics of blood cells and learn to recognize them when viewed microscopically. The cellular components of blood are illustrated in Figure 16.1.

Erythrocytes

Red blood cells (RBCs) are tiny cells with a diameter of only 7–8 μm. They have lost their nuclei during the maturation process, and their shape is like biconcave discs. RBCs contain **hemoglobin**, a red pigment that gives the red color to blood. Hemoglobin combines loosely with oxygen and carbon dioxide, enabling these respiratory gases to be transported by RBCs. Erythrocytes are the most numerous blood cells. They average about 5.4 million per cubic millimeter (mm^3) of blood in males and about 4.8 million per mm^3 in females.

Leukocytes

Leukocytes, or white blood cells (WBCs), are much larger than RBCs, and they always have a nucleus. WBCs are divided into two groups, granulocytes and agranulocytes, depending upon

---

OBJECTIVES

After completing the laboratory session, you should be able to:
1. Describe the components and general functions of blood.
2. Describe the cellular components of blood and recognize them when viewed microscopically.
3. Perform a differential white cell count and blood typing.
4. Explain the basis of blood groups, blood typing, transfusions of compatible blood, and erythroblastosis fetalis.
5. Define all terms in bold print.

Human blood carries materials to and from body cells. It consists of **plasma**, a fluid carrier, and **blood cells**, also called **formed elements**, that are suspended in the plasma. Plasma forms 55% of the blood volume, while blood cells compose the remaining 45%. About 92% of the plasma is water, and most of the remaining 8% consists of plasma proteins that give blood its viscous (sticky) character. Blood cells consist of three major types: **erythrocytes**, red blood cells that transport oxygen and carbon dioxide; **leukocytes**, white blood cells that fight infections;

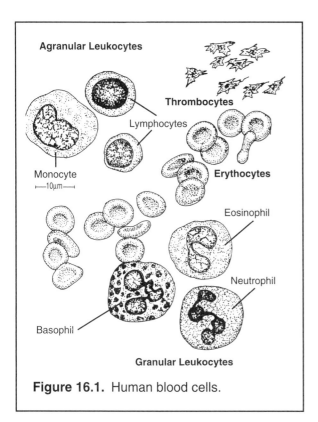

Figure 16.1. Human blood cells.

the presence or absence of cytoplasmic granules. Whenever blood cells are to be examined microscopically, they are stained with Wright's blood stain to make it easier to identify the different types of WBCs. This stains the different types of cytoplasmic granules different colors, and stains all nuclei purple.

Leukocytes are much less numerous in blood than RBCs. WBC counts range from 5,000 to 8,000 per mm^3. Leukocytes play a vital role in the body by combating infection and providing a major defense against disease-causing organisms. WBCs move through capillary walls and wander among surrounding tissue cells by amoeboid movement. It is within tissues that most of their battles against disease organisms occur. Some WBCs engulf bacteria and damaged cells and destroy them by digestion in food vacuoles. Others secrete chemicals that destroy or immobilize disease organisms.

Granulocytes

Neutrophils have a nucleus with 2–5 lobes and tiny lavender-staining granules that cause the cytoplasm to appear pale lavender. They are the most numerous leukocytes, forming 60–70% of the total WBC count. Neutrophils fight disease by removing disease-causing organisms and damaged cells by phagocytosis. The numbers of neutrophils increase rapidly during acute bacterial infections.

Eosinophils are easily recognized by their bilobed or U-shaped nucleus and red-staining cytoplasmic granules. They form only 2–4% of the total WBC count. Eosinophils neutralize histamine, a chemical released during allergic reactions, and they destroy parasitic worms.

Basophils are unique in having either a lobed or U-shaped nucleus and large, blue-staining cytoplasmic granules. They are the least numerous of the leukocytes, forming only 0.5–1% of the total WBC count. Basophils release **histamine** during allergic reactions. Histamine dilates blood vessels, increasing the blood flow to affected areas. Basophils also release **heparin**, a chemical that inhibits the formation of blood clots.

Agranulocytes

Lymphocytes are the smallest white blood cells. They are slightly larger than red blood cells. They are easily recognized by their large spherical nucleus surrounded by a small amount of cytoplasm lacking granules. Lymphocytes form 20–25% of the total WBC count, and they play a vital role in immunity. There are two types of lymphocytes. **T-lymphocytes** directly attack and destroy virus-infected cells and tumor cells, while **B-lymphocytes** produce and release into the blood antibodies that destroy disease-causing organisms and provide immunity. The number of lymphocytes increases during viral infections and antigen-antibody reactions.

Monocytes are the largest leukocytes, and are easily recognized by the presence of a large, kidney-shaped nucleus and abundant cytoplasm lacking granules. Monocytes form 3–8% of the total WBCs in blood. They destroy disease-causing organisms by phagocytizing bacteria and cells infected with viruses. An increase in the numbers of monocytes in the blood may indicate a chronic infection.

Thrombocytes

Thrombocytes or platelets are not cells but tiny, non-nucleated fragments of large cells. Thrombocytes are much smaller (2–4 μm in diameter) than RBCs, and their density ranges from 250,000 to 500,000 per mm^3 of blood. Thrombocytes play a vital role in the formation of blood clots.

Materials

Per student
Colored pencils
Compound microscope
Schilling blood chart

Per lab
Prepared slides of human blood:
   normal
   sickle-cell anemia

Assignment 1

1. Obtain a Schilling blood chart and study the appearance of blood cells in circulating blood as shown in the bottom row of the chart. Color the blood cells in Figure 16.1 as they appear on the blood chart.
2. Obtain a prepared slide of normal human blood and examine it, using your oil immersion (100×) or high-dry (40×) objective. Locate each type of blood cell. It will take careful searching to locate a basophil and an eosinophil. **Draw each type of blood cell as it appears on your slide in item 1a on Laboratory Report 16 that begins on page 367.**
3. Examine a prepared slide of blood from a patient with **sickle-cell anemia**, an inherited blood disorder. In afflicted persons, RBCs contain abnormal hemoglobin and have a sickle-like shape. Sickled erythrocytes cannot effectively transport oxygen and carbon dioxide.
4. **Complete item 1 on the laboratory report.**

DIFFERENTIAL WHITE CELL COUNT

A differential white cell count determines the percentage of each type of white blood cell in the total white cell count. It is used clinically in making a diagnosis of certain illnesses, including acute and chronic infections.

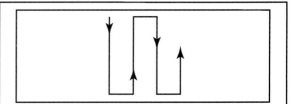

Figure 16.2. Pattern of slide movement used for slide examination when making a differential white cell count.

Materials

Per lab
Prepared slides of human blood:
   normal
   infectious mononucleosis

Assignment 2

1. Examine a prepared slide of human blood with your microscope. Use the low-power objective to locate an area where the blood cells are separated from each other. Then switch to the high-dry (40×) or oil immersion (100×) objective. Move the slide in the manner shown in Figure 16.2 as you identify and **tabulate each WBC observed in item 2a on the laboratory report** until you have tabulated 100 white blood cells.
2. Obtain a prepared slide of blood from a patient with **infectious mononucleosis**, a contagious disease caused by the Epstein-Barr virus. It occurs mostly in children and young adults and is often transmitted by kissing. Perform a differential white cell count of this slide.
3. **Complete item 2 on the laboratory report.**

BLOOD TYPING

Human blood may be classified according to the presence or absence of various **antigens** (proteins) on the surfaces of red blood cells. Because the presence of these antigens is genetically controlled, an individual's **blood type** is the same from birth to death. See Table 16.1. Blood typing most commonly tests for the presence of antigens A, B, and D (Rh).

TABLE 16.1
Percentage of ABO Blood Types

Blood Type	%U.S. Blacks	%U.S. Whites
A	25	41
B	20	7
AB	4	2
O	51	50

TABLE 16.2
Antigen-Antibody Associations

Blood Type	Antigen	Antibody
O	None	a,b
A	A	b
B	B	a
AB	AB	None
Rh $^+$	D	None
Rh $^-$	None	d*

* Antibodies are produced by an Rh$^-$ person only after Rh$^+$ red blood cells enter his or her blood.

Testing for the presence of A and B antigens determines the ABO blood group. The presence or absence of the D (Rh) antigen determines the Rh blood type. The two factors are combined in designating an individual's blood type (e.g., A,Rh$^+$; A,Rh$^-$; AB,Rh$^+$). Table 16.2 indicates the association of these **antigens** and their corresponding **antibodies** in blood.

All blood types are not compatible with each other. Therefore, knowing the blood types of both the donor and recipient in blood transfusions is imperative. The *antigens of the donor* and the *antibodies of the recipient* must be considered in blood transfusions. Transfusion of incompatible blood results in the clumping together (agglutination) of erythrocytes in the transfused blood, which may plug capillaries and result in death. Clumping of erythrocytes occurs whenever the *antigens of the donor* and the *antibodies of the recipient*, that are designated by the same letter, are brought together. See Table 16.2. The antibodies of the donor are so diluted in the recipient's blood that their effect is inconsequential.

The Rh Factor

About 85% of Caucasians and 99–100% of Chinese, Japanese, African blacks, and Native Americans possess the D (Rh) antigen and are therefore Rh$^+$. If an Rh$^-$ individual receives a single transfusion of Rh$^+$ blood, antibodies are produced against the D (Rh) antigen. However, no clumping of cells occurs, because of the gradual increase in these antibodies and the loss of erythrocytes containing the D (Rh) antigen. If the individual who produces these antibodies receives a second transfusion of Rh$^+$ blood, however, clumping will occur, and death may result.

Antibodies against the D (Rh) antigen may be produced by an Rh$^-$ woman carrying an Rh$^+$ fetus, as the result of "leakage" of fetal erythrocytes into the maternal blood. The accumulation of antibodies in maternal blood is gradual enough so that no complications result during the first pregnancy with an Rh$^+$ fetus. In subsequent pregnancies with Rh$^+$ fetuses, however, the maternal antibodies may diffuse into the fetal blood and destroy the fetal erythrocytes. This pathological condition, **erythroblastosis fetalis**, may be fatal to the fetus. The mother suffers no consequences unless she subsequently receives a transfusion of Rh$^+$ blood.

Typing Procedure

Testing for the presence of A, B, and D antigens in blood is done by adding a drop of **antiserum** containing a specific antibody to a drop of blood. If the corresponding antigen is present on red blood cells in the drop of blood, the antibody will cause the RBCs to agglutinate, forming clumps of RBCs that are visually detectable.

Since testing for the presence of the D antigen requires a temperature of 50 °C, you will use a slide-warming box and a special typing plate to allow simultaneous testing for all three antigens: A, B, and D. See Figure 16.3. In addition, each typing station has the other materials needed for the typing procedure.

Your instructor may chose to have you use either simulated blood or your own blood for blood typing. Follow your instructor's directions carefully.

Caution: *If you use your own blood, you must be careful to avoid contact with blood from other students, because their blood may contain the*

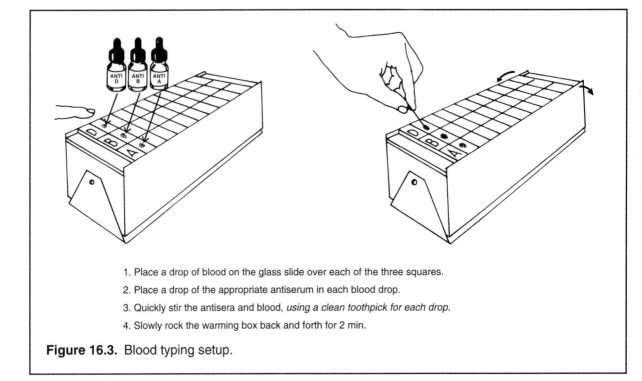

1. Place a drop of blood on the glass slide over each of the three squares.
2. Place a drop of the appropriate antiserum in each blood drop.
3. Quickly stir the antisera and blood, *using a clean toothpick for each drop.*
4. Slowly rock the warming box back and forth for 2 min.

Figure 16.3. Blood typing setup.

human immunodeficiency virus (HIV) or hepatitis viruses. Typing of your own blood may be done safely if you rigidly follow the safety precautions, typing procedures, and your instructor's directions, because you will then be in contact only with your own blood.

Your instructor has set up blood-typing stations that are provisioned with the needed supplies.

Safety Precautions

1. Avoid contact with the blood of another student.
2. Before piercing your finger, wash your hands with soap and water and disinfect the fingertip with an alcohol wipe (70% alcohol).
3. Do not remove a sterile lancet from its container until you are ready to use it. Do not touch its tip or lay it down.
4. Use a lancet *once* and *immediately* place it in the bioazard sharps container provided. *Never place a used lancet on the tabletop or in a wastebasket.*
5. Place all other materials in contact with blood, such as microscope slides, alcohol wipes, toothpicks, and paper towels, in a biohazard bag *immediately* after use.

6. When finished, wash the tabletop around the typing box with a suitable disinfectant, such as 10% household bleach. Wash your hands with soap and rinse with a disinfectant.
7. Perform the typing procedure under the direct supervision of your instructor. Follow your instructor's directions at all times.

Materials

Per typing station
Alcohol pads
Biohazard bag
Biohazard sharps container
Blood typing box with typing plate
Blood typing antisera (anti-A, anti-B, anti-D)
Household bleach, 10%
Lancets, sterile and disposable
Microscope slides, new
Toothpicks, flat

Assignment 3

1. Turn on the slide-warming box for at least 5 min prior to use.
2. Place a clean glass slide across the typing plate and allow it to warm for 3 min.

3. Cleanse your fingertip with an alcohol pad. After it has dried, pierce it with a sterile lancet. *Immediately place the used lancet in the biohazard sharps container.*
4. Place a drop of blood from your fingertip on your glass slide over each of the three squares on the typing plate.
5. *Being careful not to touch the drops of blood with the droppers*, add a drop of anti-D antiserum to the drop of blood in the D square, a drop of anti-B serum to the drop of blood in the B square, and a drop of anti-A serum to the drop of blood in the A square.
6. Quickly mix the antiserum and blood in each square with a *different* toothpick. (Using the same toothpick for more than one square will invalidate the results.)
7. Rock the warming box back and forth for exactly 2 min and then read the results. The presence of tiny red granules in the D square indicates the presence of the D antigen.

Clumping of RBCs in the B square indicates the presence of the B antigen, and clumping in the A square indicates presence of the A antigen. See Figure 16.4
8. Place your slide, alcohol pad, and anything else in contact with your blood in the biohazard bag.
9. **Complete item 3 on the laboratory report.**

Assignment 4

Your instructor has set up several microscopes as a mini-practicum so that you can check your recognition of blood cells. Your task is to identify the blood cells indicated by the microscope pointers. **Complete item 4 on the laboratory report.**

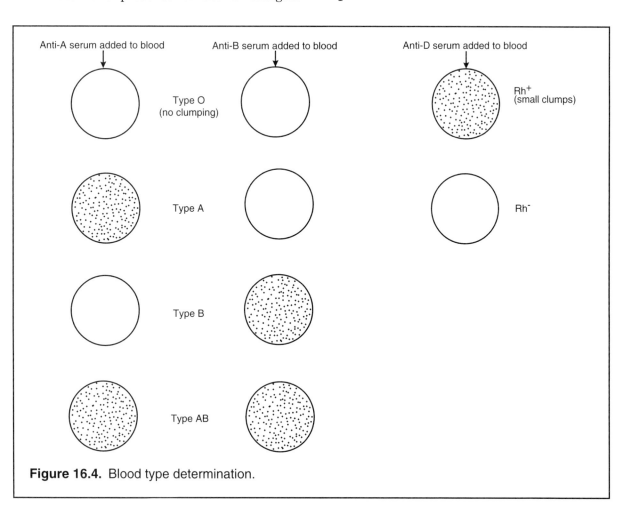

Figure 16.4. Blood type determination.

17

Gas Exchange

OBJECTIVES

After completion of the laboratory session, you should be able to:
1. Identify the parts of the respiratory system on a human torso model or on charts and describe the function of each part.
2. Describe the mechanics of breathing.
3. Determine the capacities of your lungs.
4. Define all terms in bold print.

Body cells require a continuous supply of oxygen and the constant removal of carbon dioxide. These needs are met by the interaction of the respiratory and circulatory systems. The respiratory system brings air into the lungs, where oxygen diffuses into the blood and carbon dioxide from the blood diffuses into the air in the lungs. The air in the lungs is then exhaled, and the carbon dioxide is dispersed in the atmosphere. In this way, breathing enables the continuous exchange of oxygen and carbon dioxide between air and blood in the lungs. The circulatory system transports the oxygen taken up by blood in the lungs to the body cells, and brings carbon dioxide to the lungs for exchange.

THE RESPIRATORY SYSTEM

Refer to Figure 17.1 as you read this section, and label the parts indicated.

The nasal cavity is divided into left and right portions by the **nasal septum**, a midline partition composed of bone and cartilage. The **hard** and **soft palates** separate the nasal and oral cavities. The nasal cavity warms and filters the air as it enters the respiratory system. This is accomplished as the air passes over the warm, moist, mucus-producing membrane that lines the nasal cavity. Foreign particles tend to be trapped in mucus that is carried to the pharynx by beating cilia and swallowed. The turbinates, or **nasal conchae**, which are shelf-like protuberances from the lateral walls of the nasal cavity, play an important role by increasing the surface area of the nasal cavity that is in contact with the passing air.

The **pharynx** serves as a passageway for both air and food. From the pharynx, air enters the **larynx** through an opening called the **glottis**.

The larynx is a cartilaginous box that contains the vocal cords. The **epiglottis**, a small cartilaginous projection from the upper part of the larynx, flops over to close the glottis when you swallow, thus preventing food and liquids from entering the larynx.

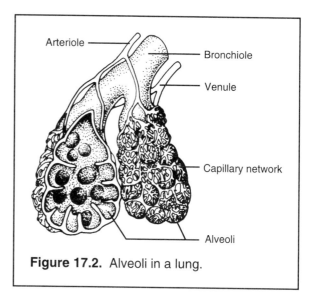

Figure 17.2. Alveoli in a lung.

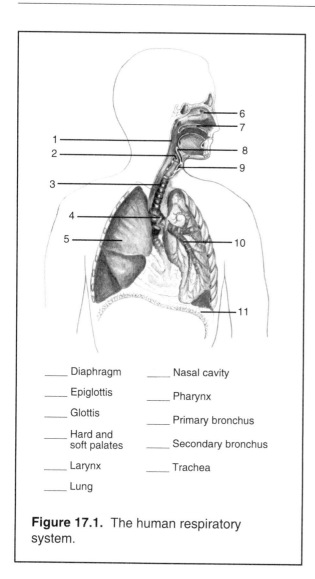

____ Diaphragm
____ Epiglottis
____ Glottis
____ Hard and soft palates
____ Larynx
____ Lung
____ Nasal cavity
____ Pharynx
____ Primary bronchus
____ Secondary bronchus
____ Trachea

Figure 17.1. The human respiratory system.

The larynx is attached to the upper end of the **trachea**, or windpipe, which is located just anterior to the esophagus. The trachea descends to about midway in the thorax, where it divides to form the **primary bronchi** that enter each lung.

Like the nasal cavity, the larynx, trachea, and bronchi are lined with **pseudostratified ciliated columnar epithelium**, whose goblet cells produce a thin layer of mucus over the epithelium. Airborne particles (e.g., dust, pollen, and bacteria) are trapped in the mucus. The beating cilia of the epithelium then move the mucus and entrapped particles upward to the pharynx, where it is swallowed. In this way foreign materials are removed from the air passages, which helps to prevent respiratory infections.

In the lungs, the primary bronchi divide to form the smaller **secondary bronchi**, which lead to the still smaller **bronchioles** and finally to the **alveoli**. The alveoli are microscopic sacs arranged in clusters at the end of each tiny air duct, and they are formed of simple squamous epithelium. The exchange of oxygen and carbon dioxide occurs between the air in the alveoli and the blood in the capillaries that surround the alveoli. Each lung has about 300 million alveoli, which increase the respiratory surface of the lung to about 75 m^2. See Figure 17.2.

The outer surface of each lung is covered by a thin, tightly adhering membrane, the **visceral pleura**, while the inner surface of the thoracic cavity is lined by a similar membrane, the **parietal pleura**. Fluid secreted by these membranes into the **pleural cavity**, the very thin space between the membranes, reduces friction as the lungs expand and contract during breathing.

The **diaphragm** is the sheet-like muscle separating the thoracic and abdominopelvic cavities. It is the primary muscle involved in breathing.

Materials

Per lab
Anatomical charts
Corrosion preparation of mammalian lungs
Human torso model

Prepared slides of:
    trachea, x.s.
    lung tissue, normal
    lung tissue, emphysematous

Assignment 1

1. Label Figure 17.1. Color-code the nasal cavity, pharynx, larynx, trachea, and primary bronchi.
2. Locate the parts of the respiratory system on the human torso model and anatomical charts.
3. Observe the corrosion preparation of mammalian lungs that shows how the bronchi divide into smaller and smaller air passages that ultimately terminate in alveoli.
4. **Complete items 1a–1c on Laboratory Report 17 that begins on page 371.**
5. Examine a prepared slide of trachea, x.s. Using the 4× objective, observe one of the cartilaginous rings that supports the tracheal wall and holds the trachea open. Cartilaginous rings also support the bronchi. Now focus on the inner surface of the trachea and switch to the 40× objective to observe the ciliated cells and goblet cells of the epithelial lining. **Complete item 1d on the laboratory report**.
6. Place your fingers on your larynx (Adam's apple). Is its cartilaginous wall hard or soft? Can you feel the vibrations of your vocal cords when speaking? What happens to your larynx when you swallow? **Complete items 1e–1h on the laboratory report**.
7. Examine prepared slides of normal and emphysematous lung tissue, using the 4× objective. Locate the alveoli in the normal lung tissue and note their size and the thickness of their walls. In emphysematous lungs, the alveolar walls rupture, producing large air spaces in the lungs that reduce the respiratory surface area. Locate such areas on the slide of emphysematous lung tissue. **Complete item 1 on the laboratory report.**

BREATHING MECHANICS

The air passages into the lungs are always open. Thus, atmospheric air and air in the lungs are always joined by air in the connecting air passages. Air moves into and out of the lungs because of changes in the pressure of air in the lungs, since atmospheric air pressure is constant at any given elevation.

The **diaphragm** is the primary muscle involved in breathing, but the **intercostal muscles** located between the ribs also participate. Recall that the diaphragm is a sheet-like muscle separating the thoracic and abdominal cavities.

Inspiration (inhalation) results from the contraction of the diaphragm, which increases the volume of the thoracic cavity and simultaneously decreases the air pressure in the lungs. When this happens, air flows into the lungs because the atmospheric air pressure is greater than the air pressure in the lungs.

Expiration (exhalation) results from the relaxation of the diaphragm, which decreases the volume of the thoracic cavity and increases the air pressure in the lungs. Air flows out of the lungs because the air pressure in the lungs is greater than the atmospheric air pressure.

Materials

Per lab
Breathing mechanics model

Assignment 2

1. Examine the breathing mechanics model. See Figure 17.3. The balloons represent the lungs, and the glass tubing represents the trachea and primary bronchi. The glass jar corresponds to the thoracic wall, and the rubber sheet simulates the diaphragm. Note that the balloons are in an enclosed space, representing the thoracic cavity.
2. Observe what happens when the rubber sheet is pulled downward and pushed upward. Determine how this works.
3. **Complete item 2 on the laboratory report.**

LUNG CAPACITY IN HUMANS

Lung capacities vary among males and females, primarily because of variations in the size of the thoracic cavity and the lungs. Lung capacities

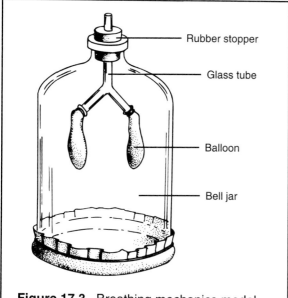

Figure 17.3. Breathing mechanics model.

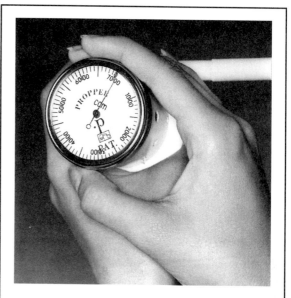

Figure 17.5. Propper spirometer. Slip on a sterile mouthpiece and rotate the dial face to zero before exhaling through the spirometer.

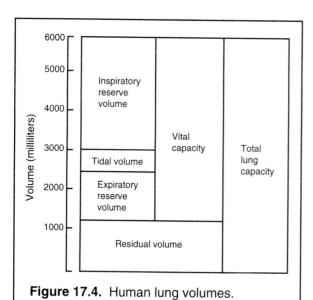

Figure 17.4. Human lung volumes.

pieces in the biohazard bag after use. **Do not inhale through the spirometer.**

Materials

Per student group
Propper spirometer
Spirometer mouthpieces, sterile and disposable

Per lab
Biohazard bag
Household bleach, 10%

Assignment 3

Determine your lung volumes as described below and record your results in item 3 on the laboratory report.

1. The volume of air exchanged between the atmosphere and the lungs during normal quiet breathing is the **tidal volume (TV)**. Determine it as described below.
   a. Rotate the dial of the spirometer so that the needle is at zero. See Figure 17.5.
   b. Disinfect the stem of the spirometer and add a sterile mouthpiece. Place the mouth-

also vary with age. Average lung capacities are shown in Figure 17.4.

In this section, you will use a spirometer to determine your lung volumes. **Caution**: *Wipe the spirometer stem with an alcohol swab or a paper towel soaked in 10% household bleach before and after using the spirometer, and use a fresh, sterile, disposable mouthpiece. Place all used mouth-*

piece in your mouth, keeping the spirometer dial upward.

c. Inhale through your nose and exhale through the spirometer for five normal quiet-breathing cycles. Record the dial reading on the spirometer and divide by 5 to determine your tidal volume.

2. The amount of air that can be forcefully exhaled after a maximum inhalation is the **vital capacity (VC)**. It is often used as an indicator of respiratory function.

   a. Rotate the dial face of the spirometer to place the needle at zero.

   b Take two deep breaths and exhale completely after each one. Then take a breath as deeply as possible and exhale through the spirometer. A slow, even expiration is best. Record the reading.

   c Repeat steps a and b two more times, resetting the dial face after each measurement.

   d. Record the average of the three measurements as your vital capacity. Compare your vital capacity with those shown in Table 17.1.

3. The volume of air that can be exhaled *after* a normal tidal volume expiration is the **expiratory reserve volume (ERV)**.

   a. Rotate the dial face of the spirometer so that the needle is at 1,000. This is to compensate for the space on the dial between 0 and 1,000.

   b. *After* a normal quiet expiration, forcefully exhale as much air as possible through the spirometer. Be sure not to take an extra breath at the start. Subtract 1,000 from the reading to determine your expiratory reserve volume. Repeat to obtain three replicates.

4. Based on your determinations above, calculate your inspiratory reserve volume. IRV = VC − (TV + ERV).

5. The volume of air exchanged during 1 min of quiet breathing is the **respiratory minute volume (RMV)**.

   a. Count the number of breathing cycles during a 3-min period and divide by 3 to get the number of cycles per minute.

   b. Calculate the RMV as follows. RMV = TV × breathing cycles per minute.

6. **Complete item 3 on the laboratory report.**

TABLE 17.1
Expected Vital Capacities (ml) for Adult Males and Females*

Males

Height (inches)	Age in Years					
		30	40	50	60	70
60	3,885	3,665	3,445	3,225	3,005	2,785
62	4,154	3,925	3,705	3,485	3,265	3,045
64	4,410	4,190	3,970	3,750	3,530	3,310
66	4,675	4,455	4,235	4,015	3,795	3,575
68	4,940	4,720	4,500	4,280	4,060	3,840
70	5,206	4,986	4,766	4,546	4,326	4,106
72	5,471	5,251	5,031	4,811	4,591	4,371
74	5,736	5,516	5,296	5,076	4,856	4,636

Females

Height (inches)	Age in Years					
		30	40	50	60	70
58	2,989	2,809	2,629	2,449	2,269	2,089
60	3,198	3,018	2,838	2,658	2,478	2,298
62	3,403	3,223	3,043	2,863	2,683	2,503
64	3,612	3,432	3,252	3,072	2,892	2,710
66	3,822	3,642	3,462	3,282	3,102	2,922
68	4,031	3,851	3,671	3,491	3,311	3,131
70	4,270	4,090	3,910	3,730	3,550	3,370
72	4,449	4,269	4,089	3,909	3,729	3,549

* Data from Propper Mfg. Co., Inc.

18

Digestion

OBJECTIVES

After completion of the laboratory session, you should be able to:

1. Describe the parts of the digestive system and state the functions of each.
2. Identify the parts of the digestive tract on a human torso model or chart.
3. Identify the parts of a tooth on a model or chart.
4. Identify the four layers of the small intestine on a microscope slide and describe the function of each.
5. Describe the role of enzymes in digestion and indicate the molecular end products of digestion.
6. Explain the effects of pH and temperature changes on the activity of pancreatic amylase.
7. Define all terms in bold print.

The cells of the body require a continuous supply of nutrients derived from the food we eat. Most food molecules are too large to be absorbed into the blood, so they must be broken down into smaller, absorbable nutrient molecules by digestion. The major functions of the **digestive system** are:

1. Ingestion of food
2. Movement of food through the digestive tract
3. Digestion of food
4. Absorption of nutrients
5. Elimination of undigestable materials

THE DIGESTIVE SYSTEM

The digestive system consists of the digestive tract and associated organs that aid the digestive process. The digestive tract (alimentary canal), through which food materials pass from mouth to anus, is about 9 meters long. The functions of its major divisions are shown in Table 18.1. Refer to Figures 18.1, 18.2, and 18.3 as you read the following sections.

Oral Cavity

The mouth contains a number of organs that assist the digestive process. The roof of the mouth is formed of the anterior **hard palate** and the posterior **soft palate**, and it separates the oral from the nasal cavity. This arrangement allows you to breathe while eating. The posterior finger-like extension of the soft palate is the **uvula**, which contracts upward when contacted by food during swallowing.

The **palatine tonsils** are located on each side of the base of the tongue. They have nothing to

TABLE 18.1
Major Functions of the Digestive Tract Divisions

Structure	Major Function
Mouth	Eating, chewing, and swallowing food; digestion of starch begins here
Pharynx	Carries food to esophagus
Esophagus	Carries food to stomach by peristalsis
Stomach	Mixes gastric juice with food to form chyme; protein digestion begins here
Small intestine	Mixes chyme with bile and intestinal and pancreatic juices; digestion and absorption of nutrients completed here
Large intestine	Decomposition of undigested materials by bacteria; reabsorption of water to form feces
Anus	Defecation

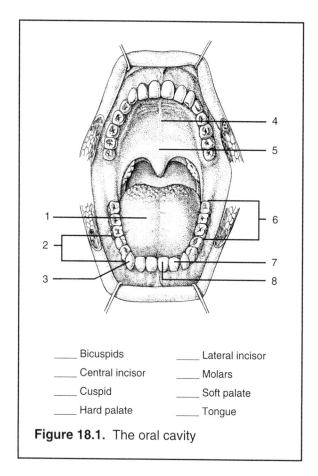

_____ Bicuspids   _____ Lateral incisor

_____ Central incisor   _____ Molars

_____ Cuspid   _____ Soft palate

_____ Hard palate   _____ Tongue

Figure 18.1. The oral cavity

do with digestion, but sometimes become enlarged and painful when infected, and make swallowing difficult.

During chewing, the **teeth** break the food into smaller pieces, and the **tongue** manipulates the food and mixes it with **saliva**. Saliva is produced by three pairs of **salivary glands** and discharged into the mouth through a number of salivary ducts. The **parotid glands** are located just in front of the ears and posterior to the angle of the jaw. If you have ever had the mumps, you are well acquainted with these glands, since they swell and hurt when infected with the mumps virus. The **sublingual glands** are located in the anterior floor of the mouth under the tongue, and the **submandibular glands** are located posterior to them. Saliva cleans and lubricates the mouth and helps to hold the food together when swallowing. It also contains an enzyme that begins the digestion of starch.

Teeth

There are 32 teeth in a complete set of permanent teeth. See. Figure 18.1. From front to back on each side of each jaw, they are **central in-**cisor, **lateral incisor**, **cuspid**, first and second **bicuspids** (premolars), and first, second, and third **molars**. The third molars (wisdom teeth) often become impacted because of the evolutionary shortening of the jaws.

Figure 18.2 shows the basic structure of a tooth. A tooth consists of two major parts: a **crown** projecting above the bone, and a **root** embedded in bone. Most of a tooth is composed of **dentin**, but the dentin of the crown is covered by a layer of **enamel**, the hardest substance in the body. The **pulp cavity** is the hollow interior of a tooth; it contains nerves and blood vessels that enter through **root canals**. The mucous membrane that covers the jaw bones and surrounds the bases of the crowns of the teeth is the **gingiva** (gum).

Swallowing

When food is pushed posteriorly by the tongue into the **pharynx**, a swallowing reflex is set in motion that causes the larynx to move upward

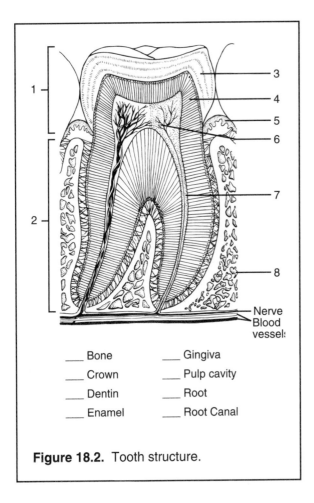

___ Bone   ___ Gingiva

___ Crown   ___ Pulp cavity

___ Dentin   ___ Root

___ Enamel ___ Root Canal

Figure 18.2. Tooth structure.

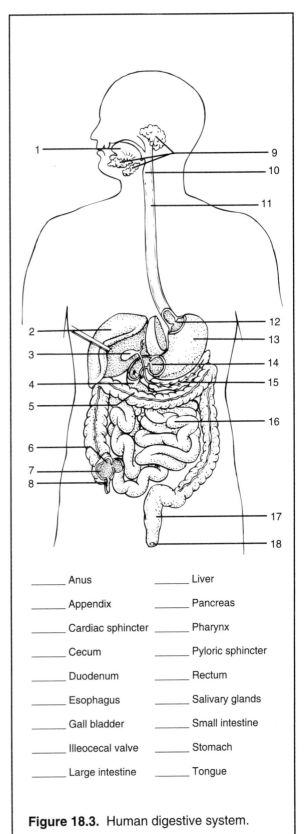

___ Anus   ___ Liver

___ Appendix   ___ Pancreas

___ Cardiac sphincter ___ Pharynx

___ Cecum   ___ Pyloric sphincter

___ Duodenum   ___ Rectum

___ Esophagus   ___ Salivary glands

___ Gall bladder   ___ Small intestine

___ Illeocecal valve ___ Stomach

___ Large intestine ___ Tongue

Figure 18.3. Human digestive system.

and the epiglottis to flop over the glottis, so that the food passes into the esophagus and not into the larynx.

Esophagus to Anus

Food is moved through the digestive tract (Figure 18.3) by **peristalsis**, the wavelike contraction of the muscles in its walls. The **esophagus** carries food from the pharynx to the **stomach**. The **cardiac sphincter**, a circular muscle located at the esophagus–stomach junction, opens to allow the passage of food and closes to prevent regurgitation. In the stomach, food is mixed with **gastric juice** and converted to a semiliquid mass called **chyme**. Enzymes in gastric juice begin the digestion of proteins and certain fats. Chyme is then released in small amounts into the small intestine. The **pyloric sphincter** controls the passage of chyme into the small intestine.

In the small intestine, chyme is mixed with

bile, pancreatic juice, and intestinal juice. Bile and pancreatic juice enter the **duodenum**, the first portion of the small intestine. The bile and pancreatic ducts join to form a common opening into the duodenum.

Bile is secreted by the **liver**, the large gland in the upper right portion of the abdominal cavity. It is temporarily stored in the **gall bladder**. Bile emulsifies fats to facilitate fat digestion. **Pancreatic juice** is produced by the **pancreas**, a pennant-shaped gland located between the stomach and the duodenum. **Intestinal juice** is secreted by the inner lining (mucosa) of the small intestine. Enzymes in pancreatic juice and intestinal juice act sequentially to complete the digestion of food. Digestion of food and absorption of nutrients into the blood are completed in the small intestine.

The nondigestible material passes into the **large intestine** via the **ileocecal valve**, another sphincter muscle. The **appendix** is a small vestigial appendage attached to the pouch-like **cecum**, the first portion of the large intestine. The large intestine (colon) includes ascending, transverse, descending, and sigmoid regions. It ends with the **rectum**, a muscular portion that expels the feces through the **anus**. Decomposition of nondigestible materials by bacteria and the reabsorption of water are the major functions of the large intestine.

Materials

Per student
Colored pencils

Per lab
Anatomical charts
Human torso model
Human head model, midsagittal section
Tooth model

Assignment 1

1. Label Figures 18.1, 18.2, and 18.3. Color-code the organs to help you learn their locations.
2. Locate the parts of the digestive system on the models and anatomical charts.
3. **Complete item 1 on Laboratory Report 18 that begins on page 375.**

HISTOLOGY OF THE SMALL INTESTINE

Examine the structure of the small intestine in cross section in Figure 18.4. Note the four layers of the intestinal wall.

Peritoneum. The outer protective membrane that lines the coelom and covers the digestive organs.

Muscle layers. Outer longitudinal and inner circular layers whose contractions mix food with digestive secretions and move the food mass by peristalsis.

Submucosa. Connective tissue containing blood vessels and nerves serving the digestive tract.

Mucosa. The inner epithelial lining that secretes intestinal juice and absorbs nutrients.

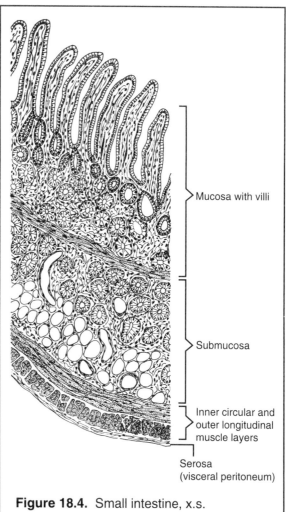

Mucosa with villi

Submucosa

Inner circular and outer longitudinal muscle layers

Serosa (visceral peritoneum)

Figure 18.4. Small intestine, x.s.

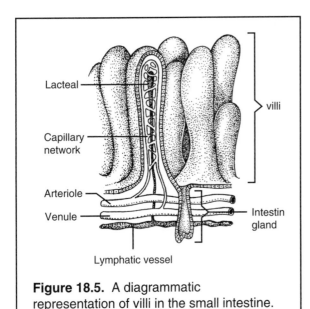

Figure 18.5. A diagrammatic representation of villi in the small intestine.

Locate the **villi**, finger-like extensions of the mucosa that project into the lumen of the small intestine. Villi increase the surface area of the mucosa, which facilitates the absorption of nutrients. See Figure 18.5.

Materials

Per student
Colored pencils
Compound microscope

Per lab
Prepared slides of small intestine, x.s.

Assignment 2

1. Color-code the four layers of the small intestine in Figure 18.4.
2. Examine a prepared slide of small intestine, x.s., at 40×. Locate the four layers of the small intestine as shown in Figure 18.4. Note the villi that greatly increase the surface area of the intestinal mucosa. Focus on the columnar epithelial lining. Are goblet cells present?
3. **Complete item 2 on the laboratory report.**

DIGESTION AND ENZYMES

The conversion of large, nonabsorbable food molecules into small, absorbable nutrient molecules occurs through the action of digestive enzymes. The role of digestive enzymes is to speed up the hydrolysis of food molecules. In **enzymatic hydrolysis**, digestive enzymes catalyze (speed up) reactions in which the addition of water molecules breaks chemical bonds in food molecules, forming smaller nutrient molecules.

$$\text{Nonabsorbable food molecules} \xrightarrow[\text{H}_2\text{O}]{\text{Digestive enzymes}} \text{Absorbable nutrient molecules}$$

Many different digestive enzymes are required to complete the digestion of food, since a particular enzyme acts on only a single type of food molecule. Saliva, gastric juice, pancreatic juice, and intestinal juice contain specific digestive enzymes that act on specific food molecules. Table 18.2 shows the absorbable end products of enzymatic hydrolysis.

Digestion of Starch

In this section, you will assess the effects of temperature and pH on the action of **pancreatic amylase**. This enzyme accelerates the hydrolysis of **starch**, a polysaccharide, to **maltose**, a disaccharide and reducing sugar. See Figure 18.6. The experiments are best done by groups of four students, with each group assigned a different temperature and with all groups sharing their results.

You will dispense solutions from dropping bottles and use a number of test tubes. To avoid contamination, the droppers from the dropping bottles must not touch other solutions. All glassware must be clean and rinsed with distilled water.

TABLE 18.2
End Products of Digestion

Food	End Products
Carbohydrates	Monosaccharides
Proteins	Amino acids
Fats	Fatty acids and monoglycerides

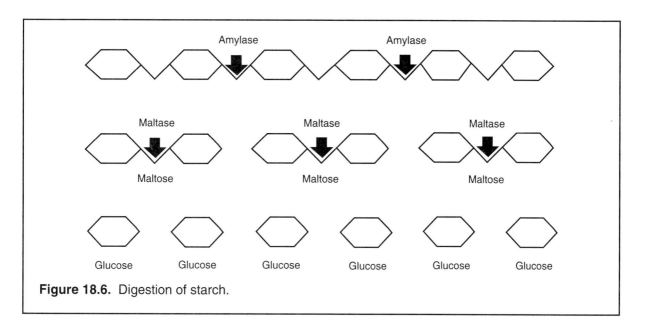

Figure 18.6. Digestion of starch.

You will dispense "droppers" and "drops" of solutions. As noted earlier, a "dropper" means *one dropper full* of solution (about 1 ml). When dispensing solutions by the "drop," hold all droppers at the same angle to dispense equal-sized drops. All drops should land squarely in the bottom of the test tube. If they run down the side of the tube, your results may be affected.

Thoroughly mix the solutions placed in a test tube. If a vortex mixer is not available, shake the tube vigorously from side to side to mix the contents.

You will use the iodine test for starch and Benedict's test for reducing sugars to determine if digestion has occurred. See Figure 18.7. Examine the demonstration of positive and negative tests prepared by your instructor and use these as standards for comparison.

Iodine Test

1. Add 3 drops of iodine solution to 1–2 droppers of solution to be tested.
2. A blue-black color indicates the presence of starch.

Benedict's Test

1. Add 5 drops of Benedict's solution to 1 dropper of solution to be tested, and heat the mixture to near boiling in a water bath for 3 min.
2. A light green, yellow, orange, or red color indicates the presence of a reducing sugar, in that order of increasing concentration.

Materials

Per student group
Beaker, 400 ml
Boiling chips
Glass marking pen
Hot plate
Test tubes, 18
Test-tube racks, submersible type
Dropping bottles of:
  Benedict's solution
  buffers, pH 5, 8, 11
  iodine solution (IKI)
  pancreatic amylase, 0.1%
  soluble starch, 0.1%

Per lab
Water baths, 5, 37, 70 °C
Celsius thermometers, 1 per water bath

Assignment 3

1. Label nine test tubes 1-9.
2. Add the pH buffer solutions to the test tubes as follows.

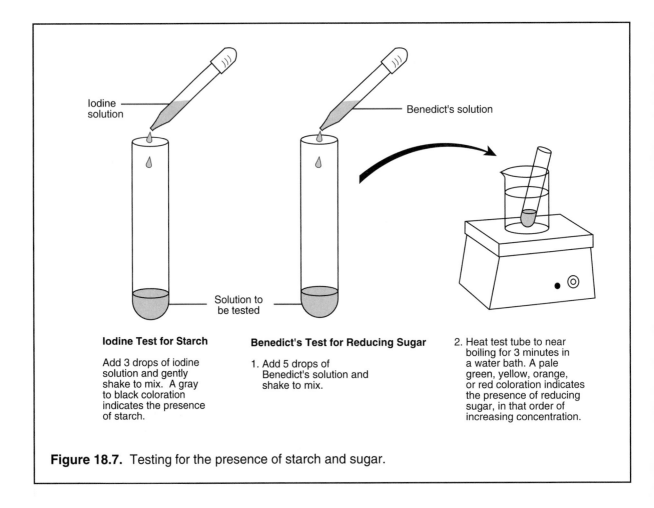

Iodine Test for Starch

Add 3 drops of iodine solution and gently shake to mix. A gray to black coloration indicates the presence of starch.

Benedict's Test for Reducing Sugar

1. Add 5 drops of Benedict's solution and shake to mix.

2. Heat test tube to near boiling for 3 minutes in a water bath. A pale green, yellow, orange, or red coloration indicates the presence of reducing sugar, in that order of increasing concentration.

Figure 18.7. Testing for the presence of starch and sugar.

Tube 1: 2 droppers pH 5
Tube 2: 2 droppers pH 5
Tube 3: 3 droppers pH 5
Tube 4: 2 droppers pH 8
Tube 5: 2 droppers pH 8
Tube 6: 3 droppers pH 8
Tube 7: 2 droppers pH 11
Tube 8: 2 droppers pH 11
Tube 9: 3 droppers pH 11

3. Add 5 drops of pancreatic amylase to tubes 1, 3, 4, 6, 7, and 9. Shake the tubes to mix well.

4. Place the test tubes in a test-tube rack and place the rack in the water bath at your assigned temperature for 10 min.

5. Add 1 dropper of starch solution to tubes 1, 2, 4, 5, 7, and 8. Shake the tubes to mix well. Leave the tubes in the water bath for 15 min. Table 17.3 summarizes the contents of each tube.

6 After 15 min, remove the test-tube rack and tubes and return to your workstation.

7. Number another set of nine tubes 1B–9B.

8. Pour half the liquid in tube 1 into tube 1B, pour half the liquid in tube 2 into tube 2B, and so on until you have divided the liquid equally between the paired tubes. You now have nine pairs of tubes with the members of each pair containing liquid of identical composition.

9. Test all the original nine tubes (1–9) for the presence of starch by adding 3 drops of iodine solution to each tube. See Figure 18.7. **Record your results in the table in item 3a on the laboratory report.**

10. Test all of the B tubes (1B–9B) for the presence of maltose, a reducing sugar, as follows. Add 5 drops of Benedict's solution to each tube, and place the tubes in a water bath (400-ml beaker half full of water) on your hot plate. Place some boiling chips in the beaker and heat the tubes to near boiling for 3–4 min. Watch for any color change in the solution.

TABLE 18.3
Summary of Tube Contents for the Starch Digestion Experiment

Tube	Droppers of Solution				
	Buffers				
	pH 5	*pH 8*	*pH 11*	*Starch*	*Drops of Amylase*
1	2			1	5
2	2			1	
3	3				5
4		2		1	5
5		2		1	
6		3			5
7			2	1	5
8			2	1	
9			3		5

Record your results in item 3a on the laboratory report.

11. Exchange results with other groups using different temperatures. **Record their results in item 3a on the laboratory report.**

12. **Complete item 3 on the laboratory report.**

19

Neural Control

OBJECTIVES

After completion of the laboratory session, you should be able to:
1. Recognize neurons when viewed microscopically and correlate neuron structure and function.
2. Dissect a sheep brain and identify its major parts.
3. Compare a sheep brain with a model of a human brain.
4. Perform and explain the mechanisms of the patellar and photopupil reflexes.
5. Measure reaction time and identify its determining variables.
6. Define all terms in bold print.

Humans possess the most highly developed nervous system of all animals, and it may be subdivided into two major components. The **central nervous system (CNS)** is composed of the **spinal cord** and **brain**. The **peripheral nervous system (PNS)** is composed of **cranial nerves**, which emanate from the brain and extend to the head and certain internal organs, and the **spinal nerves**, which extend from the spinal cord to all parts of the body except the head. Of course, well-developed **sensory receptors** are evident in the peripheral nervous system.

In this exercise, you will investigate the structure and function of the nervous system. The nervous system provides rapid coordination and control of body functions by transmitting **impulses** along neuron processes. The impulses may originate in either the brain or receptors, and they are carried to **effectors** (muscles and glands) where they initiate a particular action.

NEURONS

Neural tissue is composed of two basic types of cells. **Neurons**, or nerve cells, transmit impulses. Neuroglial cells provide structural support for neurons and prevent contact between neurons except at certain sites. Although neurons may be specialized in structure and function in various parts of the nervous system, they have many features in common. Figure 19.1 shows the basic structure of neurons, and Table 19.1 identifies the characteristics of the three functional types of neurons.

The **cell body** is an enlarged portion of the neuron that contains the nucleus. Two types of neuron processes or fibers extend from the cell body. **Dendrites** receive impulses from receptors or other neurons, and carry impulses *toward* the cell body. **Axons** carry impulses *away*

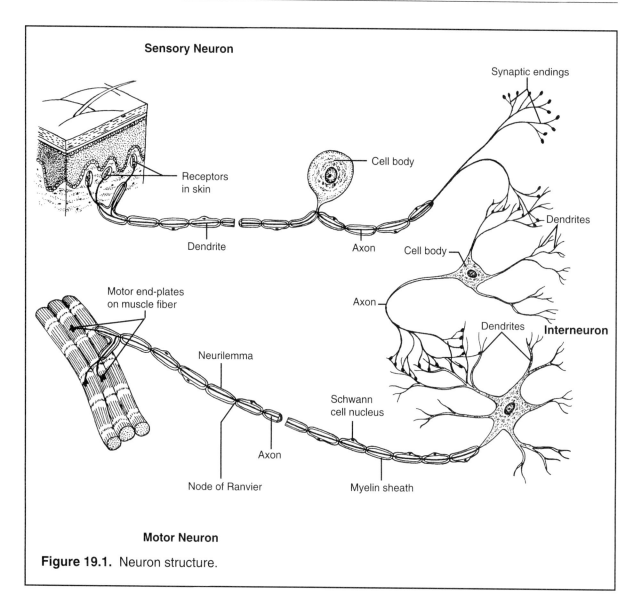

Sensory Neuron

Synaptic endings

Cell body

Receptors in skin

Dendrites

Dendrite

Axon

Cell body

Axon

Dendrites

Interneuron

Motor end-plates on muscle fiber

Neurilemma

Schwann cell nucleus

Axon

Node of Ranvier

Myelin sheath

Motor Neuron

Figure 19.1. Neuron structure.

from the cell body. A neuron may have many dendrites, but only one axon is present.

Neuron processes of the peripheral nervous system are enclosed by a covering of **Schwann cells**. In larger processes, the multiple wrappings of Schwann cells form an inner **myelin sheath**, a fatty, insulating material, while the outer layer of Schwann cells constitutes the **neurilemma**. The minute spaces between Schwann cells, where the neuron process is exposed, are called **nodes of Ranvier**. Impulses are transmitted more rapidly by myelinated

TABLE 19.1
Functional Types of Neurons

Neuron Type	Structure	Function
Sensory	Long dendrite, short axon	Carry impulses from receptors to the CNS
Interneuron	Short dendrites, short or long axon	Carry impulses within the CNS
Motor	Short dendrites, long axon	Carry impulses from the CNS to effectors

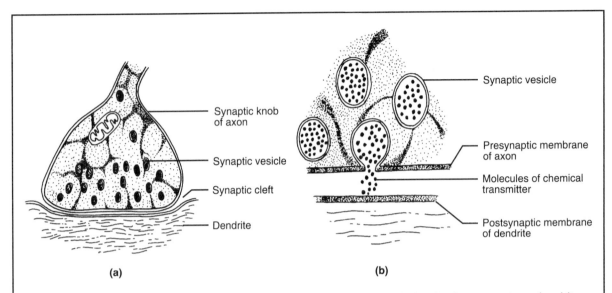

Figure 19.2. The synaptic junction. (a) Relationship of the synaptic knob of an axon to a dendrite. Synaptic vesicles migrate from the cell body to the synaptic knob. (b) Synaptic transmission. An impulse moving down the axon causes a synaptic vesicle to release a neurotrasmitter that either stimulates or inhibits the formation of an impulse in the dendrite.

nerve fibers than by unmyelinated fibers. Schwann cells provide a pathway for the regeneration of neuron processes, and are essential for the regrowth of nerve cells after injury. Schwann cells are absent in the central nervous system (CNS), but another type of neuroglial cell forms the myelin sheath of CNS neurons.

The junction of an axon tip of one neuron and a dendrite or cell body of another neuron is called a **synapse**. See Figure 19.2. Impulses passing along the axon cause the release of a **neurotransmitter** from the axon tip, or **synaptic knob**, into the **synaptic cleft**, the minute space between the axon tip, and the dendrite or cell body of a neighboring neuron. The neurotransmitter binds with receptor sites on the postsynaptic membrane of the receiving neuron, producing either stimulation or inhibition of impulse formation, in this neuron, depending on the type of neurotransmitter involved. Immediately thereafter, an enzyme breaks down or inactivates the neurotransmitter, which prevents continuous stimulation or inhibition of the postsynaptic membrane. Transmission of neural impulses across synapses is always in one direction, from axon to dendrite, because only axon tips can release neurotransmitters to activate the adjacent neuron.

Numerous substances are either known or suspected to be neurotransmitters. Among known neurotransmitters, **acetylcholine** and **norepinephrine** are stimulatory neurotransmitters, while **glycine** and **gamma aminobutyric acid (GABA)** are inhibitory neurotransmitters.

THE BRAIN

The **brain** is the control center of the nervous system. It is enclosed within the cranium and covered by the **meninges**, three layers of protective membranes. **Cerebrospinal fluid** within the meninges provides an additional cushion around the brain, to absorb shocks. Twelve pairs of cranial nerves are attached to the brain. All but one pair innervate structures of the head and neck; vagus nerves innervate the internal organs. Refer to Figures 19.3-19.7 as you study the major parts of the brain.

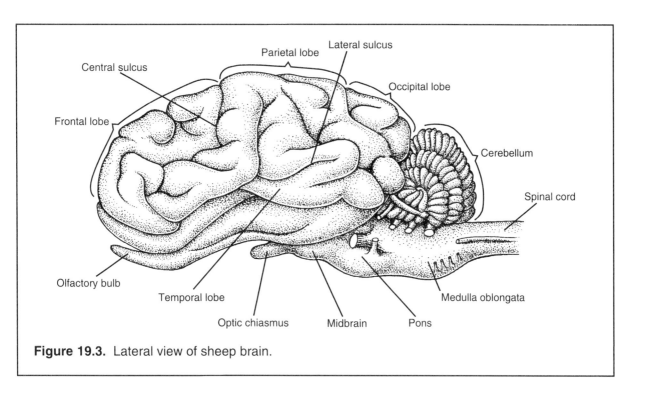

Figure 19.3. Lateral view of sheep brain.

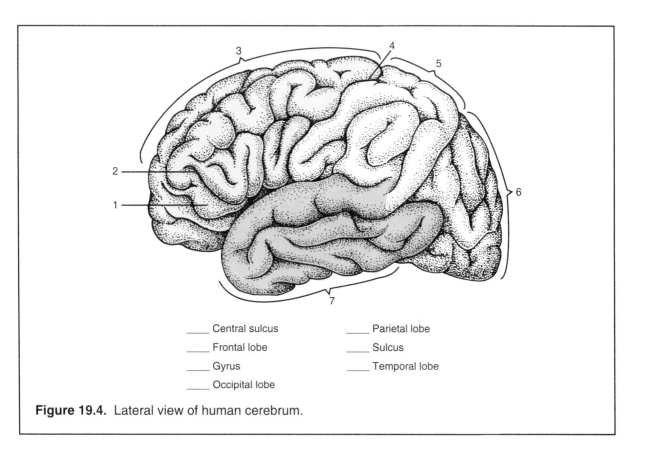

_____ Central sulcus	_____ Parietal lobe
_____ Frontal lobe	_____ Sulcus
_____ Gyrus	_____ Temporal lobe
_____ Occipital lobe	

Figure 19.4. Lateral view of human cerebrum.

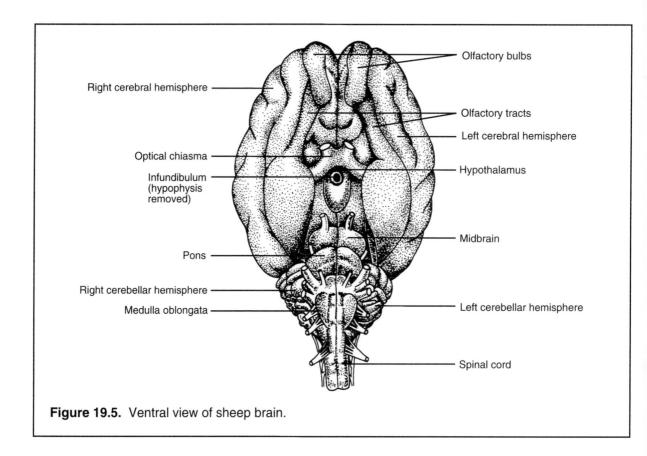

Figure 19.5. Ventral view of sheep brain.

Cerebrum

The **cerebrum** is the largest part of the brain. It consists of right and left **cerebral hemispheres** that are separated by a median **longitudinal fissure**. A mass of neuron fibers, the **corpus callosum**, enables impulses to pass between the two hemispheres of the cerebrum. The outer portion of the cerebrum, the **cerebral cortex**, is composed of neuron cell bodies and unmyelinated neuron processes. Its surface area is increased by numerous **gyri** (ridges) and **sulci** (grooves). The cerebrum initiates voluntary actions and interprets sensations. In humans, it is the seat of will, memory, and intelligence.

Each hemisphere of the cerebrum is divided into four lobes by fissures (deep grooves). Table 19.2 indicates the location and major functions of these lobes.

TABLE 19.2
Location and Function of the Cerebral Lobes

Lobe	Location	Function
Frontal	Anterior to the central sulcus	Voluntary muscular movements; intellectual processes
Parietal	Between frontal and occipital lobes	Interprets sensations from skin; speech interpretation
Temporal	Inferior to frontal and parietal lobes	Hearing; interprets auditory sensations
Occipital	Posterior part of cerebrum	Vision; interprets visual sensations

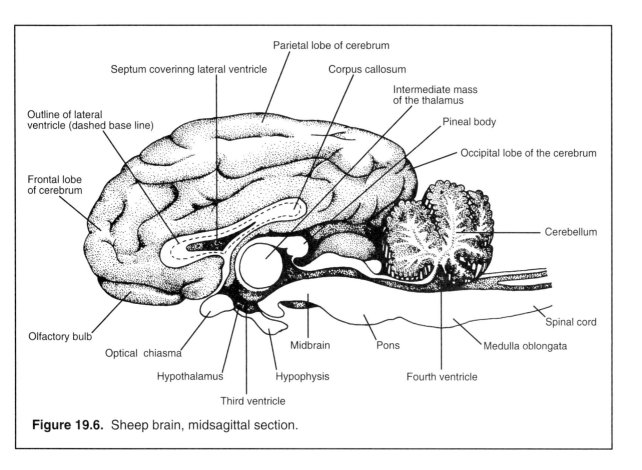

Parietal lobe of cerebrum

Septum coverinng lateral ventricle

Corpus callosum

Intermediate mass of the thalamus

Pineal body

Outline of lateral ventricle (dashed base line)

Occipital lobe of the cerebrum

Frontal lobe of cerebrum

Cerebellum

Olfactory bulb

Spinal cord

Optical chiasma

Midbrain

Pons

Medulla oblongata

Hypothalamus

Hypophysis

Fourth ventricle

Third ventricle

Figure 19.6. Sheep brain, midsagittal section.

Cerebellum

The **cerebellum** lies just below and posterior to the occipital lobe of the cerebrum. It is divided into left and right hemispheres by a shallow fissure. Muscle tone and muscular coordination are subconsciously controlled by the cerebellum.

Brain Stem

The **brain stem** is composed of a number of structures located between the cerebrum and the spinal cord. A midsagittal section of the brain stem is required in order to locate the various parts (Figure 19.6). A major function of the brain stem is the linking of higher and lower brain areas, but the individual portions also have some specific functions.

The **thalamus**, which is located at the upper end of the brain stem, consists of two lateral globular masses joined by an isthmus of tissue called the **intermediate mass**. It provides an uncritical awareness of sensations such as pain and pleasure, and is a relay station between lower brain centers and the cerebrum.

The **hypothalamus** is located just below the thalamus. It plays a major role in homeostasis of the body by controlling such functions as appetite, sleep, body temperature, and water balance. A major endocrine gland, the **hypophysis** or **pituitary gland**, is attached to the ventral wall of the hypothalamus by a short stalk, the **infundibulum**.

The **midbrain** is a small area between the thalamus and pons that is associated with certain visual reflexes. It contains the **pineal body**.

The **pons** is a rounded bulge on the ventral side of the brain stem. It is an important connecting pathway for higher and lower brain centers.

The **medulla oblongata** lies between the pons and the spinal cord. It is the lowest part of

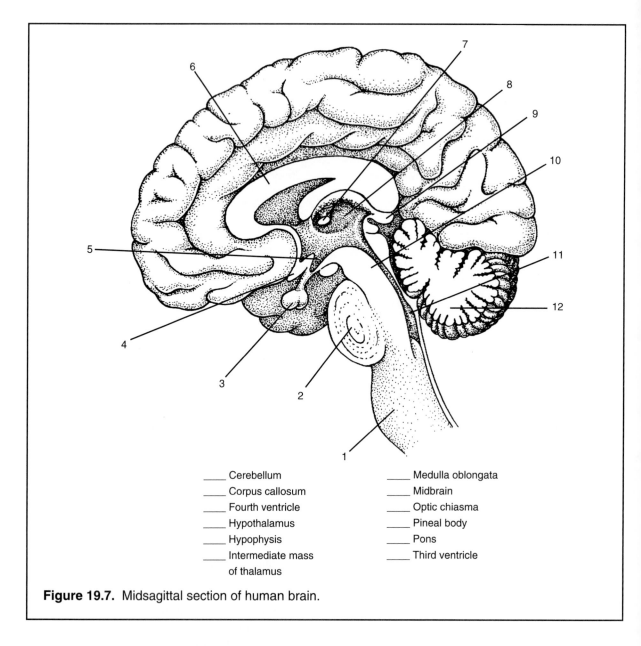

____ Cerebellum

____ Corpus callosum

____ Fourth ventricle

____ Hypothalamus

____ Hypophysis

____ Intermediate mass
of thalamus

____ Medulla oblongata

____ Midbrain

____ Optic chiasma

____ Pineal body

____ Pons

____ Third ventricle

Figure 19.7. Midsagittal section of human brain.

the brain, and all impulses passing between the brain and spinal cord must pass through it. In addition, the medulla oblongata controls heart rate, blood pressure, and breathing.

Materials

Per student group
Colored pencils
Dissecting instruments and pans

Per lab
Model of human brain
Sheep brains, whole and sectioned

Prepared slides of:
giant multipolar neurons
nerve, x.s.

Assignment 1

1. Add arrows to Figure 19.1 to indicate the direction of impulse transmission in motor and sensory neurons. Color-code the three types of neurons.
2. **Complete items 1a and 1b on Laboratory Report 19 that begins on page 377.**

3. Examine a slide of giant multipolar neurons. **Draw a neuron and label the cell body, nucleus, and neuron processes**, **in item 1c on the laboratory report**.

4. Examine a slide of nerve, x.s. **Make a drawing of your observations in item 1c on the laboratory report.** Note how the axons are arranged in bundles separated by connective tissue. Label an axon, myelin sheath, and connective tissue.

Assignment 2

1. Study and color-code Figures 19.3, 19.5, and 19.6 until you are familiar with the structures of the sheep brain.

2. Label Figures 19.4 and 19.7 and locate these parts on a model of a human brain. Color-code the cerebral lobes in Figure 19.4 and the parts of the brain in Figure 19.7.

3. Obtain an entire sheep brain for study. Note the meninges, if present, and remove them with scissors. Locate the major parts of the brain shown in Figures 19.3 and 19.5. Observe the ridges and furrows that increase the surface area of the cerebrum. Is there an advantage in this?

4. Obtain half of a sheep brain that has been sectioned along the longitudinal fissure. Locate the structures shown in Figure 19.6. Note that the hypothalamus forms the floor of the third ventricle and that the hypophysis (pituitary gland) projects ventrally from it. The **pineal body** is a remnant of a third eye found in primitive reptiles. Its function in some mammals may be to control seasonal reproductive activity based on photoperiod variation.

5. Examine the demonstration coronal sections of a sheep brain and locate the two **lateral ventricles**. Cerebrospinal fluid is secreted from blood vessels in each of the four ventricles of the brain. It circulates through the ventricles and within the meninges covering the brain and spinal cord, and it is subsequently reabsorbed back into the blood.

Note that parts of the brain in a coronal section appear either white or gray. **White matter** consists mostly of myelinated neuron processes, and **gray matter** consists mostly of neuron cell bodies and unmyelinated neuron processes. Which type of neural tissue composes the cerebral cortex? the corpus callosum?

6. **Complete item 2 on the laboratory report.**

THE SPINAL CORD

The **spinal cord** is located in the vertebral canal, and is covered by the meninges, as is the brain. It serves as a pathway for impulses between the brain, different levels of the spinal cord, and the **spinal nerves**. In humans, there are 31 pairs of spinal nerves that extend from the spinal cord and lead to all parts of the body except the head. Each spinal nerve joins the spinal cord to form ventral and dorsal roots. The **ventral root** contains fibers of motor nerves only. The **dorsal root** contains fibers of sensory nerves only, and the cell bodies of the sensory neurons are located in the **dorsal root ganglion**. See Figure 19.8. Unlike the cerebrum, the gray matter of the spinal cord is located interiorly, while the white matter is found exteriorly. What causes the difference in appearance of gray and white matter?

Materials

Per lab
Chart or model of the spinal cord
Prepared slides of cat spinal cord, x.s.

Assignment 3

1. Label Figure 19.8.

2. Examine a prepared slide of cat spinal cord, x.s., at 40× and observe the gross features. Compare your slide with Figure 19.8. Use 100× to locate the cell bodies of neurons in the gray matter, and examine them at 400×. What do the neuron fibers look like in the white matter?

3. Add arrows to Figure 19.8 to show the path of impulses.

4. **Complete item 3 on the laboratory report.**

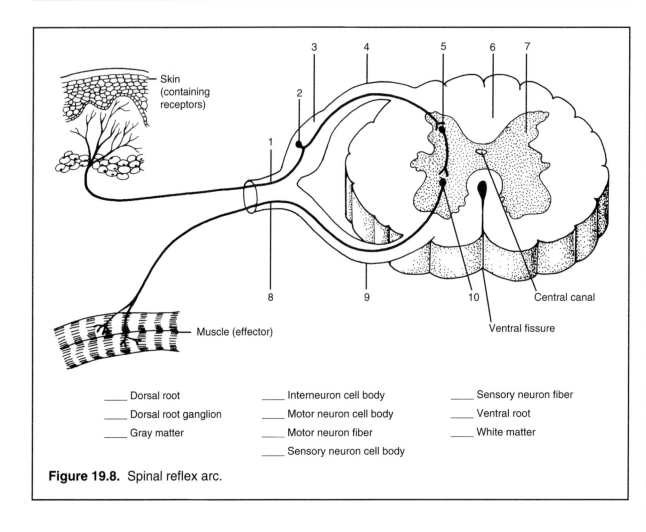

_____ Dorsal root

_____ Dorsal root ganglion

_____ Gray matter

_____ Interneuron cell body

_____ Motor neuron cell body

_____ Motor neuron fiber

_____ Sensory neuron cell body

_____ Sensory neuron fiber

_____ Ventral root

_____ White matter

Figure 19.8. Spinal reflex arc.

REFLEXES

Reflexes are involuntary responses (they require no conscious act) to specific stimuli, and they are important mechanisms for maintaining the well-being of the individual. For example, coughing and sneezing are respiratory reflexes controlled by the brain. Spinal reflexes do not involve the brain.

Simple reflexes require no more than three neurons to produce a reaction to a stimulus, and this is accomplished through a **reflex arc**. See Figure 19.8. A reflex arc consists of (1) a receptor that forms impulses on stimulation, (2) a sensory neuron that carries the impulses to the brain or spinal cord, (3) an interneuron that receives the impulses and transmits them to a motor neuron, (4) a motor neuron that carries impulses to an effector, and (5) an effector that performs the action.

Materials

Per student pair
Reflex hammer
Laboratory lamp
Penlight

Patellar Reflex

Physicians use reflex tests to assess the condition of the nervous system. The **patellar reflex** (Figure 19.9) is commonly used for this purpose. When the patellar tendon is struck just below the kneecap with a reflex hammer, the reflex action is a slight, instantaneous contraction of the large muscle (quadriceps femoris) on the front of the thigh that extends the lower leg. Striking the patellar tendon causes the quadriceps femoris muscle to be slightly stretched for an instant. This stretching causes impulses to be formed and carried along a sensory neuron to the spinal cord, where they are passed to a motor neuron that carries the impulses back to the muscle, thereby causing a weak, brief contraction. Work in pairs to perform the reflex test.

Photopupil Reflex

The photopupil reflex enables a rapid adjustment of the size of the pupil of the eye to the existing light intensity, and it is coordinated by the brain. This reflex is most easily observed in persons with light-colored eyes. When a bright light stimulates the retina of the eye, impulses are carried to the brain by sensory neurons. In the brain, the impulses are transmitted to interneurons and on to motor neurons that carry impulses to the muscles of the iris, causing them to contract. Contraction of the iris muscles decreases the size of the pupil and controls the amount of light entering the eye.

Assignment 4

1. Perform the patellar reflex as follows.
   a. The subject should sit on the edge of a table with his on her legs hanging over the edge but not touching the floor.
   b. Strike the patellar tendon (see Figure 19.9) with the small end of the reflex hammer and observe the response.
   c. Divert the subject's attention by having the subject interlock the fingers of both hands and pull the hands against each other while you strike the patellar tendon again. Is the response the same as before? If not, how do you explain the difference?
   d. Test both legs and record the results.
   e. Add arrows to Figure 19.9 to show the path of impulses.
   f. **Complete items 4a and 4b on the laboratory report.**
2. Perform the photopupil reflex as follows.
   a. Have the subject sit with eyes closed, facing a darkened part of the room for 1–2

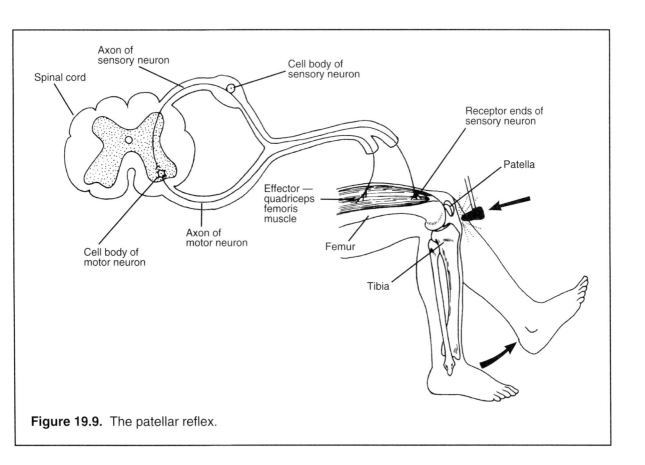

Figure 19.9. The patellar reflex.

min. When the subject opens his or her eyes, note any change in pupil size.

b. While the subject is looking into the darkened area, shine a desk lamp in his or her eyes (from about 3 ft away). Note any change in pupil size.

c. To observe the effect of unilateral stimulation, have the subject look into the darkened area and, while you observe the subject's right eye, shine a penlight in the left eye. Note any response. Then observe the left eye while stimulating the right eye.

d. **Complete item 4 on the laboratory report**.

REACTION TIME

The nervous system controls and coordinates our reactions to thousands of stimuli each day. Some of these reactions are reflexes, but many are **voluntary reactions**, responses that are consciously initiated. **Reaction time** is the time interval from the instant of stimulation to the instant of a voluntary response. All responses result from the formation of impulses by stimulation and the transmission of these impulses along neurons to effectors that bring about the response.

In reflexes, impulses flow over predetermined "automated" neural pathways involving very few neurons, and they do not require processing by the cerebral cortex. In contrast, voluntary reactions involve a greater number of neurons and synapses, and require processing of impulses by the cerebral cortex. Therefore, reflexes have much shorter response times than voluntary reactions.

The reaction time for a voluntary response is the sum of the times required for:

1. A receptor to form impulses in response to a stimulus.
2. Transmission of impulses to an integration center of the cerebral cortex.
3. Processing the impulses in the integration center.
4. Transmission of impulses to effectors.
5. Response by the effectors.

Do you think people differ in their reaction times to the same stimulus? In this section, you will test the null hypothesis that there is no difference in the reaction times of different persons in responding to the same stimulus with the same predetermined response.

Measuring Reaction Time

You will measure reaction time using a reaction-time ruler and the following procedure.

1. The subject sits on a chair or stool with the experimenter standing facing the subject.
2. The experimenter holds the *release end* of the reaction-time ruler between thumb and forefinger at about eye level or higher. See Figure 19.10.
3. The subject places the thumb and forefinger of his or her dominant hand about an inch apart and on each side of the *thumb line* at the lower end of the ruler. The subject's attention is focused on the ruler at the thumb line.
4. When the subject says he or she is ready, the experimenter, within 10 sec, releases the ruler. The subject, seeing the falling ruler, catches it between thumb and forefinger as quickly as possible.
5. The reaction time is read in milliseconds at the upper edge of the thumb and recorded.
6. The test is repeated 5 times and the average reaction time is calculated. If any reaction time is grossly different, discard it and repeat the test to obtain 5 results that are fairly consistent.

Materials

Per student pair
Reaction-time ruler

Assignment 5

1. Have your partner measure your reaction time 5 times and **record the 5 results in item 5a on the laboratory report.**
2. Calculate your average reaction time. Write it on the board for the class tabulation. **Complete items 5b–5f on the laboratory report.**
3. Do you think practice and learning will decrease your reaction time? Repeat the reaction-time test 20 times without recording the reaction time. Then repeat the test 5 times and record your reaction times.
4. **Complete item 5 on the laboratory report.**

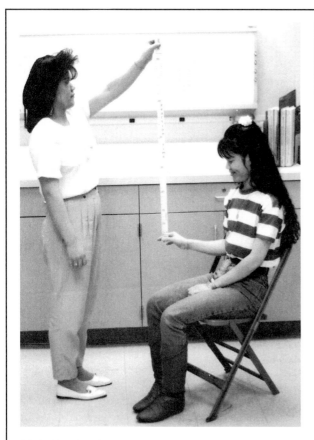

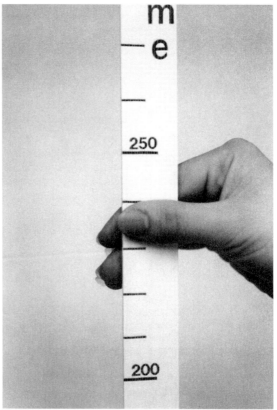

(a) Subject holds thumb and forefinger 1 inch apart, with the upper margin of the thumb at the thumb line of the reaction time ruler. When ruler is released, the subject closes the thumb and forefinger as quickly as possible in order to grasp ruler.

(b) Read the time in milliseconds at the upper edge of the thumb. In this photo, upper edge of the thumb is at 235 msec.

Figure 19.10. Measuring visual reaction time.

Assignment 6

Your instructor has set up a sheep brain with numbered pins indicating certain structures as a mini-practicum. Your task is to identify the structures indicated by the pins. **Complete item 6 on the laboratory report.**

20

Sensory Perception

OBJECTIVES

After completion of the laboratory session, you should be able to:
1. Describe the basic structure and function of the eye and ear and identify the parts on charts and models.
2. Explain the basic characteristics of sensory perception as determined by the tests performed in this exercise.
3. Define all terms in bold print.

Sensations result from the interaction of three components of the nervous system.

1. **Sensory receptors** generate impulses upon stimulation.
2. **Sensory neurons** carry the impulses to the brain or spinal cord, and **interneurons** carry the impulses to the sensory interpretive centers in the brain.
3. The **cerebral cortex** interprets the impulses as sensations.

The nature of a sensation is determined by the part of the brain receiving the impulses rather than by the type of receptors being stimulated. For example, the auditory center interprets all impulses it receives as sound sensations, regardless, of their origin.

In this exercise, you will study the structure and function of the eye and the ear, and perform tests that will demonstrate certain characteristics of sensory perception.

THE EYE

The eyes contain the receptors for light stimuli, and they are well protected by the surrounding skull bones and the eyelids. The eyelids and the anterior surface of the eye are covered by a mucous membrane, the **conjunctiva**, which contains many blood vessels and pain receptors except in the area where it covers the cornea of the eye. The **lacrimal gland**, located over the upper, lateral portion of the eye, produces tears that cleanse the surface of the conjunctiva and keep it moist.

Structure of the Eye

Refer to Figure 20.1 as you study this section.

The eye is a hollow ball, roughly spherical in shape. Its wall is composed of three distinct layers.

The outer layer is composed of (1) the fibrous **sclera**, which forms the white portion of the eye, and (2) the transparent **cornea**, which forms the anterior bulge where light enters the eye.

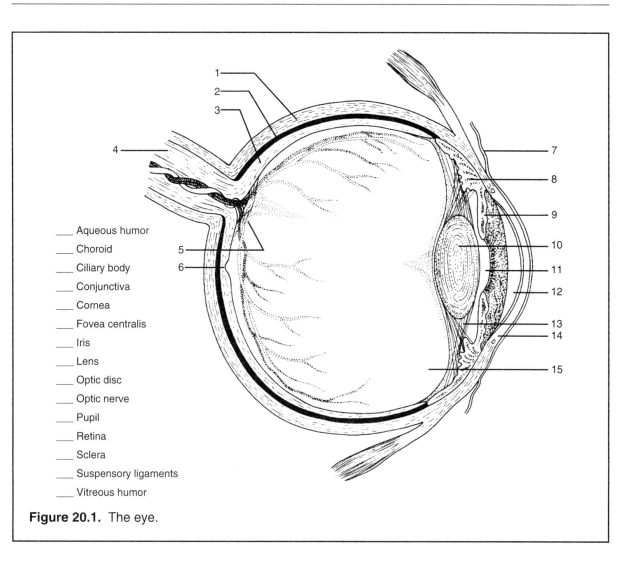

_____ Aqueous humor

_____ Choroid

_____ Ciliary body

_____ Conjunctiva

_____ Cornea

_____ Fovea centralis

_____ Iris

_____ Lens

_____ Optic disc

_____ Optic nerve

_____ Pupil

_____ Retina

_____ Sclera

_____ Suspensory ligaments

_____ Vitreous humor

Figure 20.1. The eye.

The middle layer consists of three parts. The **choroid** coat, a black layer, absorbs excess light that passes through the retina, and contains the blood vessels that nourish the eye. Anteriorly, the **ciliary body** forms a ring of muscle that controls the shape of the **lens**, which is suspended from the ciliary body by the **suspensory ligaments**. The **iris** controls the amount of light entering the eye by controlling the size of the **pupil**, the opening in the center of the iris.

The **retina** composes the inner layer of the wall of the eye. It contains the photoreceptors: **rods** for black-and-white vision and **cones** for color vision. The neuron fibers coalesce at the **optic disc**, where they enter the **optic nerve**, which carries impulses to the brain. The optic disc is known as the blind spot since it has no receptors. A tiny depression just lateral to the op-

tic disc, the **fovea centralis**, contains densely packed cones for sharp, direct vision.

The interior of the eye behind the lens is filled with **vitreous humor**, a transparent, jelly-like substance that helps hold the retina in place and gives shape to the eye. A watery fluid, the **aqueous humor**, fills the space between the lens and cornea.

Materials

Per student group
Colored pencils
Dissecting instruments and pan
Beef eye, preferably fresh

Per lab
Eye model

1 Label Figure 20.1 and color-code the sclera, cornea, choroid, ciliary body, iris, retina, and lens.
2. Locate the parts of the eye on an eye model.
3. **Complete items 1a and 1b on Laboratory Report 20 that begins on page 381.**
4. Dissect a beef eye, following these procedures.
   a. Examine the external surface of the eye. Locate the optic nerve and trim away any remnants of the extrinsic eye muscles and conjunctiva. Is the curvature of the cornea greater than the rest of the eyeball? What is the shape of the pupil?
   b. Use a sharp scalpel to make a small incision in the sclera about 0.5 cm from the edge of the cornea. See Figure 20.2. Holding the eye firmly but gently, insert scissors into this incision and cut through the sclera around the eyeball while holding the cornea upward. The fluid that exudes when making this cut is aqueous humor.
   c. Now gently lift off the anterior part of the eye and place it on the dissecting pan with its inner surface upward. In preserved eyes, the lens often remains with the cornea, but in fresh eyes it usually remains with the vitreous humor.
   d. Examine the anterior portion. Locate the ciliary body, a thickened black ring, and the iris. There may be a little aqueous humor remaining next to the inner surface of the cornea.
   e. Observe the vitreous humor and the attached lens in the posterior part of the eye. Pour the vitreous humor onto the dissecting pan and note its jelly-like consistency.
   f. Use a dissecting needle to separate the periphery of the lens from the ciliary body on the surface of the vitreous humor. Hold the lens to your eye and look through it across the room. What is unusual about the image? Place it on this printed page. Does it magnify the print?
   g. Look at the interior of the posterior portion of the eye to see the thin, beige retina that now is wrinkled since the vitreous humor has been removed. Note that the retina easily separates from the underlying choroid but is attached at the blind spot,

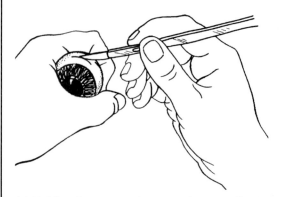

(a) Holding the eye as shown, make a small cut through the sclera with a sharp scalpel about 0.5 cm from the edge of the cornea.

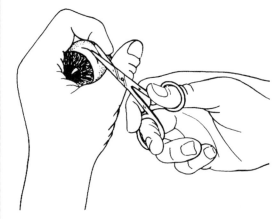

(b) Insert the point of a scissors into the cut and continue the cut through the wall around the eye

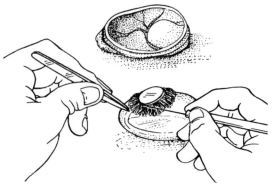

(c) After pouring the vitreous humor onto the dissecting pan, carefully separate the lens from the ciliary body with a dissecting needle.

Figure 20.2. Three steps in dissecting a beef eye.

the junction of the optic nerve with the retina.

h. Note the iridescent portion of the choroid that aids dim-light vision by reflecting light that has passed through the retina back through the retina. This iridescence causes animals' eyes to "shine" at night, reflecting light back to a light source.

i. Dispose of the eye as directed by your instructor.

5. **Complete item 1 on the laboratory report.**

Function of the Eye

Light waves reflecting from objects are bent as they pass through the cornea. They continue through the pupil and lens, and the lens focuses the light rays on the retina. The lens accommodates for near and distance vision as its shape is changed by muscles in the ciliary body. Photoreceptors in the retina form impulses that are transmitted via the optic nerve to the visual center in the brain, where they are interpreted as visual images.

Materials

Per lab
Meter sticks
Color-blindness test plates (Ishihara)
Astigmatism charts
Snellen eye charts

Assignment 2

Perform the following visual tests to learn more about visual sensations. Work in pairs. If you wear corrective lenses, perform the tests with and without them and note any differences in your results.

Blind Spot. No rods or cones are located at the junction of the retina and optic nerve. This site is known as the optic disc or blind spot. It can be located by using the following procedure.

1. Hold Figure 20.3 about 50 cm (20 in.) in front of your eyes.
2. Cover your left eye and focus with the right eye on the cross. You will be able to see the dot as well.

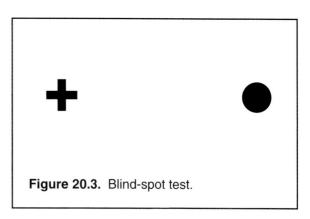

Figure 20.3. Blind-spot test.

TABLE 20.1
Age and Accommodation

| Age | Near Point | |
	Inches	Centimeters
10	3.0	7.5
20	3.5	8.8
30	4.5	11.3
40	6.8	17.0
50	20.7	51.8
60	33.0	82.5

3. Slowly move the figure toward your eyes while focusing on the cross, until the dot disappears.
4. Have your partner measure and record the distance from your eye to the figure at the point where the dot disappears.
5. Test the left eye in a similar manner, but focus on the dot and watch for the cross to disappear.
6. **Complete item 2a on the laboratory report.**

Near Point. The shortest distance from your eye that an object is in sharp focus is called the **near point**. The shorter this distance, the more elastic the lens and the greater the eye's ability to accommodate for changes in distance. Elasticity of the lens is greatest in infants, and it gradually decreases with age. See Table 20.1. Accommodation is minimal after 60 yrs of age, a condition called presbyopia. How does this relate to the common usage of bifocal lenses by older persons?

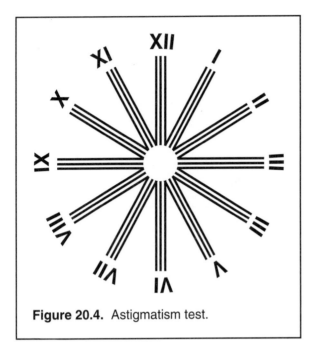

Figure 20.4. Astigmatism test.

1. Hold this page in front of you at arm's length. Close one eye, focus on a word in this sentence, and slowly move the page toward your face until the image is blurred. Then move the page away until the image is sharp. Have your partner measure the distance between your eye and the page.
2. Test the other eye in the same manner.
3. **Complete item 2b on the laboratory report.**

Astigmatism. This condition results from an unequal curvature of either the cornea or the lens, which prevents light rays from being focused with equal sharpness on the retina.

1. Cover one eye and focus on the circle in the center of Figure 20.4. If the radiating lines appear equally dark and in sharp focus, no astigmatism exists. If astigmatism exists, record the number of the lines that appear lighter in color or blurred.
2. Test the other eye in the same manner.
3. **Complete item 2c on the laboratory report.**

Acuity. Visual acuity refers to the ability to distinguish objects in accordance with a standardized scale. It may be measured using a Snellen eye chart. If you can read the letters that are designated to be read at 20 ft at a distance of 20 ft, you have 20/20 vision. If the smallest letters that you can read at 20 ft are those designated to be read at 30 ft, you have 20/30 vision.

1. Stand 20 ft from the Snellen eye chart on the wall, while your partner stands next to the chart and points out the lines to read.
2. Cover one eye and read the lines as requested. Record the rating of the smallest letters read correctly.
3. Test the other eye in the same manner.
4. **Complete item 2d on the laboratory report.**

Color Blindness. The sensation of color vision depends on the degree to which impulses are formed by the three types of cones (receptors for red, green, and blue light) in the retina. The most common type of color blindness is red-green color blindness, which is caused by a deficit in cones stimulated by either red or green light. People with such a deficit have difficulty distinguishing reds and greens. A totally colorblind person sees everything as shades of gray.

1. Your partner holds the color-blindness test plates about 30 in. from your eyes in good light and allows you 5 sec to view each plate and give your response. Your partner records your responses in item 2e on your laboratory report.
2. **Complete item 2e on the laboratory report.**

THE EAR

The ear contains not only receptors for sound stimuli, but also receptors involved in maintaining equilibrium. For ease of study, the ear may be subdivided into the external ear, middle ear, and internal ear. Refer to Figure 20.5 as you study this section.

External Ear

The **external ear** includes the **auricle** or **pinna**, the flap of cartilage and skin commonly called the "ear," and the **external auditory canal** that leads inward through the temporal bone to the **tympanic membrane**, or eardrum.

Middle Ear

The **middle ear**, a small cavity in the temporal bone, is connected to the pharynx by the

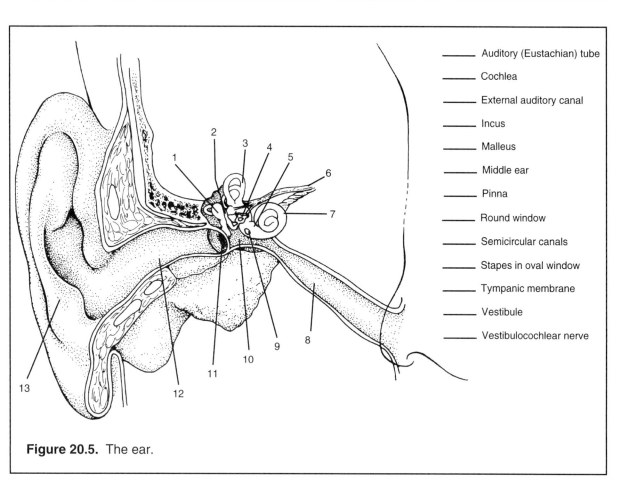

Figure 20.5. The ear.

_____ Auditory (Eustachian) tube
_____ Cochlea
_____ External auditory canal
_____ Incus
_____ Malleus
_____ Middle ear
_____ Pinna
_____ Round window
_____ Semicircular canals
_____ Stapes in oval window
_____ Tympanic membrane
_____ Vestibule
_____ Vestibulocochlear nerve

auditory (eustachian) tube. It is filled with air that enters or leaves via the eustachian tube, depending on the air pressure at each end of the tube. Three small bones, the **ear ossicles**, form a lever system that extends from the eardrum to the inner ear. In sequence, they are (1) the **malleus** (hammer) (label 1), which is attached to the ear drum, (2) the **incus** (anvil) (label 2), and (3) the **stapes** (stirrup) (label 4), which is inserted into the **oval window** of the inner ear.

Inner Ear

The **inner ear** consists of a complex of interconnecting tubes and chambers that are embedded in the temporal bone and are filled with fluid. The inner ear is subdivided into three major parts.

1. The **cochlea** is coiled like a snail shell and contains the receptors for sound stimuli.
2. The **vestibule** is the enlarged portion at the base of the cochlea. The stirrup is inserted into the oval window of the vestibule, and the round window is covered by a thin membrane. These two windows are involved in the transmission of sound stimuli. In addition, the vestibule contains receptors for static equilibrium that inform the brain of the position of the head.
3. The three **semicircular canals** contain receptors for dynamic equilibrium that inform the brain when the head is turned or when the entire body is rotated.

The Hearing Process

The auricle channels sound waves into the auditory canal, and as they strike the eardrum, it vibrates at their frequency. This vibration causes a comparable movement of the ear ossicles (hammer, anvil, and stirrup), which, in turn, transfer their movements to the fluid in the vestibule and cochlea of the inner ear. The movement of the fluid is made possible by the thin membrane of the round window, which moves *out and in* in synchrony with the *in and out* movement of the stirrup in the oval window, since the fluid cannot be compressed. The movement of the fluid in the cochlea stimulates the sound receptors, which send impulses to the auditory center in the brain, where they are interpreted as sound sensations.

Tone or pitch is determined by the particular receptors that are stimulated and the part of the brain that receives the impulses. Loudness is determined by the frequency of the impulses that reach the brain, which, in turn depends on the intensity of the vibrations that produce these impulses.

There are two kinds of hearing loss. **Conduction deafness** results from damage that prevents sound vibrations from reaching the inner ear. It is usually correctable by surgery or hearing aids. **Nerve deafness** is caused by damage to the sound receptors or neurons that transmit impulses to the brain. Nerve deafness usually results from exposure to loud sounds, and is not correctable.

Materials

Per student group
Colored pencils
Cotton for ear plugs
Meter sticks
Pocket watches, spring wound
Tuning forks

Per lab
Model of the ear

Assignment 3

1. Label and color-code Figure 20.5.
2. Locate the structures shown in Figure 20.5 on the ear model or chart.
3. **Complete item 3 on the laboratory report.**

Assignment 4

Watch-Tick Test. This is a simple test to detect hearing loss at a single sound frequency. It requires a quiet area. Work in teams of three students. One student is the subject, another moves the watch, and the third measures and records distances.

1. Have the subject sit in a chair and plug one ear with cotton. The subject is to look straight ahead and indicate by hand signals when the first and last ticks are heard.
2. Start with the watch about 3 ft laterally from the ear being tested and move it *slowly* toward the ear until the subject indicates the first tick is heard. Measure and record the distance from ear to watch.
3. Start with the watch close to the ear and *slowly* move it away from the ear until the last tick is heard. Measure and record the distance from ear to watch.
4. Calculate the average of the two measurements.
5. Test the hearing of the other ear in the same manner.
6. **Complete item 4a on the laboratory report.**

Rinne Test. A tuning fork is used in this test, which distinguishes between nerve and conduction deafness *in cases in which some hearing loss exists.* Work in pairs.

1. Have the subject sit in a chair and plug one ear with cotton. The subject is to indicate by hand signals when the sound is heard or not heard.
2. Strike the tuning fork against the heel of your hand to set it in motion. *Never strike it against a hard object.*
3. Hold the tuning fork 6–9 in. away from the ear being tested with the edge of the tuning fork toward the ear as shown in Figure 20.6.
4. The sound will be heard initially by persons with normal hearing and those with minimal hearing loss. As the sound fades, a point will be reached at which it will no longer be heard. (The subject is to indicate this point by a hand signal.) When this occurs, place the end of the tuning fork against the temporal bone behind the ear. See Figure 20.6. *Where a slight hearing loss exists* and the sound reappears, some conduction deafness is present.
5. Persons with a severe hearing loss will not hear the sound in step 4, or will hear it only briefly. If the sound reappears when the end of the tuning fork is placed against the temporal bone, conduction deafness is evident. If it does not reappear, nerve deafness exists.
6. **Complete item 4b on the laboratory report.**

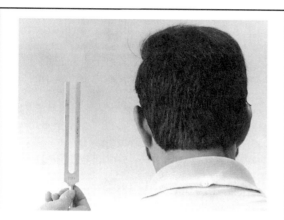

(a) Hold a vibrating tuning fork 6-9 in. from the ear with the edge of the tuning fork toward the ear.

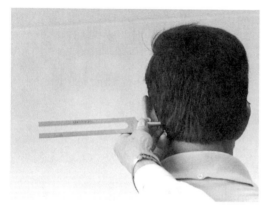

(b) When the sound is no longer heard, place the end of the tuning fork against the temporal bone behind the ear.

Figure 20.6. Rinne test.

Static Balance. The inner ear is not the only receptor involved in the maintenance of equilibrium. Pressure and touch receptors in the skin, stretch receptors in the muscles, and light receptors in the eyes also are involved. The brain constantly receives impulses from these receptors and subconsciously initiates any necessary corrective motor actions. The following test is a simple way to observe how this interaction functions. Work in groups of two to four students.

1. Use a meter stick to draw a series of vertical lines on the chalkboard about 5 cm (2 in.) apart. Cover an area about 1 m wide. This will help you detect body movements.

2. Have the subject remove his or her shoes and stand in front of the lined area facing you.

3. With feet together and arms at the side, the subject is to try to stand perfectly still for 30 sec while you watch for any swaying motion. Record the degree of swaying motion as slight, moderate, or great.

4. Repeat the test with the subject's eyes closed. Record the degree of movement. Try it again to see if extending the arms laterally helps the subject to maintain balance. What happens when the subject stands on only one foot with the hands at the side and with the eyes closed?

5. **Complete item 4 on the laboratory report.**

SKIN RECEPTORS

Human skin contains receptors for touch, pain, pressure, hot, and cold stimuli. The following tests will enable you to detect certain characteristics of sensations involving some of these receptors. Work in pairs to perform these experiments.

Materials

Per student group
Beakers, 400 ml, 3
Celsius thermometer
Clock or watch with second hand
Coins
Dividers
Hot plate
Metric ruler

Per lab
Crushed ice

Assignment 5

Distribution of Touch Receptors. For you to perceive two simultaneous stimuli as two touch sensations, the stimuli must be far enough apart to stimulate two touch receptors that are separated by at least one unstimulated touch receptor. This characteristic can be used to determined the density of touch receptors in the skin.

1. With the subject's eyes closed, touch his or

her skin with one or two points of the dividers. The subject reports the sensation as either "one" or "two." Start with the points of the dividers close together and gradually increase the distance between the points until the subject reports a two-point stimulus as a two-touch sensation about 75% of the time. Measure and record the distance between the tips of the dividers as the minimum distance evoking a two-point sensation.

2. Use the procedure in step 1 to determine the minimum distance giving a two-point sensation on the (1) inside of the forearm, (2) back of the neck, (3) palm of the hand, and (4) tip of the index finger.

3. **Complete item 5a on the laboratory report.**

Adaptation to Stimuli. Your nervous system has the ability to ignore stimuli or impulses so that you are not constantly bombarded with insignificant sensations.

1. Have the subject rest a forearm on the top of the table with the palm of the hand up.

2. With the subject's eyes closed, place a coin on the inner surface of the forearm. The subject is to indicate awareness of the presence of the coin and also the instant the sensation disappears. Record the time between these two events as the adaptation time.

3. Repeat the test using several coins stacked up to make a heavier object that increases the intensity of the stimulus. Determine the adaptation time.

4. **Complete items 5b and 5c on the laboratory report.**

Intensity of Sensations. The intensity of a sensation is usually proportional to the intensity of the stimulus. This occurs because receptors generate more impulses when the strength of the stimulus is increased, and the brain interprets the arrival of more impulses as a greater sensation.

1. Fill three 400-ml beakers with ice water, tap water, and warm water (about 50 °C), respectively).

2. Place your index finger in the water in each beaker, in sequence, and note the sensations. Can you recognize the temperature differences?

3. Use some small beakers to prepare water with slight differences in temperature, in order to determine the smallest difference that is detectable. Record this differential.

4. Now return to the original three beakers. Place your index finger in the ice water and note the sensation. Then immerse your whole hand and note the sensation. Repeat this procedure with the warm water.

5. **Complete items 5d–5f on the laboratory report.**

6. Use the three beakers of ice water, tap water, and warm water as in the previous experiment. Be sure that the warm water has not cooled.

7. Place one hand in ice water and the other in warm water. Note the sensation. Does it change with time? Explain.

8. After 2 min, place both hands in the beaker of tap water. Note the sensation. Does it change with time?

9. **Complete item 5 on the laboratory report.**

TASTE

The taste of food not only increases the appeal of food, but also increases the flow of saliva. In fact, just the thought of delicious food will stimulate salivation. Try it and see. What we usually refer to as taste is a combination of both taste and smell. This is why food "loses its taste" when you have a bad head cold that prevents airborne molecules of food from reaching your odor receptors in the roof of your nasal cavity.

Taste buds, receptors for taste, are located on the sides of very small projections, **papillae**, on the upper surface of the tongue. There are only four basic tastes—sweet, sour, bitter, and salt—and the distribution of taste buds sensitive to these tastes is not uniform, as shown in Figure 20.7.

As shown in Figure 20.8, the sensory **taste cells** of a taste bud are embedded in epithelial tissue on the tongue, with only their taste hairs protruding through a tiny **taste pore**. When the substances capable of activating the taste cells contact the taste hairs, the taste cells are stimulated, forming impulses that are carried by neurons to the brain.

Work in pairs to perform the following

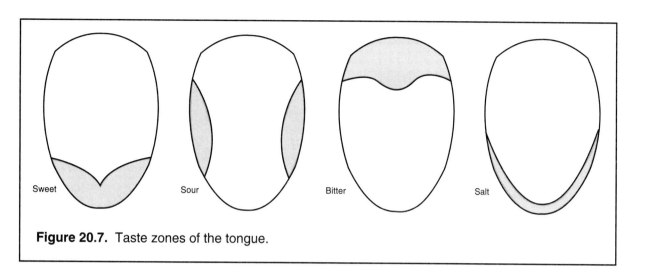

Figure 20.7. Taste zones of the tongue.

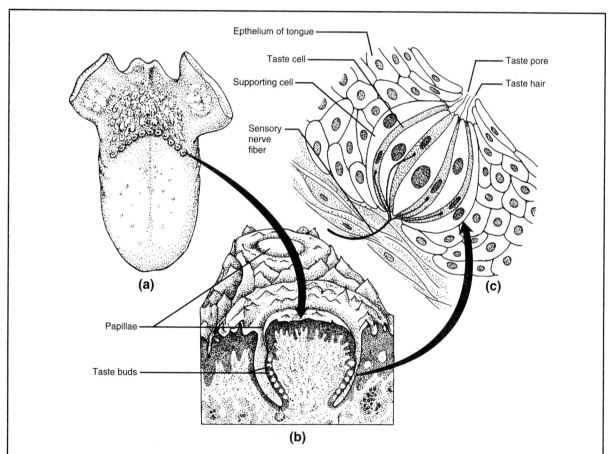

Figure 20.8. (a) Papillae containing taste buds are located on the upper surface of the tongue. (b) Taste buds are located along the outer margins of papillae on the tongue. (c) A taste bud consists of a bulb-shaped cluster of taste cells and supporting cells embedded in ephithelial tissue. The hair-like tips of taste cells protrude slightly through a taste pore.

experiments and alternate the roles of subject and experimenter.

Materials

Per student pair
Cotton swabs, 6
Facial tissues, 6
Paper cups, 2
Sucrose granules
Dropping bottles of:
   10% sodium chloride (salt)
   10% sucrose (sweet)
   10% vinegar (sour)

Assignment 6

1. Verify the locations of the sweet, sour, and salt taste receptors, as shown in Figure 20.7, as follows.
   a. Obtain 2 paper cups, several facial tissues, and 6 cotton swabs.
   b. Have the subject blot the upper surface of the tongue with a facial tissue, and stick the tongue out.
   c. Place several drops of sucrose solution on a cotton swab. Swab the tip of the tongue. Can the subject detect a sweet taste without withdrawing the tongue into his/her mouth? Is there any difference in the taste sensation after withdrawing the tongue?
   d. Have the subject rinse his/her mouth with water and blot the tongue and stick it out as before. Now swab an area where taste buds for sweet should be absent. Can the subject detect a sweet taste without withdrawing the tongue? Is there any difference in the taste sensation after withdrawing the tongue?
   e. Use the above procedure to verify the distribution of taste receptors for sour and salt.
   f. **Complete item 6a on the laboratory report.**
2. Determine if taste buds can detect molecules that are not in solution.
   a. Obtain several facial tissues and about half of a teaspoon of sugar granules on a folded paper towel.
   b. Have the subject blot and stick out his/her tongue as before.
   c. Sprinkle a number of sugar granules on the "sweet zone" of the tongue. Can the subject detect a sweet taste without withdrawing the tongue into his/her mouth? After 5 sec, have the subject withdraw the tongue. Is there any difference in the sensation?
   d. **Complete item 6 on the laboratory report.**

21

Support and Movement

OBJECTIVES

After completion of the laboratory session, you should be able to:

1. Identify the major bones and types of articulations in a human skeleton.
2. Identify the parts of a typical long bone.
3. Describe and identify the sexual differences in male and female pelvic girdles.
4. Describe and identify the types of levers formed by skeletal muscles and bones.
5. Identify the action of selected skeletal muscles.
6. Contrast the microscopic appearance of striations in relaxed and contracted skeletal muscle tissue.
7. Describe the roles of actin and myosin in muscle contraction.
8. Define all terms in bold print.

Support and movement of the body result from the interaction of the skeletal and muscle systems. These two body systems are so closely associated, both structurally and functionally, that we will consider them together in this exercise.

The skeletal system consists mainly of **bones** that join together, forming articulations (joints). At movable articulations, the bones are held together by **ligaments**, bands of dense connective tissue. In addition, there are many associated **cartilages**. The skeletal system provides: (1) protection for vital organs, (2) support for the body, (3) sites for muscle attachment, (4) a storehouse for calcium salts, and (5) formation of blood cells.

The muscle system consists of **skeletal muscles** that are attached to bones by **tendons**, bands or cords of inelastic, dense, fibrous connective tissue. Skeletal muscles are specialized for contraction (shortening), and they always span across a joint formed by bones. Thus, contraction of a skeletal muscle causes movement of one of the bones at the joint. In this way, the contraction of skeletal muscles enables body movements.

THE SKELETON

The skeleton consists of two major subdivisions. The **axial skeleton** is composed of the skull, vertebral column, ribs, and sternum. The **appendicular skeleton** consists of the bones of the upper extremities and the pectoral girdle and the bones of the lower extremities and the pelvic girdle. See Figure 21.1.

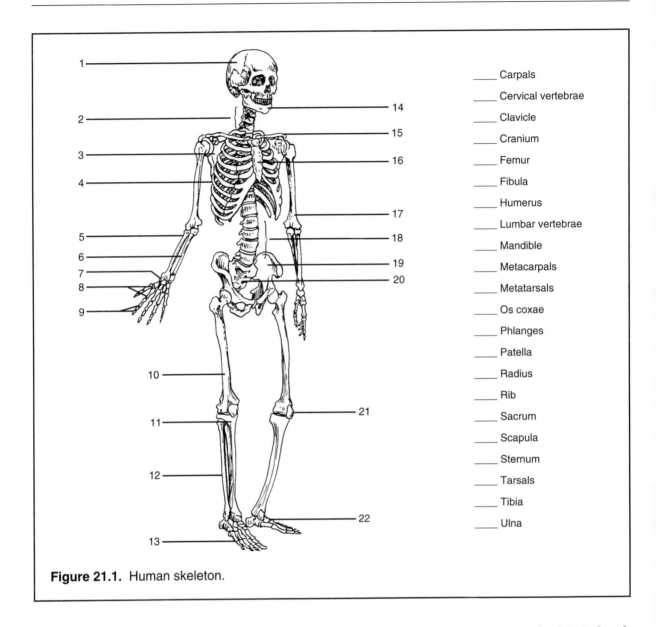

Figure 21.1. Human skeleton.

Column labels (with blanks):

_____ Carpals
_____ Cervical vertebrae
_____ Clavicle
_____ Cranium
_____ Femur
_____ Fibula
_____ Humerus
_____ Lumbar vertebrae
_____ Mandible
_____ Metacarpals
_____ Metatarsals
_____ Os coxae
_____ Phlanges
_____ Patella
_____ Radius
_____ Rib
_____ Sacrum
_____ Scapula
_____ Sternum
_____ Tarsals
_____ Tibia
_____ Ulna

Axial Skeleton

The **skull** is composed of the **cranium** (8 fused bones encasing the brain), 13 fused **facial bones**, and the movable **mandible** (lower jaw).

The **vertebral column** consists of vertebrae separated by **intervertebral discs**, which are composed of fibrocartilage. Vertebrae are subdivided as follows:

Cervical vertebrae: 7 vertebrae of the neck
Thoracic vertebrae: 12 vertebrae of the thorax, to which ribs are attached
Lumbar vertebrae: 5 large vertebrae of the lower back
Sacrum: a bone formed of 5 fused vertebrae

Coccyx: the tailbone, formed of 3–5 fused, rudimentary vertebrae

There are 12 pairs of **ribs**. The first 10 pairs are joined to the **sternum** (breastbone) by costal cartilages to form the thoracic cage. The last 2 pairs of ribs are short and unattached anteriorly. They are called floating ribs.

Appendicular Skeleton

The **pectoral girdle** supports the upper extremities. It consists of a **clavicle** (collarbone) and **scapula** (shoulder blade) on each side of the body. The clavicle is attached to the sternum on one end and to the scapula on the other. The scapula is supported by muscles that allow mobility for the shoulder.

The **humerus** (upper arm bone) articulates with the scapula at the shoulder and with the **ulna** and **radius** at the elbow. The wrist is composed of eight **carpal bones** that lie between the (1) ulna and radius and (2) **metacarpals**, the bones of the hand. The **phalanges** are the bones of the fingers and thumb.

The **pelvic girdle** consists of two **coxal bones** (hipbones) that join together anteriorly at the pubic symphysis and are fused posteriorly to the sacrum. This provides a sturdy support for the lower extremities.

The **femur** (thighbone) articulates with a coxal bone at the hip and with the **tibia** (shinbone) at the knee. The **patella** (kneecap) is embedded in the tendon anterior to the knee joint. The smaller bone of the lower leg is the **fibula**. Both tibia and fibula articulate with the **tarsal bones** forming the ankle and posterior part of the foot. The anterior foot bones are five **metatarsals**, and the toe bones are the **phalanges**.

Articulations

The bones of the skeleton are attached to each other by ligaments in such a way that varying degrees of movement occur at the joints. Articulations are categorized according to the degree of movement that is possible.

1. **Immovable joints** are rigid, such as those that occur between the skull bones.
2. **Slightly movable joints** allow a little movement, such as those between vertebrae.
3. **Freely movable joints** are the most common and allow the broad range of movement noted in the appendages. There are several types:
   a. **Hinge joints** allow movement in one direction only.
   b. **Ball-and-socket joints** allow angular movement in all directions plus rotation.
   c. **Gliding joints** occur where bones slide over each other, e.g., wrist and ankle bones.
   d. **Pivot joints** enable rotation in only one axis, e.g., between the first and second cervical vertebrae.

Materials

Per student
Colored pencils
Pipe cleaners

Per lab
Articulated human skeletons

Assignment 1

1. Label Figure 21.1 and color the axial skeleton blue.
2. Using Figure 21.1, identify the bones of an articulated skeleton and learn their recognition features.
3. **Complete items 1a-1d on Laboratory Report 21 that begins on page 385.**
4. Locate examples of each type of articulation on the articulated skeleton.
5. **Complete item 1 on the laboratory report.**

Sexual Differences of the Pelvis

The structure of the pelvic girdle is different in males and females, primarily because the female pelvis is adapted for childbirth. Table 21.1 lists some of the major differences between male and female pelvic girdles. Compare these characteristics with Figure 21.2. Since the characteristics of a pelvis may have considerable variations from the ideal, sex determination of a pelvis is based on a combination of characteristics rather than on a single characteristic.

Calculation of a **pelvic ratio** is helpful in determining the sex of a pelvic girdle. Two measurements are required. The ratio is calculated by dividing the first measurement by the second.

1. The distance between the tips of the ischial spines
2. The distance between the inner surface of the pubic symphysis and the upper, inner surface of the sacrum

$$\text{Pelvic ratio} = \frac{\text{first measurement}}{\text{second measurement}}$$

Females usually have a ratio of 1.0 or more, while males usually have a ratio of 0.8 or less.

TABLE 21.1
Comparison of the Male and Female Pelvis

Characteristic	Male	Female
General Structure	Not tilted forward; narrower and longer; heavier bones	Tilted forward; broader and shorter; lighter bones
Acetabula	Larger and closer together	Smaller and farther appart
Pubic angle	Less than 90°	Greater than 90°
Sacrum	Longer and narrower	Shorter and wider
Coccyx	More curved; less movable	Straighter; more movable
Pelvic brim	Narrower and heart shaped	Wider and oval shaped
Ischial spines	Longer, sharper, closer together, and project more medially	Shorter, blunt, farther apart, and project more posteriorly

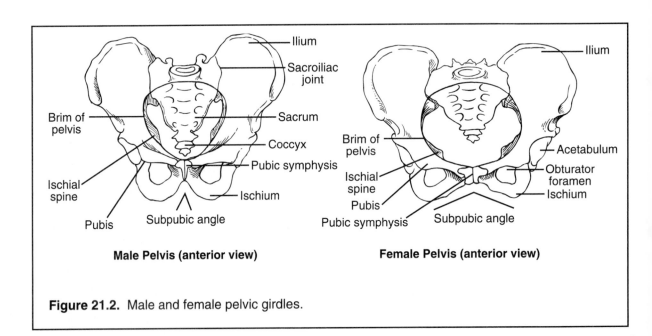

Figure 21.2. Male and female pelvic girdles.

Fetal Skeleton

During development, bones (except skull bones) of a fetus are first formed of hyaline cartilage. After this, ossification centers appear within the cartilage, and the formation of bone gradually replaces the cartilage. At the time of birth, the bones still have a large cartilage content. Skull bones form within membranes rather than in cartilage, but they also are incompletely formed at birth. The incompletely formed fetal skeleton provides greater flexibility during birth and the ability to grow after birth. Skeletal growth is complete at about 25 yrs of age.

Materials

Per student group
Dividers or calipers
Measuring tape
Metric ruler
Pipe cleaners

Per lab
Articulated skeleton
Fetal skeleton
Pelvic girdles, male and female

Assignment 2

1. Study the differences of the male and female pelvic girdles noted in Table 21.1 and Figure 21.2.
2. Examine the male and female pelvic girdles provided, and compare their distinguishing characteristics with Table 21.1 and Figure 21.2.
3. Use dividers or calipers to make the two measurements needed to calculate the pelvic ratio. Calculate the pelvic ratio for both male and female pelvic girdles.
4. Determine the pelvic ratio for the pelvic girdle of the articulated skeleton and examine its other features to determine the gender of the skeleton.
5. **Complete item 2 on the laboratory report.**

Assignment 3

1. Examine the fetal skeleton. Note the incomplete ossification of the bones. Observe how the brain of a baby is incompletely protected by cranial bones because of the spaces (fontanels) between the skull bones.
2. **Complete item 3 on the laboratory report.**

MACROSCOPIC BONE STRUCTURE

Figure 21.3 depicts the structure of a typical long bone, a human humerus. A long bone is characterized by a shaft of bone, the **diaphysis**, which extends between two enlarged portions forming the ends of the bone, the **epiphyses**. The articular surface of each epiphysis is covered by an **articular cartilage** (hyaline cartilage), which reduces friction in the joints and protects the ends of the bone. The rest of the bone is covered by the **periosteum**, a tough, tightly adhering membrane containing tiny blood vessels that penetrate into the bone. Bone deposition by the periosteum contributes to the growth in diameter of the bone. Larger blood vessels and nerves enter the bone through a channel called a **foramen**.

A longitudinal section of the bone reveals that the epiphyses consist of **cancellous** (spongy) **bone** covered by a thin layer of **compact bone**, while the diaphysis is formed of heavy, compact bone. **Red marrow** fills the spaces in the spongy bone. It forms red and white blood cells and platelets. The **medullary cavity** is filled with fatty **yellow marrow**. In immature bones, an **epiphyseal disc** of cartilage is located between the diaphysis and each epiphysis; this is the site of linear growth. A mature bone lacks this cartilage, since it has been replaced by bone, and only an **epiphyseal line** of fusion remains.

Materials

Per student group
Beef femur or tibia, fresh and split lengthwise
Colored pencils
Dissecting instruments and pan

Per lab
Beef knee joint split lengthwise
Human femur, split

Assignment 4

1. Label Figure 21.3. Color-code spongy bone red, medullary cavity yellow, articular cartilage blue, and epiphyseal disc green. **Complete item 4a on the laboratory report.**
2. Examine the split human femur. Locate the spongy and compact bone, epiphyseal line, and medullary cavity.
3. Obtain a split beef femur. Locate the medullary cavity, compact bone, spongy bone, epiphyseal disc, and articular cartilage.
4. Remove some of the yellow marrow to observe the size of the medullary cavity. Does yellow marrow contain fat? Is red marrow well supplied with blood?
5. Feel the articular cartilage. How is it suited for its function?
6. Use a scalpel to cut off a small piece of periosteum. Is the periosteum loosely attached? Is it weak or strong? **Complete items 4b–4f on the laboratory report.**

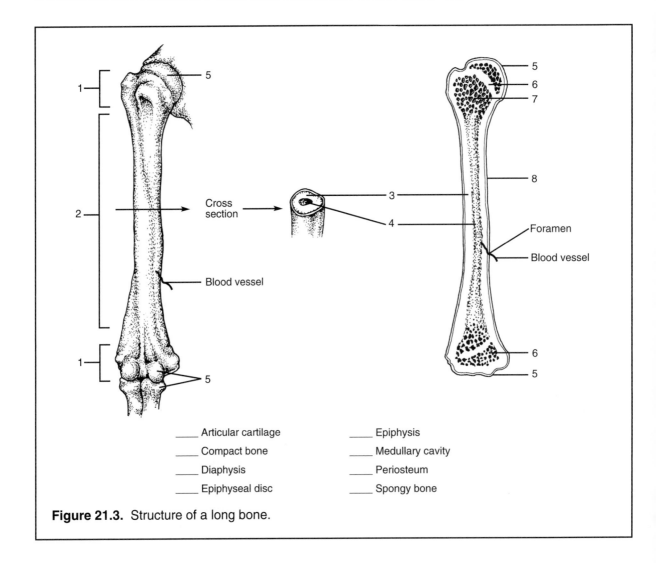

Figure 21.3. Structure of a long bone.

_____ Articular cartilage
_____ Compact bone
_____ Diaphysis
_____ Epiphyseal disc

_____ Epiphysis
_____ Medullary cavity
_____ Periosteum
_____ Spongy bone

7. Examine the split beef knee joint. Note how the ligaments hold the bones together. Examine the ligaments. **Complete item 4 on the laboratory report.**

SKELETAL MUSCLES

Skeletal muscles span a joint and are attached at each end to different bones by **tendons**. The end of a muscle that moves during contraction is the **insertion**, while the stationary (nonmovable) end is the **origin**.

Skeletal muscles and bones of the skeleton are arranged to form a system of levers that work together to produce a variety of movements. A lever consists of a rigid rod that moves about a fixed point called a **fulcrum**. Two opposing

forces act on a lever. The **resistance** in the weight to be moved by the **force** applied at a specific point on the lever. In the arrangement of muscles and bones, the rigid rod of the lever is formed by a *bone* and the fulcrum is a *joint*. The resistance may be the weight of an arm, leg, or body part, and it may also include an additional weight, such as a weight held in a hand. The force causing movement is always the **contraction force** of a muscle applied at the point of insertion of the muscle on a bone.

The three types of levers formed in the body are diagramed in Figure 21.4. Each type is characterized by the relative positions of the resistance (R), fulcrum (F), and contraction force (CF). In a **first-class lever**, the fulcrum is located between the contraction force and resistance. In a **second-class lever**, the resistance

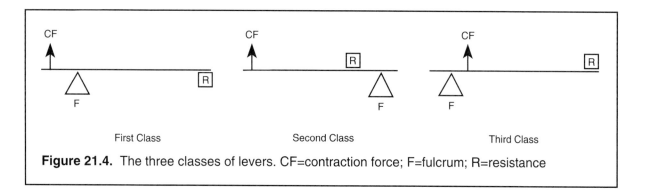

Figure 21.4. The three classes of levers. CF=contraction force; F=fulcrum; R=resistance

First Class   Second Class   Third Class

is located between the contraction force and the fulcrum. In a third-class lever, the contraction force is located between the fulcrum and the resistance. The nearer the contraction force is located to the fulcrum, the lower is the mechanical advantage and the greater is the speed of movement of the body part.

Skeletal muscles are arranged in **antagonistic groups**, meaning that opposing muscles move the same body part in opposite directions. This arrangement is necessary because muscles can only contract. Here are two examples of antagonistic muscle actions.

Flexor muscles decrease the angle of a joint, while **extensor muscles** increase the angle of a joint.

Abductor muscles move a body part away from the midline of the body, while **adductor muscles** move a body part toward the midline of the body.

The biceps and triceps muscles of the upper arm are examples of antagonistic muscles. The biceps muscle flexes the forearm while the triceps is relaxed, and the triceps extends the forearm while the biceps is relaxed. See Figure 21.5 and note the action of these muscles and the types of levers involved.

Assignment 5

1. Study Figures 21.4 and 21.5 until you know the characteristics of each type of lever.
2. Place your left hand on your upper right arm and note the contraction and relaxation of the biceps and triceps muscles as you flex and extend your right forearm.
3. In Figure 21.6, record on each diagram the lo-

cation of the resistance (R), fulcrum (F), and contraction force (CF). Then record the type of movement (flexion, extension, abduction, adduction) and lever class for each diagram.
4. **Complete item 5 on the laboratory report.**

MYOFIBRIL ULTRASTRUCTURE AND CONTRACTION

The striations of a skeletal muscle fiber result from the arrangement of actin and myosin myofilaments within each myofibril. See Figure 21.7. When a myofibril is relaxed, the thicker **myosin myofilaments** are the main components of, and only occur within, the dark-colored A bands of the myofibril. The thinner **actin myofilaments** are the only myofilaments in the light-colored I bands, but they extend into the A bands from the **Z lines.** Z lines form the boundaries of a **sarcomere**, the contractile unit of a myofibril.

Mechanics of Contraction

Contraction of a muscle fiber results from the interaction of actin and myosin in the presence of calcium and magnesium ions. Adenosine triphosphate (ATP) supplies the required energy (see chapter 8).

When a muscle fiber is activated by a neural impulse, the cross-bridges of the myosin myofilaments attach to active sites on the actin myofilament and bend to exert a power stroke that pulls the actin filaments toward the center of the A band. After the power stroke, the cross-bridges separate from the first actin active sites, attach to the next active sites, and produce an-

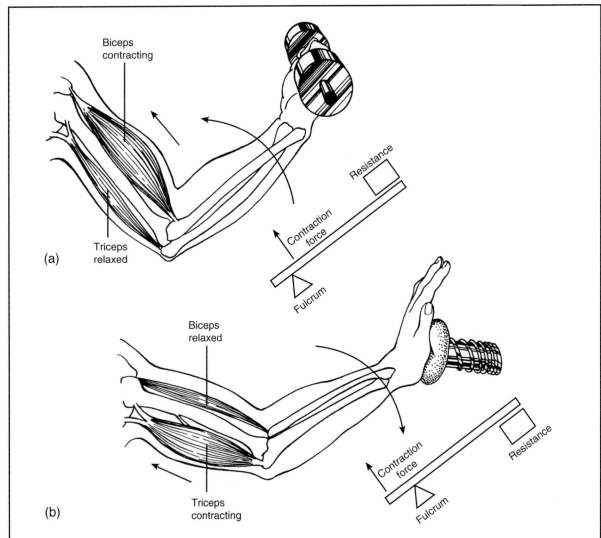

Figure 21.5. Antagonistic functions of biceps and triceps. (a) Contraction of the biceps flexes the forearm via a third-class lever. (b) Contraction of the triceps extends the forearm via a first-class lever.

other power stroke. This process is repeated until maximum contraction is attained.

Experimental Muscle Contraction

In this section, you will induce muscle fibers to contract, using ATP and magnesium (Mg^{++}) and potassium (K^+) ions. Your instructor has prepared 2-cm segments of muscle fibers from a glycerinated rabbit psoas muscle. The segments have been placed in a petri dish of glycerol and teased apart to yield thin strands consisting of very few muscle fibers. The strands should be no more than 0.2 mm thick.

Materials

Per student
Microscopes, compound and dissecting
Dissecting instruments
Microscope slides and cover glasses
Plastic ruler, clear and flat
Glycerinated rabbit psoas muscle, 3–6 strands

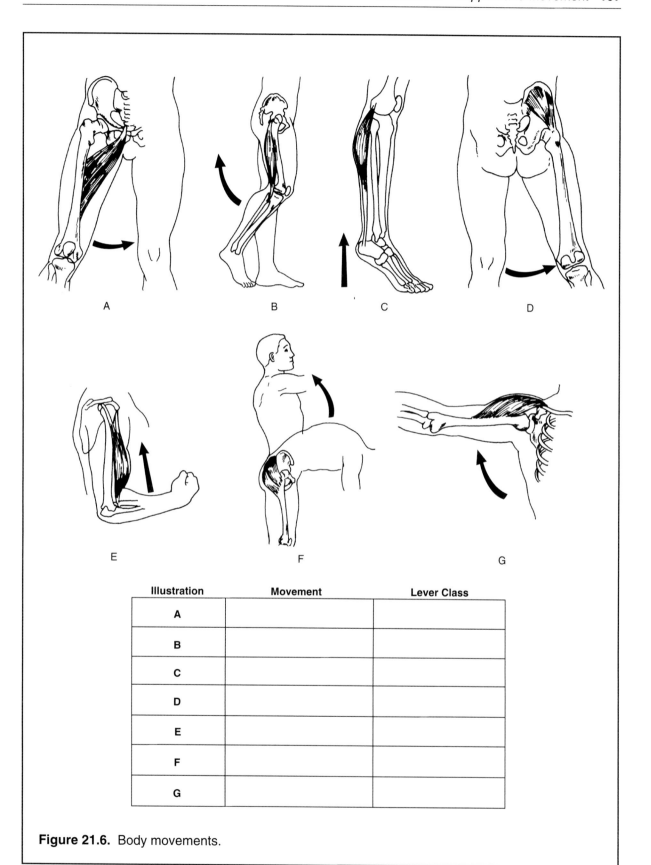

Illustration	Movement	Lever Class
A		
B		
C		
D		
E		
F		
G		

Figure 21.6. Body movements.

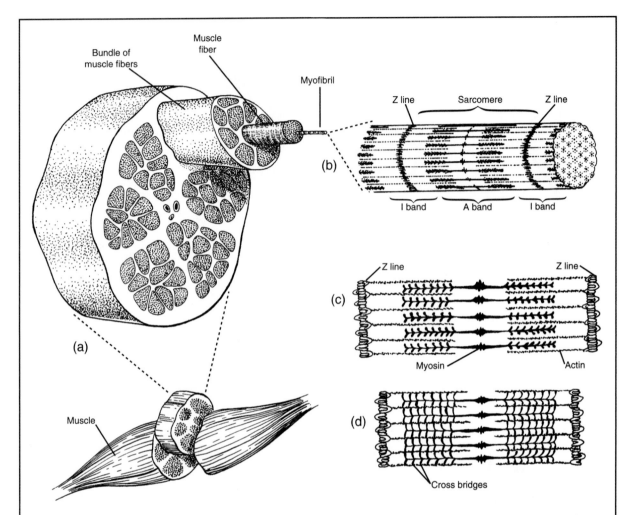

Figure 21.7. Structure of skeletal muscle. (a) The arrangement of muscle fibers within a muscle. (b) The ultrastructure of a myofibril showing the relationship of actin and myosin filaments. Note how the arrangment of actin and myosin differs in (c) a relaxed state and (d) a partially contracted state.

Dropping bottles of:
ATP, 0.25%, in triple-distilled water
glycerol
magnesium chloride, 0.001 M
potassium chloride, 0.05 M

Assignment 6

1. Place a few muscle strands in a small drop of glycerol on a microscope slide and add a cover glass. Observe them with the high-dry or oil-immersion objective. **Draw the pattern of striations of the relaxed muscle fibers in item 6a of the laboratory report.**
2. Place three to five of the thinnest strands on another slide in just enough glycerol to moisten them. Arrange the strands straight, parallel, and close together.
3. Use a dissecting microscope to measure the length of the relaxed strands by placing the slide on a clear plastic ruler. **Record their lengths in item 6b on the laboratory report.** Note the width of the strands.
4. While observing through the dissecting microscope, add one drop from each of the

solutions: ATP, magnesium chloride, and potassium chloride. Note any changes in length or width of the strands.

5. Remeasure and **record the length of the contracted strands in item 6b on the laboratory report.**

6. Place a few contracted strands in a small drop of glycerol on another slide and add a cover glass. Observe the strands with the high-power or oil-immersion objective. **Draw the pattern of striations in item 6a on the laboratory report.**

7. **Complete item 6 on the laboratory report.**

Assignment 7

Your instructor has set up several disarticulated human bones as a mini-practicum. Your task is to identify the bones. **Complete item 7 on the laboratory report.**

22

Excretion

OBJECTIVES

After completion of the laboratory session, you should be able to:
1. Identify the components of the urinary system and their functions.
2. Identify the parts of a dissected kidney.
3. Describe the basic function of the kidney.
4. Perform a urinalysis using a "dipstick."
5. Identify normal and abnormal components of urine and correlate abnormal components with disease conditions.
6. Define all terms in bold print.

Metabolic processes produce waste products that are harmful to body cells, so the wastes must be removed from body fluids. The principal metabolic wastes are carbon dioxide and nitrogenous wastes that accumulate in the blood. They must be removed as fast as they accumulate, so that their concentrations in the blood remain within normal limits. Carbon dioxide is removed by the respiratory system. **Nitrogenous wastes** are removed by the **urinary system** during the formation of **urine** as the blood is cleansed by the kidneys.

The primary nitrogenous wastes are ammonia, urea, and uric acid. When amino acids are deaminated by the liver, the amine groups tend to form **ammonia (NH$_3$)**, a substance toxic to body cells. To get rid of the ammonia, the liver converts it into **urea**. Urea is the most abundant nitrogenous waste in urine. A relatively small amount of **uric acid** is also present in urine as a result of the breakdown of nucleic acids.

The process of urinary excretion not only removes nitrogenous wastes from the body, but also regulates the concentration of ions, water, and other substances in body fluids.

THE URINARY SYSTEM

The urinary system consists of (1) a pair of **kidneys** that remove nitrogenous wastes, excess ions, and water from the blood to form **urine**; (2) a pair of **ureters** that carry urine by peristalsis from the kidney to the urinary bladder; (3) a **urinary bladder** that serves as a temporary storage container for urine; and (4) a **urethra** that carries urine from the bladder during urination. See Figure 22.1.

The Kidney

The basic structure of a kidney is shown in coronal section in Figure 22.2. The outer portion is the **cortex**, which contains vast numbers of capillaries and **nephrons**. The inner **medulla** con-

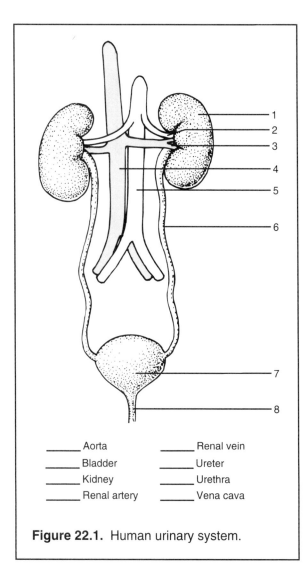

_____ Aorta   _____ Renal vein

_____ Bladder   _____ Ureter

_____ Kidney   _____ Urethra

_____ Renal artery   _____ Vena cava

Figure 22.1. Human urinary system.

tains the **renal pyramids**, which are composed mainly of **collecting tubules**. The tip (papilla) of each pyramid is inserted into a funnel-like **calyx** that unites with the **renal pelvis**.

The Nephron

The nephron is the functional unit of the kidney. Each human kidney contains about 1 million nephrons. A nephron consists of (1) a **Bowman's (glomerular) capsule** that surrounds an arteriole capillary tuft, known as a **glomerulus**, and (2) a tortuous, thin-walled **tubule** that leads to a collecting tubule. The nephron tubule consists of three parts. The **proximal convoluted tubule** leads from the nephron capsule to a U-shaped **loop of**

Henle that is contiguous with a **distal convoluted tubule**. Many renal tubules are joined to a single collecting tubule.

Materials

Per student group
Colored pencils
Dissecting kit and pan
Microscope, compound
Sheep kidney, preferably fresh, mid-coronally sectioned

Per lab
Model of human urinary system
Model of human kidney, mid-coronal section
Prepared slides of kidney cortex
Sheep kidneys, triple injected and preserved

Assignment 1

Complete item 1 on Laboratory Report 22 that begins on page 389.

Assignment 2

1. Label and color-code parts of the urinary system in Figure 22.1 and parts of the kidney and nephron in Figure 22.2.
2. Study the models of the urinary system and kidney. Locate the parts shown in Figures 22.1 and 22.2.
3. Examine the demonstration whole sheep kidney. Note its distinctive shape and locate the ureter and renal blood vessels.
4. Obtain a sheep kidney that has been coronally sectioned and locate the parts shown in Figure 22.2a. Correlate the structure of the parts with their functions as previously described.
5. Examine the sectioned triple-injected kidney set up under a dissection microscope. Locate the glomeruli and nephron capsules.
6. Examine a prepared slide of kidney cortex. Locate a glomerulus and nephron capsule. Note the thin walls of the tubules. How many cells thick is the tubule wall?
7. **Complete item 2 on the laboratory report.**

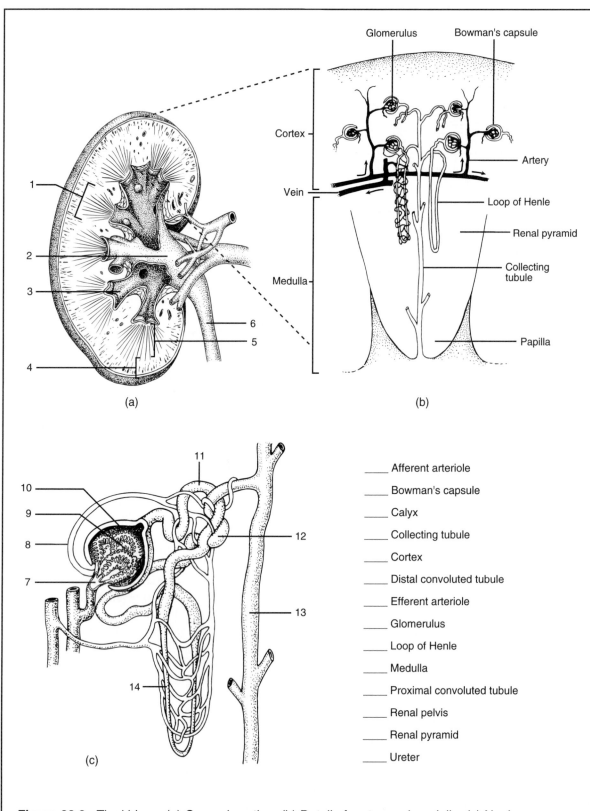

Figure 22.2. The kidney. (a) Coronal section. (b) Detail of cortex and medulla. (c) Nephron.

Urine Formation

Blood is carried from the aorta to the kidneys by renal arteries, and returned from the kidneys by renal veins into the vena cava. While passing through the kidneys, the concentration of nitrogenous wastes and other substances in the blood is controlled by (1) the partial removal of diffusible materials in the glomerular blood by **filtration** into Bowman's capsules, (2) **selective reabsorption** of useful substances from the tubules back into the blood, and (3) **secretion** of certain ions from the capillary blood into the tubules.

The renal artery branches into progressively smaller arteries, leading ultimately to the afferent arterioles supplying blood to the glomeruli. The **afferent arteriole** leading to a glomerulus is larger in diameter than the **efferent arteriole** that carries blood away from the glomerulus. This causes an increase in blood pressure in the glomerulus and accelerates the filtration of materials from the glomerular blood into the Bowman's capsule. Thus, some of all diffusible materials pass from the glomerulus into Bowman's capsule.

The fluid in the capsule, known as the **filtrate**, flows through the nephron tubule, which is enveloped by a capillary network. The tubule selectively reabsorbs needed materials—especially water, nutrients, and mineral ions—from the filtrate into the capillary blood as the filtrate moves through the tubule. At the same time, certain excess ions (especially H^+ and K^+) may be secreted from the capillary blood into the filtrate. Both **tubule reabsorption** and **tubule**

TABLE 22.1
Volume (liters) of Blood, Filtrate, and Urine Formed in Humans in a 24-Hr Period

Blood flow through the kidneys	1,700
Filtrate removed from blood	180
Urine formed	1.5

secretion involve active and passive transport mechanisms. The remaining fluid flows from the nephron tubule into a collecting tubule of a renal pyramid, continues into a calyx, and moves on into the renal pelvis. This fluid, now called **urine**, leaves the kidney via the ureter and is carried to the urinary bladder by peristalsis.

Assignment 3

1. Study Table 22.1. Note the relationships among the daily volumes of blood circulated through the human kidneys, the filtrate removed, and the urine formed. **Complete items 3a-3d on the laboratory report.**
2. Study Table 22.2. Note how the concentration of selected substances varies in different parts of the nephron and blood vessels. Use Figure 22.2 to orient yourself to the anatomical association of the structures involved.

   Note that the concentration of urea is constant in columns 1, 2, and 3 of Table 22.2. This indicates that urea is a small molecule that easily passes from the glomerulus into

TABLE 22.2
Concentration (mg/100ml) of Dissolved Substances in Selected Blood Vessels and Regions of the Nephron

| | 1 | 2 | 3 | 4 | 5 |
Materials in Water Solution	Afferent Arteriole	Efferent Arteriole	Bowman's Capsule (Filtrate)	Collecting Tubule (Urine)*	Renal Vein
Urea	30	30	30	2,000	25
Uric acid	4	4	4	50	3.3
Inorganic salts	720	720	720	1,500	719
Protein	7,000	8,000	0	0	7,050
Amino acids	50	50	50	0	48
Glucose	100	100	100	0	98

* Trace amounts are not included.

Bowman's capsule. How do you explain the change in urea concentration in columns 4 and 5?

3. **Complete item 3 on the laboratory report.**

URINALYSIS

In humans, the composition of urine is commonly used to assess the general functioning of the body. Diet, exercise, and stress may cause variations in the composition and concentration of the urine but significant deviations usually result from malfunctions of the body. See Table 22.3.

Materials

Per student
Collecting cup, plastic

TABLE 22.3
Urine Components Evaluated by Urinalysis

Component	Normal	Abnormal*
Color	Straw to amber	Pink, red-brown, or smoky urine may indicate blood in the urine. The higher the specific gravity, the darker is the color. Nearly colorless urine may result from excessive fluid intake, alcohol ingestion, diabetes insipidis, or chronic nephritis.
Turbidity	Clear to slightly turbid	Excess and persistent turbidity may indicate pus or blood in the urine.
pH	4.8–8.0; Average = 6.0	Acid urine may result from a diet high in protein (meats and cereals) or a high fever; alkaline urine results from a vegetarian diet or bacterial infections of the urinary tract.
Specific (ADH) gravity	1.003–1.035	Low values result from a deficiency of antidiuretic hormone (ADH) or kidney damage that impairs water reabsorption. High values result from diabetes mellitus or kidney disease, allowing proteins to enter filtrate.
Blood or hemoglobin	Absent	Presence of intact RBCs may result from lower urinary tract infections, kidney disease allowing RBCs to enter filtrate, lupus, or severe hypertension. Presence of hemoglobin occurs in extensive burns and trauma, hemolytic anemia, malaria, and incompatible transfusions.
Protein	Absent or trace	Proteins are present in kidney diseases allowing proteins to enter the filtrate, and they may be present in fever, trauma, anemia, leukemia, hypertension, and other nonrenal disorders. They may occur due to excessive exercise and high-protein diets.
Glucose	Absent or trace	Presence usually indicates diabetes mellitus.
Ketones	Absent or trace	Presence results from excessive fat metabolism as in diabetes mellitius and starvation.
Bilirubin	Small amounts	Excess may indicate liver disease (hepatitis or cirrhosis) or blockage of bile ducts.
Nitrite (bacteria)	Absent	Presence indicates a bacterial urinary tract infection.
Leukocytes (pus)	Absent	Presence indicates a urinary tract infection.

* Only a few causes of abnormal values are noted.

Multistix 9 SG reagent strips, 6
Set of test tubes of simulated urine samples

Per lab
Biohazard bag

Assignment 4

In this section, you will use "dip sticks" to analyze several simulated urine samples in order to determine if they are normal or if their characteristics suggest possible disease. Table 22.3 indicates normal values and some abnormal characteristics. Study it before proceeding.

1. Obtain six Multistix 9 SG reagent strips (dip sticks) and a color chart that indicates how to read the results. Study the color chart carefully to be sure that you understand the time requirements for reading the dip sticks and the location and interpretation of each reagent band. Note: The sequence of the reagent bands from the tip to the handle of a dip stick matches the sequence of specific tests listed from top to bottom on the color chart.
2. Obtain six numbered test tubes containing simulated urine samples. Place them in a test-tube rack at your workstation.

3. Analyze the urine samples one at a time by dipping a dip stick completely into the sample so that all reagent bands are immersed. Remove the dip stick and place it on a paper towel with the reagent bands facing upward. Read your results after 1 min.
4. If you wish to test your own urine, obtain a plastic collecting cup and a Multistix 9 SG reagent strip from your instructor and perform the test in the restroom. Dispose of the urine in a toilet, place the collecting cup in the biohazard bag, and read your results. Then place the dip stick in the biohazard bag.
5. **Record your results in item 4a on the laboratory report. Compare your results with Table 22.3 and complete item 4 on the laboratory report.**

Assignment 5

Your instructor has set up a sectioned sheep kidney and a model of the urinary system with selected structures indicated by numbered pins or tags as a mini-practicum. Your task is to identify the structures. **Complete item 5 on the laboratory report.**

23 Reproduction

OBJECTIVES

After completion of the laboratory session, you should be able to:
1. Describe the structure and function of the male and female reproductive systems.
2. Identify parts of the reproductive systems on models and charts.
3. Describe the process of gametogenesis.
4. Identify the stages of sperm and egg development when viewed microscopically.
5. Describe the mode of action and relative effectiveness of birth control methods.
6. Define all terms in bold print.

The human male and female reproductive systems are specially adapted for their roles in gametic sexual reproduction. **Gametes** are sex cells—sperm and eggs—and are produced by **gonads**, the sex glands. Sperm are produced by **testes**, the male sex glands, and eggs are produced by **ovaries**, the female sex glands. During copulation (sexual intercourse), sperm are deposited in the female reproductive tract, and they quickly move into the uterus and oviducts in search of an egg. In humans as in all mammals, **internal fertilization**, the fusion of sperm and egg, usually occurs in an oviduct. Fertilization results in the formation of a **zygote**, the first cell of a potential child. As the embryo develops, it is carried into the uterus, where **internal development** takes place, leading to the birth of a baby in about 280 days.

MALE REPRODUCTIVE SYSTEM

Locate the organs described here in Figure 23.1. The male gonads are paired testes that are held in the sac-like **scrotum**. This arrangement holds the testes outside the body cavity and at a temperature of 94–95 °F, which is necessary for the production of viable **spermatozoa**. Muscles in the wall of the scrotum relax or contract to position the testes farther from or closer to the body, and in this way regulate the temperature of the testes.

A testis contains numerous **seminiferous tubules** that produce the spermatozoa. **Interstitial cells** are located between the tubules, and secrete **testosterone**, the male hormone responsible for the sex drive and the development of the sex organs and secondary sexual characteristics. The secretion of **interstitial cell stimulating hormone (ICSH)** by the hypophysis (pituitary gland) activates the interstitial cells.

The **penis**, the male copulatory organ, contains

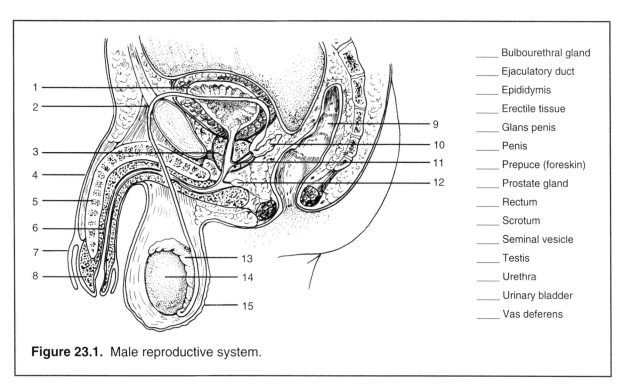

Figure 23.1. Male reproductive system.

1 _____
2 _____
3 _____
4 _____
5 _____
6 _____
7 _____
8 _____
9 _____
10 _____
11 _____
12 _____
13 _____
14 _____
15 _____

_____ Bulbourethral gland
_____ Ejaculatory duct
_____ Epididymis
_____ Erectile tissue
_____ Glans penis
_____ Penis
_____ Prepuce (foreskin)
_____ Prostate gland
_____ Rectum
_____ Scrotum
_____ Seminal vesicle
_____ Testis
_____ Urethra
_____ Urinary bladder
_____ Vas deferens

three cylinders of spongy **erectile tissue** that fill with blood during sexual excitement to produce an erection. A circular fold of tissue, the **prepuce** (foreskin) covers the **glans penis**. For hygienic reasons, the prepuce of male babies is often removed by a surgical procedure called **circumcision**.

Mature, but inactive, sperm are carried down the seminiferous tubules to the **epididymis**, a long, coiled tube on the surface of the testis. Sperm are stored here until they are propelled through the reproductive tract by wavelike contractions during **ejaculation**.

The **bulbourethral glands** open into the urethra below the prostate gland and secrete an alkaline liquid that neutralizes the acidity of the urethra prior to ejaculation.

At the male orgasm, sperm pass from the epididymis into the **vas deferens**, a duct that exits the scrotum and enters the body cavity via the inguinal canal. It continues across the surface of the urinary bladder to join with the **ejaculatory duct** within the **prostate gland**, which encircles the urethra just below the bladder. Alkaline secretions from the **seminal vesicles** are mixed with the sperm just before the vas deferentia enter the prostate gland, where sperm-

activating prostatic secretions are added. Muscular contractions force **semen**, the mixture of sperm and glandular secretions, out through the urethra.

FEMALE REPRODUCTIVE SYSTEM

Locate the organs described here in Figure 23.2. The external female genitalia consist of (1) two folds of skin surrounding the vaginal and urethral openings, and known as the **labia majora** (outer folds) and **labia minora** (inner folds), respectively, and (2) the **clitoris**, a nodule of erectile tissue homologous to the penis in the male. Collectively, these structures are called the **vulva**.

The **vagina** is a collapsible tube extending 4–6 in. from its external opening to the uterus. It serves as both the female copulatory organ and the birth canal. The **uterus** is a pear-shaped organ located over and posterior to the urinary bladder. The **cervix** of the uterus extends a short distance into the upper end of the vagina.

A pair of **ovaries**, the female gonads, are located laterally to the uterus, where they are

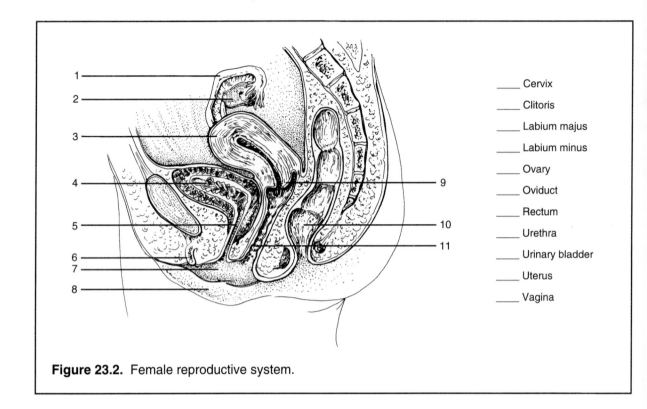

Figure 23.2. Female reproductive system.

_____	Cervix
_____	Clitoris
_____	Labium majus
_____	Labium minus
_____	Ovary
_____	Oviduct
_____	Rectum
_____	Urethra
_____	Urinary bladder
_____	Uterus
_____	Vagina

supported by ligaments. One egg is released from alternate ovaries about every 28 days. The egg is picked up by the expanded end of the **oviduct** and carried toward the uterus by beating cilia of the ciliated epithelium lining the oviduct.

Ovaries secrete two female hormones. **Estrogen** is responsible for the development of the sex organs, the female sex drive, the female secondary sex characteristics, and the monthly buildup of the uterine lining. **Progesterone** prepares the uterine lining for the implantation of an early embryo. In turn, ovarian function is controlled by hormones released from the hypophysis (pituitary gland). Consult your text for a discussion of the ovarian and uterine cycles.

Materials

Per student
Colored pencils

Per lab
Anatomical charts of male and female reproductive systems
Models of male and female reproductive systems

Assignment 1

1. Label Figures 23.1 and 23.2. Color-code the organs of the reproductive systems.
2. Locate the organs of the male and female reproductive systems on the models and charts provided.
3. **Complete item 1 on Laboratory Report 23 that begins on page 393.**

GAMETOGENESIS

The formation of gametes is called gametogenesis. It includes meiotic cell division, which reduces the chromosome number of gametes to one half that of somatic cells. For example, the diploid (2n) chromosome number in humans is 46, while the haploid (n) gametes (egg and sperm cells) contain only 23 chromosomes. If you need to review meiosis, see Exercise 9.

The patterns of gametogenesis described here and illustrated in Figures 23.3 and 23.4 are typical of mammals, although slight variations may occur among individual species.

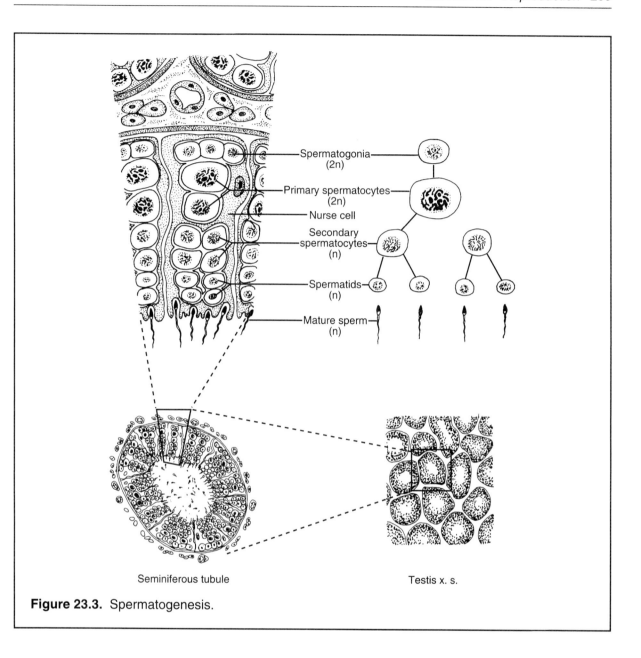

Figure 23.3. Spermatogenesis.

Spermatogenesis

Sperm formation occurs in the seminiferous tubules of the testes. **Spermatogonia** (2n) are the outermost cells of the tubule. They divide mitotically to form a **primary spermatocyte** and a replacement spermatogonium. The primary spermatocyte then divides meiotically to yield two **secondary spermatocytes** (n) after meiotic division I, and four **spermatids** (n) after meiotic division II. The spermatids attach to "nurse cells" and mature into spermatozoa. Maturation includes the loss of most of the cytoplasm of the spermatozoa and the formation of a flagellum from a centriole. The sperm head consists mostly of the cell nucleus. Sperm are carried along the seminiferous tubules and reach the epididymis in about 10 days.

Oogenesis

Prior to the birth of a female child, some **oogonia** of the germinal epithelium surrounding each ovary enlarge, become surrounded by follicular cells, and move into the ovary. These oogonia (2n) become the **primary oocytes** (2n), and

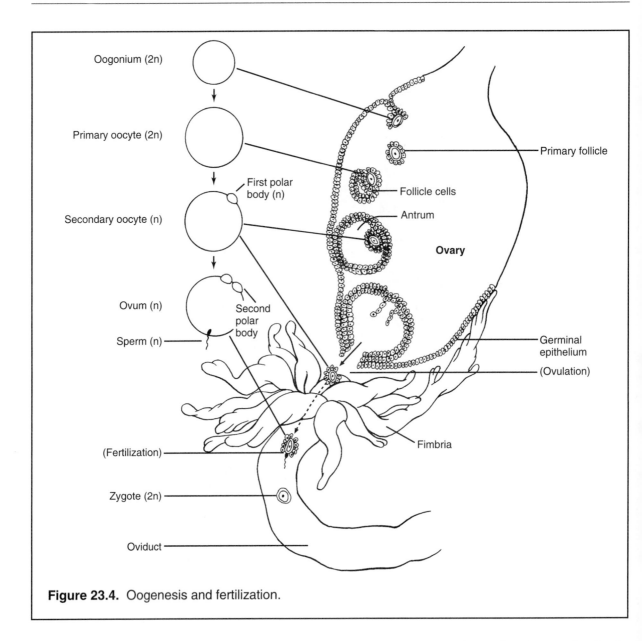

Figure 23.4. Oogenesis and fertilization.

enter the prophase of meiosis I before oogenesis is arrested. The primary oocytes remain inactive at this stage until puberty.

At puberty, **follicle stimulating hormone (FSH)** and **luteinizing hormone (LH)** secreted by the pituitary gland stimulate primary oocytes and follicle cells to undergo further growth. Each month, one follicle develops more rapidly than the others to become a mature or **Graafian follicle**, which is filled with fluid containing a large amount of **estrogen** secreted by the follicular cells. Meiotic division I proceeds to produce (1) a **secondary oocyte** (n) that receives most of the cytoplasm from the primary

oocyte and (2) a much smaller **first polar body** that remains attached to the secondary oocyte.

Ovulation occurs when the mature follicle ruptures and ejects the follicular fluid and secondary oocyte through the ovary wall. The secondary oocyte enters the oviduct and is carried toward the uterus by the ciliated epithelium that lines the oviduct. Note that while it is common to speak of the "egg" as being released by the ovary in ovulation, a secondary oocyte is actually released. After ovulation, the empty follicle becomes the **corpus luteum**, which produces progesterone to maintain the uterine lining.

No further division occurs unless the secondary oocyte is penetrated by a sperm. If this occurs, the secondary oocyte completes meiotic division II to form the **egg** (n) and another polar body. The first polar body also may complete meiosis II to form an additional polar body. Subsequently, the egg and sperm nuclei fuse to form the diploid zygote. The polar bodies disintegrate.

Materials

Per student
Compound microscope

Per lab
Prepared slides of:
    cat testis, sectioned
    cat ovary, sectioned with mature follicle
    cat ovary, sectioned with corpus luteum
    human sperm

Assignment 2

1. Examine a prepared slide of cat testis. Locate the seminiferous tubules, which produce spermatozoa, and the interstitial cells, which produce testosterone, the male hormone. Compare your observations with Figure 23.3, and locate the cells involved in spermatogenesis.
2. Examine the prepared slide of human sperm. Note their small size. About 350 million sperm are released in an ejaculation.
3. Examine a prepared slide of cat ovary. Compare your slide with Figure 23.4. Locate the germinal epithelium, a primary follicle with a primary oocyte, and a Graafian (mature) follicle containing a secondary oocyte.
4. Examine a prepared slide of cat ovary showing a corpus luteum, which secretes progesterone to maintain the uterine lining.
5. **Complete item 2 on the laboratory report.**

BIRTH CONTROL

The control of fertility is one of the major concerns of modern society, not only because of the desire to prevent unwanted pregnancies but also because of the great need to curb a human population growth rate that is out of control.

TABLE 23.1
Effectiveness of Birth Control Methods

Method	Pregnancies per 100 Sexually Active Women per Year
Abstinence	0
Tubal ligation	0
Vasectomy	0.4
Norplant	5
Oral contraceptive	5
IUD	5
Condom (high quality)	14
Diaphragm plus spermicide	16
Sponge	17
Rhythm method	24
Spermicide only	25
Withdrawal	26
Condom (poor quality)	30
Douche	60
No birth control method	90

Table 23.1 indicates the effectiveness of common birth-control methods. Descriptions of some of these methods follow.

Tubal ligation is a surgical procedure in which a small section of each oviduct is removed and the cut ends are tied.

Vasectomy is a surgical procedure in which a small section of each vas deferens is removed and the cut ends are tied.

The contraceptive pill consists of synthetic estrogens and progesterones that inhibit the development of ovarian follicles and ovulation.

An **intrauterine device (IUD)** is a metal or plastic device placed in the uterus by a physician, and it remains there for long periods of time. An IUD prevents implantation of an embryo in the uterine lining. Studies have shown that the use of an IUD increases the probability of pelvic inflammatory disease and the probability of female sterility.

A **diaphragm** is a dome-shaped device inserted into the vagina and placed over the cervix prior to sexual intercourse. A **cervical cap** is similar to a diaphragm. Both are used with spermicides.

A **sponge** is a sponge-like device that is inserted into the vagina and placed against the cervix prior to sexual intercourse. The sponge contains a spermicide.

A **condom** is a thin, tight-fitting sheath of latex or lamb intestine that is worn over the penis during sexual intercourse. It is more effective when used with a spermicide. Only latex condoms provide some protection against sexually transmitted diseases.

Spermicides are chemicals that are lethal to sperm. They are marketed as foams, jellies, creams, and suppositories.

The **rhythm method** involves abstention from sexual intercourse during a woman's fertile period, which extends from a few days before to a few days after ovulation. It requires a woman to take her temperature each morning before arising, in order to detect the 0.2–0.6 °F drop in body temperature that occurs just prior to ovulation and the 0.2–0.6 °F rise that occurs immediately after ovulation.

Withdrawal is the removal of the penis from the vagina just prior to ejaculation. It's contraceptive effectiveness is limited because some sperm are often emitted from the penis prior to ejaculation.

A **douche** is the rinsing out of the vagina after sexual intercourse. It is not very because sperm can enter the uterus within 1.5 min after being deposited in the vagina.

Materials

Per lab
Demonstration table of birth-control devices and spermicides

Assignment 3

1. Examine the various birth control methods set up as a demonstration. Read the directions for use that accompany each device or spermicide.
2. **Complete item 3 on the laboratory report.**

24

Fertilization and Development

OBJECTIVES

After completion of the laboratory session, you should be able to:

1. Describe activation and cleavage and identify activated eggs and cleavage stages when observed with the microscope.
2. Describe and identify a blastula and a gastrula.
3. Describe the formation of the germ layers and identify the germ layers in a late gastrula.
4. Indicate the adult tissues and organs formed from the germ layers.
5. Identify the extra-embryonic membranes and state their functions.
6. Correlate size and visible features with the age of human fetuses.
7. Define all terms in bold print.

The union of gametes and early embryological development are difficult to study in many animals, especially humans. By comparison, echinoderms are good subjects for such a study, because their gametes are easy to procure and minimal care is needed to maintain the adults and embryos. In addition, early embryological development in echinoderms is similar to that in chordates, including humans.

The penetration of an egg by a sperm is called **activation**. The subsequent fusion of egg and sperm nuclei is **fertilization**, and this process forms the diploid **zygote**. A series of mitotic divisions called **cleavage** begins in the zygote and produces progressively smaller cells. As cleavage progresses, a solid ball of cells, the **morula**, is formed, and continued cleavage produces the **blastula**, a hollow ball of cells. This concludes the cleavage process. The blastula is not much larger than the zygote.

Mitotic divisions continue beyond the cleavage process and transform the blastula into a **gastrula** by a process called gastrulation. This stage of early development is completed by the formation of the three **embryonic tissues** or **germ layers**.

FERTILIZATION AND DEVELOPMENT IN THE SEA URCHIN

In this portion of the exercise, you will study fertilization and early development in the sea urchin.

Procurement of Gametes

Your instructor will obtain gametes for the entire class. Several sea urchins may be needed to

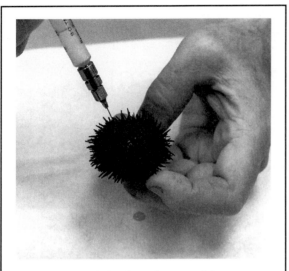

Figure 24.1. Injection of sea urchin.

Figure 24.2. Shedding female on beaker of seawater.

find a male and female, since the sex of these animals cannot be easily determined by external examination.

Materials

Per lab
Beakers, 50 ml and 100 ml
Dropping bottles
Finger bowls
Hypodermic needles, 22 gauge
Hypodermic syringes, 5 ml
Medicine droppers
Syracuse dishes
Potassium chloride solution, 0.5 M
Seawater at 20 °C
Sea urchins

Procedure

1. Inject 1 ml of the 0.5 M potassium chloride solution into each of three or four sea urchins. Insert the hypodermic needle through the membranous region around the mouth, as shown in Figure 24.1.
2. Place the urchins on paper towels with the oral (mouth) side down. Watch for the release of the gamete secretions from the aboral surface. The sperm secretion is white, and the egg secretion is pale buff in color.
3. As soon as a female starts shedding (releas-ing eggs), place her on a beaker full of cold (20°C) seawater, aboral side down, so that the aboral surface is in the water. Release of all the eggs will take several minutes. See Figure 24.2.
4. After the eggs have been released, discard the female sea urchin. Swirl the eggs and water to wash the eggs and allow them to settle to the bottom of the beaker. Pour off the water and add fresh seawater. Repeat this washing procedure twice. It will facilitate the activation process.
5. After the final washing, swirl the water to disperse the eggs. Then pour about 25 ml of seawater and eggs into each of five or six finger bowls and keep them at 20 °C until used. When the bowls are stacked to reduce evaporation, the eggs will remain viable for 2–3 days at 20 °C.
6. Allow several minutes for the sperm secretion to accumulate on the aboral surface of a male. Then remove the secretion with a medicine dropper and place it in a Syracuse dish. See Figure 24.3. Undiluted sperm secretion in a covered dish will be viable for 2–3 days at 20 °C. Prepare a sperm solution, just before use, by placing 2 drops of sperm secretion in 25 ml of seawater, and dispense this from a dropping bottle.
7. Keep gametes at 20 °C until used.

Figure 24.3. Removing sperm with a dropper.

Activation and Fertilization

Chemicals called **gamones** are released by both the sperm and eggs of sea urchins. These chemicals serve to attract sperm to the eggs and also enable the attachment of a sperm to the egg membrane. Similar substances are released by human gametes. The first sperm to attach near the **animal pole** of the egg causes activation. The activated egg then extends a **fertilization cone** through the egg membranes to engulf the

sperm head and draw it into the egg. The sperm tail remains outside the egg. A rapid release of substances from cytoplasmic vesicles into the space between the **inner** and **outer egg membranes** immediately follows, and results in the rapid inflow of fluid into this space. The accumulation of fluid pushes the outer membrane farther outward and prevents penetration of another sperm. The outer membrane is now called the **fertilization membrane**. Study Figure 24.4. Subsequently, the egg and sperm nuclei fuse to form the diploid nucleus of the zygote in the process of **fertilization**.

Early Embryology

After fertilization, the **zygote** begins a series of mitotic divisions that produce successively smaller cells. These divisions are called **cleavage**, and the first division occurs about 40–60 min after activation. Succeeding divisions occur at approximately 30-min intervals. The cells formed by the cleavage divisions are called **blastomeres**. See Figure 24.5.

The first cleavage division passes through the **animal** and **vegetal poles** of the zygote to yield cells of equal size. The second division also passes through both poles to form 4 cells of equal size.

The third division is perpendicular to the po-

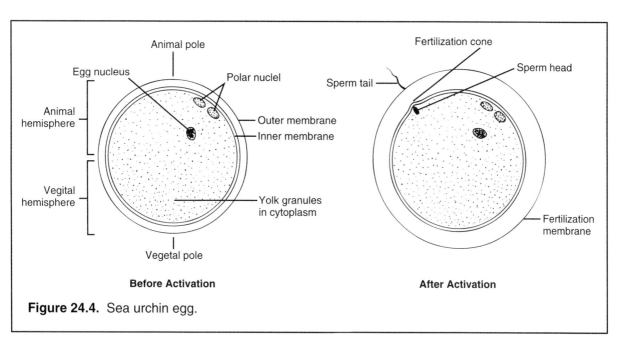

Before Activation

Animal pole

Egg nucleus

Polar nuclei

Animal hemisphere

Outer membrane

Inner membrane

Vegital hemisphere

Yolk granules in cytoplasm

Vegetal pole

After Activation

Fertilization cone

Sperm head

Sperm tail

Fertilization membrane

Figure 24.4. Sea urchin egg.

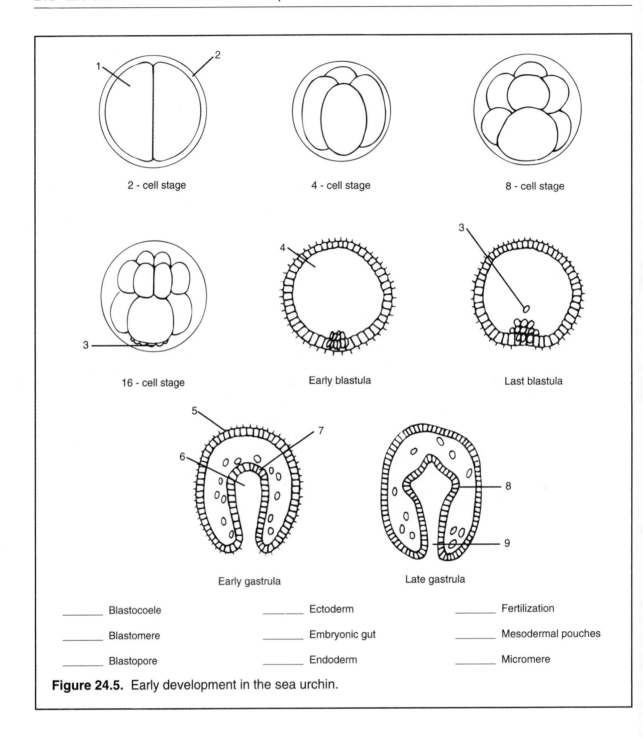

2 - cell stage

4 - cell stage

8 - cell stage

16 - cell stage

Early blastula

Last blastula

Early gastrula

Late gastrula

_____ Blastocoele	_____ Ectoderm	_____ Fertilization
_____ Blastomere	_____ Embryonic gut	_____ Mesodermal pouches
_____ Blastopore	_____ Endoderm	_____ Micromere

Figure 24.5. Early development in the sea urchin.

lar axis and forms 8 cells. The cells of the **animal hemisphere** are slightly smaller than those of the **vegetal hemisphere** because of the greater concentration of **yolk** in the vegetal hemisphere. The fourth division forms 16 cells; however, it forms 4 large cells and 4 tiny cells in the vegetal hemisphere. The tiny cells are called **micromeres**.

The **blastula** is formed about 9 hr after activation and is no longer enveloped by the fertilization membrane. The inner cavity of the blastula, the **blastocoele**, is filled with fluid.

TABLE 24.1
Examples of Tissues and Organs Formed from the Germ Layers in Humans

Endoderm	Mesoderm	Ectoderm

Linings of the
   Urinary bladder
   Digestive tract
   Respiratory tract
Liver
Pancreas
Thyroid, parathyroids,
   and thymus
	Skeleton
	Muscles
	Kidneys
	Gonads
	Blood, heart, and blood ves-
	   sels
	Reproductive organs
	Dermis of the skin
		Epidermis, including hair and nails
		Inner ear
		Lens, retina, and cornea of the eye
		Brain, spinal cord, nerves, and adrenal
		   medulla

Cilia develop on the outer surfaces of the cells, and their beating provides rotational motion of the blastula.

Continued division of the cells of the blastula results in the inward growth (invagination) of cells at the vegetal pole, led by the micromeres. The **early gastrula** consists of two cell layers. The inner cell layer, the **endoderm**, forms the **embryonic gut** (archenteron), which opens to the exterior via the **blastopore**. The outer cell layer is the **ectoderm**. In the **late gastrula**, pouches bud off the endoderm to form the **mesoderm**. All three germ layers (embryonic tissues) are now present, and all later-appearing adult tissues and organs are derived from the three germ layers. See Table 24.1. In both echinoderms and chordates, the blastopore becomes the anus, and a mouth forms later from another opening at the other end of the embryonic gut.

Materials

Per student
Colored pencils
Depression slides and cover glasses

Per lab
Glass marking pens
Medicine droppers
Toothpicks, flat
Ward's culture gum
Developing embryos at 3, 6, 12, 24, 48,
   and 96 hr
Unfertilized eggs in seawater at 20 °C
Sperm solution in dropping bottles at 20 °C
Prepared slides of sea urchin blastula, gastrula,
   and larval stages

Assignment 1

Complete item 1 on Laboratory Report 24 that begins on page 395.

Assignment 2

1. **Complete item 2a on the laboratory report.**
2. Place one drop of the egg-and-seawater mixture (8–12 eggs) in a depression slide and observe without a cover glass. Use the 10× objective. Compare the eggs with Figure 24.4. **Draw two or three eggs in the space for item 2b on the laboratory report.**
3. While the slide is on the microscope stage, add one drop of the sperm mixture at the edge of the depression, record the time, and quickly observe with the 10× objective. Note how the motile sperm cluster around the eggs. Why? Observe the rapid formation of the fertilization membrane. When most of the eggs have been activated, record the time. **Draw two or three activated eggs in the space for item 2b on the laboratory report.**
4. Use a toothpick to place a small amount of Ward's culture gum around the depression on the slide and add a cover glass. This prevents evaporation of water but allows passage of O_2 and CO_2. Write your initials on the slide with a glass-marking pen.

5. Place your slide, egg mixture, and sperm mixture in the refrigerator at 20 °C.
6. **Complete item 2c on the laboratory report.**
7. Label Figure 24.5. Color the cells as follows to distinguish the embryonic tissue:
   ectoderm—blue
   mesoderm—red
   endoderm—yellow
8. Examine your slide of activated eggs at 15–20-min intervals, and try to observe the division of the zygote. Keep the slide at 20 °C when not observing it.
9. Prepare and observe, in age sequence, slides of different stages of sea urchin development. Make only one slide at a time. Label it by age and with your initials. Keep it at 20 °C when not observing it. Examine these slides at 15–20-min intervals to observe a cell division. Compare your observations with Figure 24.5.
10. Compare the living blastula, gastrula, and larval stages with the prepared slides.
11. When finished, clean the slides, cover glasses, microscope stage, and objectives to remove all traces of seawater.
12. **Complete item 2 on the laboratory report.**

CHORDATE DEVELOPMENT

Examining the early development in a simple chordate, amphioxus, will help in understanding early development in humans. All chordates, including humans, exhibit a similar developmental pattern, and this pattern is also similar to development in the sea urchin. See Figure 24.6.

In amphioxus, gastrulation forms ectodermal and endodermal layers as in the sea urchin, but mesoderm formation is a bit different. In amphioxus, two lateral mesodermal pouches bud off of the endoderm as in the sea urchin, but the dorsal portion of the endoderm between the pouches also becomes mesoderm that subsequently forms the **notochord**. A notochord is a flexible rod that provides support for the body, and is an evolutionary forerunner of the vertebral column in vertebrates.

The formation of the notochord induces the overlying ectoderm to develop into a **neural tube**. In humans, as in other vertebrates, the neural tube subsequently develops into the brain and spinal cord. The lateral mesodermal pouches form the typical mesodermal structures, as noted in Table 24.1.

Amniote Egg

Embryos of all chordates develop in an aqueous (water) environment. Among vertebrates, female fish and amphibians lay their eggs in water and males simultaneously deposit sperm over the eggs. Such organisms utilize external fertilization and their young undergo external development. It is a giant evolutionary step from the external fertilization and external development of fish and amphibians to the internal fertilization and internal development found in humans and most mammals. This transition occurred in several stages and began long before humans appeared on Earth.

Reptiles were the first truly terrestrial vertebrates because they solved the problem of reproduction without returning to water. Their successful adaptations include internal fertilization and the **amniote egg**. The reptilian egg has a leathery shell that prevents excessive water loss and allows an exchange of oxygen and carbon dioxide between the developing embryo and the atmosphere. An adequate supply of stored nutrients (yolk and albumin) enables the development of the embryo to hatching. In addition, special protective membranes enclose the embryo.

The features of the amniote egg are shown in Figure 24.7. Note the reptilian **extra embryonic membranes**. The **amnion** surrounds the embryo and contains the **amniotic fluid** that provides the embryo with its own "private pond" in which to develop. Instead of returning to water for embryonic development, reptiles "brought the water to the embryo." The **yolk sac** envelops the yolk and absorbs nutrients for the embryo. The **allantois** is an embryonic urinary bladder, but it also spreads out against the outer membranes and serves as a gas-exchange organ. The **chorion** is the outermost membrane and encompasses all of the others.

Figure 24.7b shows how the extra-embryonic membranes are utilized in humans and most mammals. As in reptiles, the amnion envelops

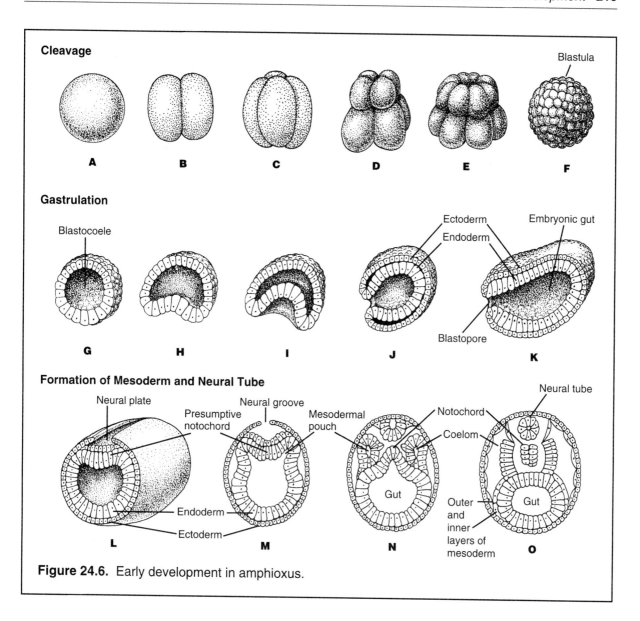

Cleavage

Blastula

A   B   C   D   E   F

Gastrulation

Blastocoele

Ectoderm

Endoderm

Embryonic gut

Blastopore

G   H   I   J   K

Formation of Mesoderm and Neural Tube

Neural plate

Presumptive notochord

Neural groove

Mesodermal pouch

Notochord

Neural tube

Coelom

Endoderm

Ectoderm

Gut

Outer and inner layers of mesoderm

Gut

L   M   N   O

Figure 24.6. Early development in amphioxus.

the embryo, providing a "private pond" of amniotic fluid for the embryo, and the chorion surrounds all other membranes. The allantois carries embryonic blood vessels to the chorion. It forms the major part of the **umbilical cord** and, along with the chorion, forms the **placenta** that attaches the embryo to the uterine lining of the mother. Because there is no yolk in the development of the mammalian embryo, the yolk sac plays a minor role as part of the umbilical cord. The amnion forms the outer covering of the umbilical cord.

The umbilical cord carries the blood of the embryo to and from the placenta. The blood of the embryo is separated from the blood of the mother by thin membranes in the placenta, where an exchange of materials occurs:

$$\text{Maternal blood} \underset{\text{wastes and } CO_2}{\overset{\text{nutrients and } O_2}{\rightleftharpoons}} \text{Embryo blood}$$

Thus, mammals, including humans, have capitalized on the extra-embryonic membranes "invented" by reptiles.

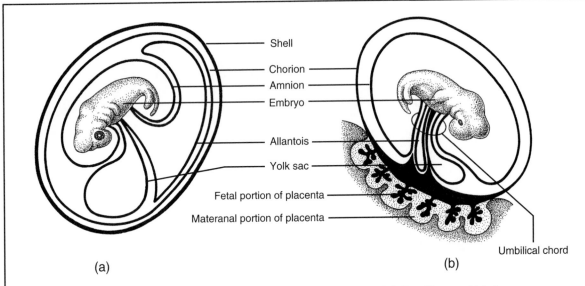

Figure 24.7. Extra-embryonic membranes of the amniote egg. (a) Reptiles and birds. (b) Humans and most mammals.

Materials

Per student
Colored pencils

Per lab
Prepared slides of:
  amphioxus, cleavage
  amphioxus, gastrula
  amphioxus, x.s., neurula stage

Assignment 3

1. Study Figure 24.6. Note the similarity of development in amphioxus with that of the sea urchin up to the gastrula. Note that (a) the mesoderm is formed from pouches that bud off the endoderm, (b) the mesoderm destined to be the notochord is located between the mesodermal pouches, and (c) the neural plate is formed by the dorsal part of the ectoderm and folds up to form the neural tube.
2. Color the embryonic tissues in Figure 24.6k, n, and o:
  ectoderm—blue
  neural tube—green
  mesoderm—red
  endoderm—yellow

3. Examine a prepared slide of amphioxus development and locate stages like those in Figure 24.6.
4. **Complete item 3 on the laboratory report.**

HUMAN DEVELOPMENT

A human embryo (blastocyst) is implanted in the uterus about 5 days after fertilization, and germ layers and extra-embryonic membranes are evident by the 14th day. See Figure 24.8. Chorionic villi (projections) attach the embryo firmly to the uterine lining, and some of these villi will become part of the embryonic portion of the placenta. The first 8 weeks of development is known as the **embryonic period**, and the remainder of pregnancy is known as the **fetal period**.

At 8 weeks, the **fetus** is recognizably human, with all organ systems in rudimentary form. It is about 3 cm ($1\frac{1}{4}''$) in length and 1 g in weight. Among the characteristics of the 8-wk fetus are: (1) The limbs are recognizable as arms and legs, and the fingers and toes are formed. (2) Bone formation begins, and internal organs continue to form. (3) The head is nearly as large as the body.

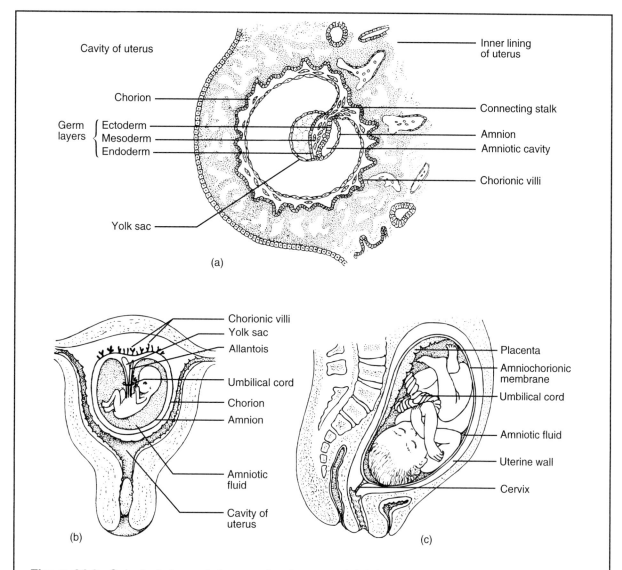

Figure 24.8. Selected stages in human development. (a) A human embryo, about 14 days old, implanted in the uterus. Note the germ layers and extra-embryonic membranes. (b) Uterus with a fetus about 10 weeks old. Note the extra-embryonic membranes and the umbilical cord. (c) A fullterm fetus with head pressed against the cervix. Note the placenta, umbilical cord, and the fetal position.

All major brain regions are present, and the eyes are far apart with the eyelids fused. (4) The cardiovascular system is functional.

Subsequent development results in a baby being born about 266 days after fertilization (conception) or 280 days from the last mensis. When born prematurely, a fetus has about a 15% chance of survival if born at 24 weeks, but nearly a 100% chance at 30 weeks.

Materials

Per student
Colored pencils

Per lab
Demonstration of pregnant cat or pig uterus
Model of a pregnant human female torso
Models of human developmental stages
Preserved human fetuses of various ages

Assignment 4

1. Study Figure 24.7 until you understand the location of the extra-embryonic membranes.
2. Study Figure 24.8a, noting the formation of the extra-embryonic membranes and germ layers. Color-code the germ layers: ectoderm—green; mesoderm—red; and endoderm—yellow.
3. Examine Figure 24.8b, noting the components of the umbilical cord, the placenta, and the amniotic fluid enveloping the fetus.
4. **Complete items 4a and 4b on the laboratory report.**
5. Examine the pregnant human torso model and compare it with Figure 24.8c. Note how the amnion and chorion are pressed together, forming the amniochorion in late stages of pregnancy. Does the fetus fill all available space?
6. Examine the pregnant cat or pig uterus, noting the amniochorion, placenta, and umbilical cord.
7. Examine the series of models showing human development. Note the progression of development with the age of the fetus.
8. Examine the series of preserved human fetuses, noting the degree of development and the age of each fetus.
9. **Complete the laboratory report.**

Part 4 Organismic Diversity

25

Monerans, Protists, and Fungi

After completion of the laboratory session, you should be able to:
1. Describe the distinguishing characteristics of monerans, protists, and fungi.
2. Identify representatives of these groups.
3. Describe the reproductive patterns of selected representatives.
4. Define all terms in bold print.

TABLE 25.1
Classification of the Human Species,
Homo sapiens

Taxonomic Category	Classification of Humans
Kingdom	Animalia
Phylum	Chordata
Class	Mammalia
Order	Primates
Family	Hominidae
Genus	*Homo*
Species	*sapiens*

The next few exercises are included to acquaint you with the variety of organisms in the biotic world. The emphasis is on the distinguishing characteristics and life cycles of the major groups of living organisms.

Taxonomy is the science of classifying organisms. Several **taxonomic categories** are used as shown in Table 25.1. The kingdom is the category containing the largest number of species (kinds of organisms), while the species category contains the fewest kinds of organisms—only one. Organisms are classified according to their degree of similarity, which indicates the degree of their evolutionary relationship. Therefore, a kingdom contains related phyla, a phylum contains related classes, and so forth.

The **phylum** category has been traditionally used by zoologists in classifying animals, while botanists have traditionally used the term **division** for this same category in classifying plants and plant-like organisms. These terms have persisted in the classification of organisms even though zoology and botany have now been unified in the single subject of biology.

The scientific name of an organism is composed of both its genus and species names. For example, *Homo sapiens* is the scientific name for humans. Note that the genus name is capitalized while the species name is not. Also, the scientific name is always printed in italics.

Biologists use a five-kingdom system of classi-

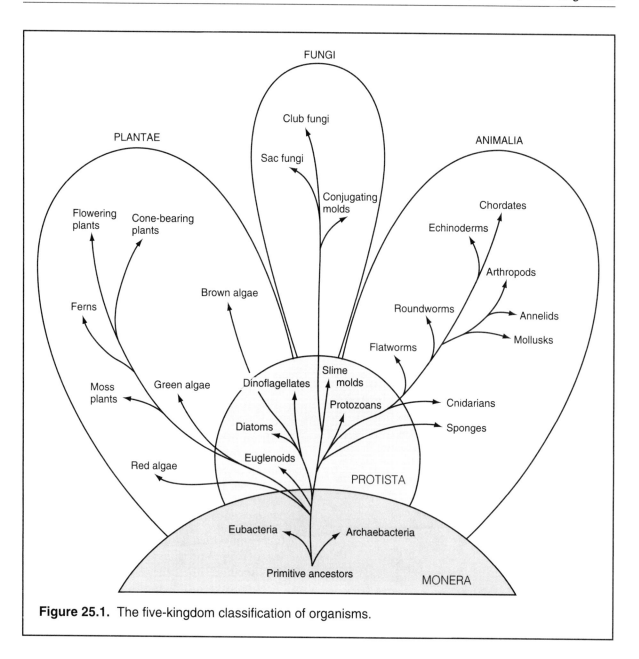

Figure 25.1. The five-kingdom classification of organisms.

fication. The five kingdoms are **Monera, Protista, Fungi, Plantae**, and **Animalia**. The major organismic groups composing each kingdom are shown in Figure 25.1. A classification of the major organismic groups is found in Appendix D, page 433.

KINGDOM MONERA

Members of the kingdom **Monera** (**Prokaryotae**) are the smallest, simplest organisms.

Most are unicellular but some are **colonial** (composed of a group of independently functioning cells). All cells are **prokaryotic,** since they lack a nucleus and other membrane-bound organelles. A somewhat rigid, nonliving cell wall is secreted external to the cell (plasma) membrane. See Figure 3.1 to review the structure of a prokaryote cell.

The kingdom Monera is divided into two subkingdoms: Eubacteria and Archaebacteria. **Eubacteria** consists of true bacteria and contains a multitude of species. **Archaebacteria** are relatively rare bacterialike prokaryotes with a unique biochemistry. They live in extreme environments, such as at the high temperatures of deep sea vents. We will consider only the Eubacteria.

Bacteria (Subkingdom Eubacteria)

Bacterial cells exhibit three characteristic shapes: **bacillus** (rodlike shape; bacilli plural); **coccus** (spherical shape; cocci plural); and **spirillum** (spiral shape; spirilla plural). See Figure 25.2 and Color Plates 1.1 to 1.3. Some species possess **bacterial flagella** (Plate 1.2) which enable movement. Bacterial flagella are distinctly different from the flagella of eukaryotic cells noted in Exercise 3. Although cell shape and the presence or absence of flagella are used in the identification of bacterial species, a determination of their biochemistry and physiology is required for identification.

Most bacteria are **heterotrophs,** meaning that they must obtain their organic nutrients from the environment. Most heterotrophs are **saprotrophs** which obtain nutrients by decaying organic matter and dead organisms. Saprotrophs provide a great service by converting organic debris into inorganic chemicals for repeated use by organisms. A few heterotrophs are **parasites** which cause disease in other organizms as they obtain organic nutrients from their cells and tissues.

Some bacteria are **autotrophs** which can produce their own organic nutrients. **Chemosynthetic bacteria** are able to oxidize inorganic chemicals, such as ammonia or sulfur, and capture the released energy to synthesize organic nutrients. **Photosynthetic bacteria** use light energy to synthesize their organic nutrients. A few photosynthetic bacteria possess *bacteriochlorophyll,* a light-capturing pigment that is different from chlorophyll, which enables them to carry on a non-oxygen producing form of photosynthesis.

Most photosynthetic bacteria belong to a group known as **cyanobacteria,** which live in freshwater, marine or moist terrestrial environments. Cyanobacteria lack chloroplasts but possess *chlorophyll a,* which enables them to carry on the same kind of oxygen-producing photosynthesis found in plants. Although cyanobacteria are usually blue-green in color, they also contain other pigments which may cause them to appear yellow, orange or red. The cells of cyanobacteria differ from other bacterial cells by being larger and either spherical or cylindrical in shape. And they often have a gelatinous coat external to the cell wall. See Plate 1.4.

Reproduction in Bacteria

Bacteria reproduce by **binary fission,** an asexual form of reproduction. See Figure 9.1 on page 68. This form of cell division enables bacteria to produce massive numbers of individuals within a brief time. When conditions are favorable, a cell may divide to form two cells every twenty minutes.

Some bacteria are able to form non-reproductive **endospores** within the bacterial cell. An endospore consists of a bacterial chromosome (DNA) and cytoplasm enveloped by a thick endospore wall. When the bacterial cell dies, the endospore remains viable and is resistant to unfavorable conditions which kill most other bacteria. When conditions become favorable, the endospore grows into a bacterial cell, and reproduction resumes by binary fission. See Plate 1.3.

A sexual process called **conjugation** also occurs in bacteria. Conjugation is the transfer of genetic material (DNA) from one cell to another, and it occurs when adjacent cells are joined by a

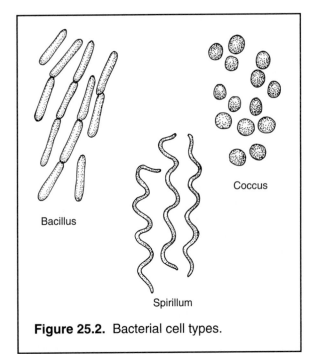

Coccus

Bacillus

Spirillum

Figure 25.2. Bacterial cell types.

tiny tube formed by their plasma membranes. The donor DNA combines with, and may replace, homologous segments in the recipient cell's DNA, increasing genetic variability. Subsequently, conjugating bacterial cells separate, and reproduction occurs by binary fission. The transfer of donor DNA and its fusion with DNA of the recipient cell is analogous to the transfer of sperm and fusion of sperm and egg DNAs in higher organisms.

Materials

Per student
Compound microscope

Per lab
Antibiotic disks
Cultures of cyanobacteria
　Gloeocapsa
　Oscillatoria
Demonstration microscope setups:
　motile bacteria (*Pseudomonas*), 1,000×
　Gloeocapsa, 1,000×
　Oscillatoria, 1,000×
Microscope slides and cover glasses
Pour plates of:
　Escherichia coli B
　Staphylococcus epidermidis
Prepared slides of bacterial types

Assignment 1

1. Obtain a prepared slide of bacterial types. There are three stained smears of bacteria on the slide, and each consists of a different morphological type. Use reduced illumination to examine each type at 400×. Bacterial cells are smaller than you probably imagine. **Draw a few cells of each type in item 1a on Laboratory Report 25 that begins on page 399.**
2. Examine the demonstration setup of motile bacteria at 1,000×. **Complete item 1b on the laboratory report.**
3. Examine the demonstration nutrient-agar plates of *Escherichia coli B* and *Staphylococcus epidermidis* on which disks containing selected antibiotics were placed prior to incubation. The antibiotic diffuses from each disk into the agar and kills susceptible bacterial

species where its concentration is lethal. The diameter of the clear no-growth area around an antibiotic disk indicates the effectiveness of the antibiotic against the bacterial species. Measure the diameter of the no-growth area of each antibiotic disk for each bacterial species. **Complete item 1c on the laboratory report.**
4. Examine the demonstration setups of live *Gloeocapsa* and *Oscillatoria* at 1,000×. Note the shape of each colony, the shape of the cells in each colony, the color of their chromoplasm, and the gelatinous sheath, if present.
5. **Complete item 1 on the laboratory report.**

KINGDOM PROTISTA

Protists are a heterogeneous group of unicellular or colonial organisms that exhibit animal-like, plant-like, or fungus-like characteristics. The major groups are probably not closely related but are products of evolutionary lines that diverged millions of years ago. Protists and all higher organisms are composed of **eukaryotic cells** that contain membrane-bound organelles.

Mitotic cell division is the most common form of reproduction among protists. However, sexual reproduction does occur. In haploid (n) protists, only the zygote is diploid, and it promptly divides by meiotic division, forming four new haploid individuals. In diploid (2n) protists, haploid gametes are formed by meiotic division, and the fusion of gametes forms a new diploid individual.

Protozoans: Animal-like Protists

The animal-like protists known as protozoans lack a cell wall and are usually motile. Most forms engulf food into vacuoles, where it is digested. Some absorb nutrients through the cell membrane, and a few are parasitic. Protozoans occur in most habitats where water is available. Water tends to diffuse into freshwater protozoans, and these protozoans possess **contractile vacuoles** that repeatedly collect and pump out the excess water to maintain their water balance. Three groups of protozoans are shown in Figure 25.3.

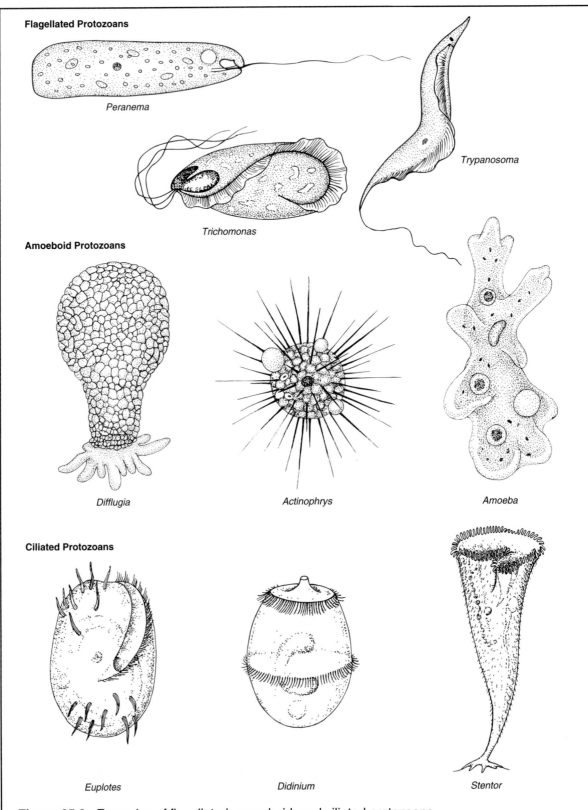

Figure 25.3. Examples of flagellated, ameoboid, and ciliated protozoans.

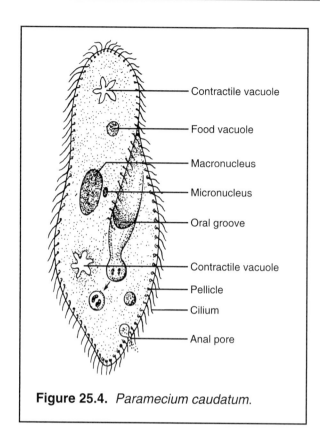

Figure 25.4. *Paramecium caudatum.*

forms of amoeboid protozoans secrete a shell for protection. Calcareous shells of foraminiferans and siliceous shells of radiolarians are abundant in ocean sediments. *Entamoeba histolytica* is a parasitic amoeboid protozoan that causes amoebic dysentery in humans. See Plates 2.4, 2.5, 3.1, and 3.2.

Ciliated Protozoans (Phylum Ciliophora)

Ciliates are the most advanced and complex of the protozoans. They are characterized by the presence of a **macronucleus,** one or more **micronuclei,** and movement by means of numerous **cilia,** hair-like processes extending from the cell. A flexible outer covering, the **pellicle,** is located exterior to the cell membrane. *Paramecium* is a common example of ciliated protozoan that also possesses **trichocysts,** tiny dart-like weapons for offense and defense, located just under the cell surface. Food organisms are swept down the **oral groove** by cilia and into food vacuoles, where digestion occurs. See Figures 25.3 and 25.4 and Plates 3.3 and 3.4.

Sporozoans (Phylum Sporozoa)

All species of sporozoans lack motility and are internal parasites of animals. *Plasmodium vivax,* a pathogen causing malaria, is a typical example. It is transmitted by the bite of *Anopheles* mosquitoes and infests red blood cells and liver cells of human hosts. Malaria has probably caused the death of more humans than any other disease. See Plate 3.5.

Plant-like Protists

The plant-like protists possess chloroplasts containing chlorophyll, and most of them have a cell wall. You will study representatives of three major groups of these protists.

Euglenoids (Division Euglenophyta)

These unicellular protists possess both plant-like and animal-like characteristics. They have **chlorophylls a** and **b** in their chloroplasts, a flagellum for movement, and an "eyespot" (stigma) that detects light intensity. A cell wall is absent. See Figure 25.5 and Plates 4.1 and 4.2.

Flagellated Protozoans (Phylum Zoomastigophora)

These primitive protozoans have one or more **flagella** that provide them with a means of movement. Food may be engulfed and digested in vacuoles, or nutrients may be absorbed through the cell membrane.

Some flagellated protozoans have established symbiotic relationships with other organisms. *Trypanosoma brucei* is a parasitic protozoan that causes African sleeping sickness. It lives in the blood and nervous system of its vertebrate host and is transmitted by the bite of tsetse flies. *Trichonympha collaris* is a mutualistic symbiont that lives in the gut of termites. It digests the cellulose (wood) to produce simple carbohydrates that can be digested or utilized by the termite. See Plates 2.1–2.3.

Amoeboid Protozoans (Phylum Sarcodina)

These protists move by means of **pseudopodia,** flowing extensions of the cell. Prey organisms are engulfed and digested in food vacuoles. Some

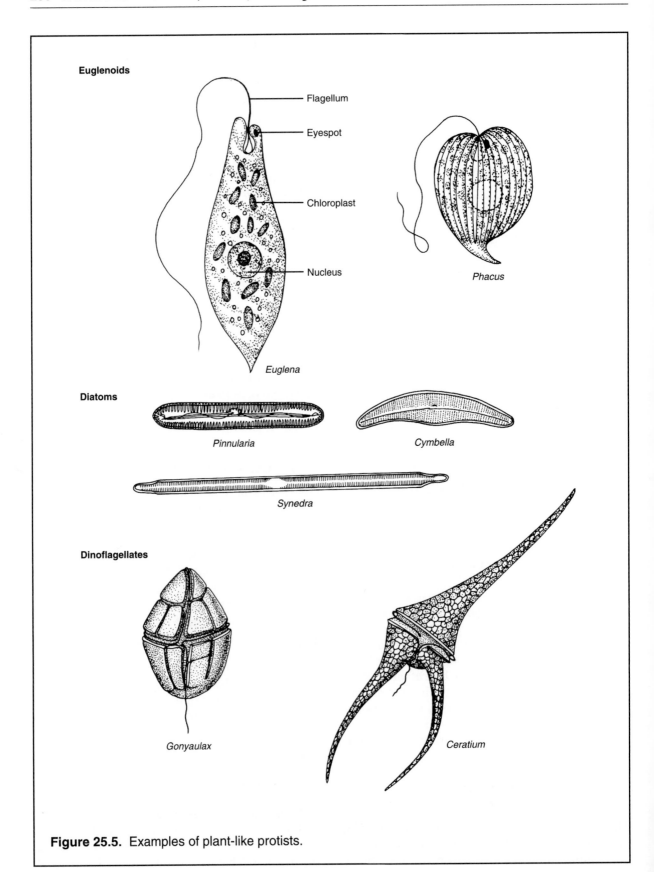

Figure 25.5. Examples of plant-like protists.

Dinoflagellates (Division Dinoflagellata)

Most of these unicellular protists are marine and photosynthetic, with **chlorophylls a** and **c** in their chloroplasts. Most have a cell wall of cellulose, and all forms have two flagella. One of these flagella lies in a groove around the equator of the cell, and the other hangs free from the end of the cell. When nutrients are abundant, certain marine species of dinoflagellates reproduce in such enormous numbers that the water turns a reddish color, a condition known as the red tide. Toxins produced by some species may cause massive fish kills and make shellfish unfit for human consumption. See Plates 4.3 and 4.4.

Diatoms (Division Bacillariophyta)

These protists are unicellular and have a cell wall made of silica, a natural glass. The cell wall consists of two halves that fit together like the top and bottom of a box. **Chlorophylls a** and **c** are found in the chloroplasts of diatoms. When diatoms die, their siliceous walls settle to the bottom of the water. In some areas, these cell walls have formed massive deposits of diatomaceous earth that are mined and used commercially in a variety of products, such as fine abrasive cleaners, toothpaste, filters, and insulation. Much of the earth's atmospheric oxygen has been produced by photosynthesis in marine diatoms. See Plate 4.5.

Slime Molds: Fungus-like Protists

The feeding stages of slime molds resemble amoeboid protozoans, but the slime molds reproduce by forming sporangia that produce spores.

Cellular Slime Molds (Division Acrasiomycota)

Members of this group live as single-celled amoeboid organisms that engulf bacteria in leaf litter and soil. Poor environmental conditions cause the cells to congregate in what is known as a "swarming stage," and which results in the formation of a sporangium containing spores. *Dictyostelium* is an example of this group.

Plasmodial Slime Molds (Division Myxomycota)

Feeding stages of plasmodial slime molds consist of strands of protoplasm streaming along in amoeboid fashion and engulfing bacteria on leaf litter, rotting wood, and the like. The large plasmodium is multinucleate. Unfavorable conditions stimulate the plasmodium to migrate and ultimately to form a sporangium with spores. *Physarum* is an example of a plasmodial slime mold. See Plates 5.1 and 5.2.

Materials

Per student
Compound microscope

Per lab
Medicine droppers
Microscope slides and cover glasses
Protoslo
Toothpicks
Cultures of:
   Euglena
   Paramecium, containing yeast cells stained with congo red
   Pelomyxa
   Physarum
   pond water
Diatomaceous earth
Demonstration setup showing effect of light on the distribution of *Euglena* in a partially shaded Petri dish
Demonstration microscope setups of:
   diatomaceous earth
   dinoflagellates
   foraminifera
   radiolaria
Termites, living
Prepared slides of:
   diatoms
   dinoflagellates
   foraminifera
   radiolaria

Assignment 2

1. Obtain a live termite and separate the abdomen from the thorax. Squeeze the abdominal contents into a drop of water on a microscope slide. Add a cover glass and examine at 100×.

Locate a multiflagellated *Trichonympha* and examine it at 400×. **Complete items 2a-2b on the laboratory report.**

2. Examine the *Pelomyxa* culture set up under a dissecting microscope. Locate a specimen on the bottom of the jar, remove it with a dropper, and place it, along with a drop of water, on a microscope slide. Examine this large, multinucleate amoeboid protozoan at 40× and 100×. Locate food and contractile vacuoles within it, and observe the flowing motion of its cytoplasm as pseudopodia form during **amoeboid movement. Complete item 2c on the laboratory report.**

3. Examine the slides of foraminifera and radiolaria "shells" set up under demonstration microscopes. How do they differ?

4. Place a drop of the *Paramecium* culture containing stained yeast cells on a microscope slide and examine it at 40× or 100× without a cover glass. Observe how *Paramecium* moves and what happens when it bumps into an object.

5. Add a drop of Protoslo to the drop of *Paramecium* culture on your slide and mix it with a toothpick. Add a cover glass and at 40× locate a *Paramecium* stuck in the Protoslo. Examine it at 100× and 400×. Locate the structures shown in Figure 25.4. Since *Paramecium* has been feeding on stained yeast cells, the red-stained yeast cells will be visible in food vacuoles. As digestion proceeds, the yeast cells will change from bright red to purple to blue.

6. **Complete item 2 on the laboratory report.**

Assignment 3

1. Make a slide of a drop of the *Euglena* culture and examine it at 100× and 400×. Note the color, manner of movement, and lack of a cell wall. **Complete item 3a on the laboratory report.**

2. Observe the distribution of *Euglena* in the partially shaded Petri dish set up as a demonstration. *Do not move the dish.* Lift the black-paper light shield to make your observations, and return it to its prior position. **Complete item 3b on the laboratory report.**

3. Examine the demonstration microscope setup of dinoflagellates. **Complete item 3c and draw 1-2 cells in item 3d on the laboratory report.**

4. Examine a prepared slide of diatoms. Note the symmetry of their cell walls. Examine the sample of diatomaceous earth and the demonstration microscope setup of diatomaceous earth. **Complete item 3 on the laboratory report.**

Assignment 4

Examine the demonstration setups of slime molds. Note the organization of the plasmodium stage and the sporangia of the reproductive stage. Compare them with Plates 5.1 and 5.2. **Complete item 4 on the laboratory report.**

Assignment 5

Make and examine several slides of pond water and observe the monerans and protists present. **Complete item 5 on the laboratory report.**

FUNGI

The **kingdom fungi** contains a large and diverse group of heterotrophic organisms that occur in freshwater, marine, and terrestrial habitats. Terrestrial fungi reproduce by **spores,** dormant reproductive cells that are dispersed by wind and that germinate to form a new fungus when conditions are favorable. Most fungi are multicellular; only a few are unicellular.

Fungi are either **saprotrophs** or **parasites.** Most species are saprotrophs and play a beneficial role in decomposing nonliving organic matter, but some saprotrophs cause serious damage to stored food products. Rusts and mildews are important fungal plant parasites. Ringworm and athlete's foot are common human ailments caused by fungi.

The vegetative (nonreproductive) body of a multicellular fungus is called a **mycelium,** and is composed of thread-like filaments known as **hyphae.** Hyphae are formed of cells joined end to end. The cells of hyphae may be separated by cell walls (septate hyphae), or the cell walls may be incomplete or absent (nonseptate hyphae). The cell walls are formed of chitin. See Plate 5.6.

Nutrients are obtained by hyphae that secrete digestive enzymes into the surrounding substrate, which is digested extracellularly. The resulting nutrients are then absorbed into the hyphae.

Spores of fungi are formed either from terminal cells of reproductive hyphae or in **sporangia,** enlarged structures at the ends of specialized hyphae. Some fungi form **fruiting bodies** that contain spore-forming hyphae.

Conjugating Molds (Division Zygomycota)

The common black bread mold, *Rhizopus stolonifer*, is an example of this group. It produces three types of hyphae. **Stolon hyphae** spread over the surface of bread as the mycelium grows. **Rhizoid hyphae** penetrate the bread to digest it and to anchor the mycelium. **Sporangiophores** are upright hyphae that form a **sporangium** at their tips. Asexual **mitospores** (spores formed by mitosis) develop within the sporangia and are released when mature. Germination of these spores forms the haploid hyphae of a new mycelium. See Figure 25.6 and Plates 5.3 and 5.4.

Asexual reproduction by mitospores occurs continuously. Sexual reproduction occurs only when opposite mating types (designated + and −) come in contact. Then, as shown in Figure 25.7, special cells become **gametes** (n) that fuse forming a **zygote** (2n). The zygote develops a resistant cell wall, forming a **zygospore,** the characteristic that gives its name to this group of fungi. Subsequently, the zygospore germinates and forms a sporangiophore whose sporangium produces both + and − haploid spores.

Sac Fungi (Division Ascomycota)

Yeasts, mildews, most molds, and cup fungi belong to the sac fungi. Many members of this group are important parasites of plants and animals, while others provide benefits to humans. Yeasts are used in brewing and winemaking and to release bubbles of CO_2 to make dough rise in the baking industry. *Penicillium* species produce antibiotics.

Asexual reproduction occurs by **budding** in

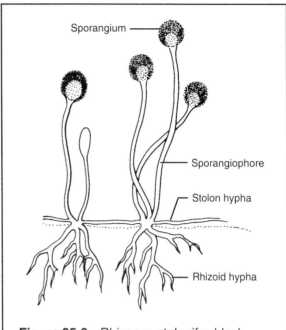

Figure 25.6. *Rhizopus stolonifer,* black bread mold.

yeasts and by **conidiospores** in mildews and molds. Conidiospores are formed by mitotic division at the tips of reproductive hyphae. See Plate 5.5. Sexual reproduction occurs as well. Some forms, like cup fungi, undergo a complex sexual process that culminates in the production of a **fruiting body** that contains the spore-forming hyphae. The tips of these hyphae enlarge to form tiny sacs, known as **asci,** within which **ascospores** are formed by meiotic division. See Plates 6.1a and 6.1b.

Club Fungi (Division Basidiomycota)

Mushrooms, puffballs, and shelf fungi belong to the club fungi. See Plates 6.2–6.4. Many forms of these fungi are beneficial saprotrophs, but some are serious plant parasites, such as rusts and smuts, that cause enormous losses in wheat, corn, and other cereal crops.

The common mushroom, *Coprinus,* is a suitable example of this group. The mushroom mycelium derives nutrients from nonliving organic substances in the soil. It is haploid and occurs in two mating types, + and −. Whenever opposite mating types come in contact with each

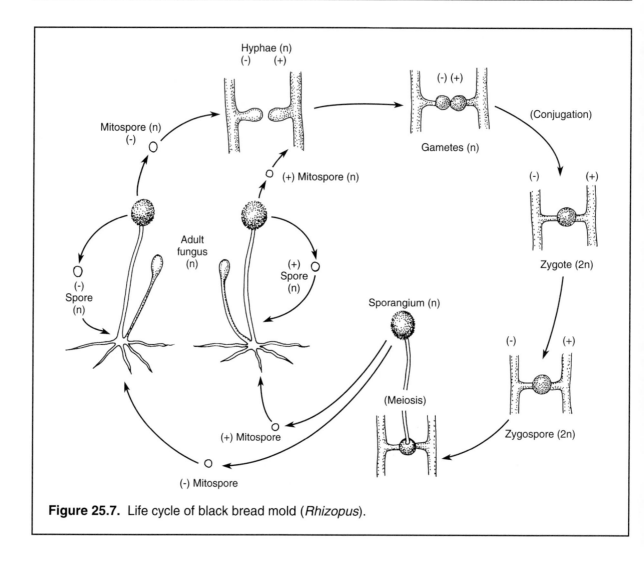

Figure 25.7. Life cycle of black bread mold (*Rhizopus*).

other, cells (analogous to gametes) from the hyphae of each mating type grow toward each other and fuse, forming a single cell containing separate nuclei of each mating type. The nuclei do not fuse; instead, mitotic division of the cell forms **dikaryotic hyphae** whose cells contain two nuclei (n + n), one of each mating type. See Figure 25.8.

Growth of many dikaryotic hyphae results in the formation of a **fruiting body,** the structure commonly called a "mushroom." Spore-forming hyphae of the gills of the mushroom develop terminal enlarged cells called **basidia.** The two nuclei in each basidium then fuse, forming a zygote nucleus (2n). By meiotic division, basidia form four **basidiospores,** two of each mating type, which are released and dispersed by air cur-

rents. Basidiospores germinate to form haploid hyphae by mitotic division.

Materials

Per lab

Demonstration cultures under dissecting microscopes:

    Penicillium on citrus fruit or agar

    Rhizopus on bread or agar

    Rhizopus mating types with zygospores on agar

Dropping bottles of methylene blue, 0.01%

Medicine droppers

Microscope slides and cover glasses

Mushrooms

Yeast culture in 5% glucose

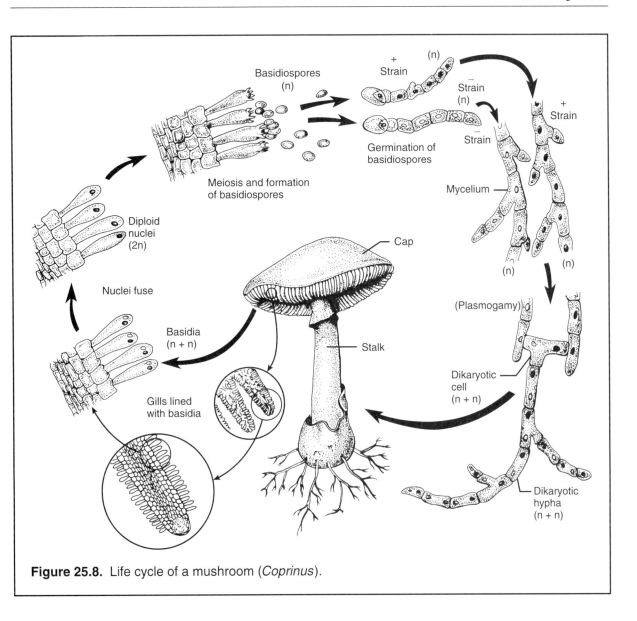

Figure 25.8. Life cycle of a mushroom (*Coprinus*).

Prepared slides of:
Rhizopus zygospores
Coprinus gills, x.s.
"Unknown" monerans, protists, and fungi

Assignment 6

1. **Complete items 6a-6c on the laboratory report.**
2. Examine the *Rhizopus* colony set up under a dissecting microscope. Note its growth pat-

tern and the location of mature and immature sporangia. Locate the three types of hyphae.
3. Examine (a) *Rhizopus* mating colonies and the gametes and zygospores that have formed, and (b) a prepared slide of *Rhizopus* zygospores. **Complete items 6d–6f on the laboratory report.**
4. Place a drop of the yeast culture on a microscope slide. Add a drop of methylene blue and a cover glass, and examine the stained cells at 100× and 400×. Note how buds form on the cells, and locate the nuclei. **Draw a few**

budding yeast cells in item 6g on the laboratory report.

5. Examine the *Penicillium* mold set up under a dissecting microscope. Note the growth pattern and the color of mature and immature conidiospores. **Complete items 6h-6j on the laboratory report.**

6. Examine a mushroom. Locate the stalk, cap, and gills.

7. Examine a prepared slide of *Coprinus* gill, x.s. Locate the hyphae forming the gill, the basidia, and basidiospores. **Complete item 6 on the laboratory report.**

Assignment 7

Your instructor has set up "unknown" monerans, protists and fungi for you to identify as a mini-practicum. **Complete item 7 on the laboratory report.**

26

Plants

OBJECTIVES

After completion of the laboratory session, you should be able to:
1. Describe the characteristics of the plant groups studied.
2. Classify members of the plant groups to division.
3. Describe the reproductive cycles of the plant groups studied.
4. Identify parts of flowers and seeds and describe their functions.
5. Define all terms in bold print.

Plants, members of the **kingdom Plantae,** are all photosynthetic autotrophs. Their photosynthetic pigments are confined to plastids, and their cell walls are composed of cellulose. Mature plant cells possess a large central vacuole.

The classification system used in this book places the algae in the kingdom Plantae, along with multicellular land plants. However, some biologists place the algae in the kingdom Protista. Algae have relationships with both kingdoms, as you will see.

ALGAE

Algae occur primarily in freshwater and marine habitats, but some forms occur in moist terrestrial areas. Algae lack vascular tissue and true leaves, stems, and roots, although complex forms may have structures that resemble them. There are three groups of algae, based on their coloration, which is due to pigments in their chloroplasts. These three groups are the green, brown, and red algae.

Green Algae (Division Chlorophyta)

Green algae may be unicellular, colonial, or multicellular. Most species occur in freshwater or marine habitats, but a few occur in moist areas on land. The presence of (1) **chlorophylls a** and **b** in chloroplasts, (2) cellulose cell walls, (3) starch as the nutrient storage form, and (4) whiplash flagella on motile cells suggests that green algae are ancestral to higher plants. See Figure 26.1 and Plates 7.1–7.5 and 8.1.

Brown Algae (Division Phaeophyta)

These multicellular algae are almost exclusively marine and are often called seaweeds because of their abundance along rocky seacoasts. Their brownish color results from the presence of the pigment **fucoxanthin** in addition to

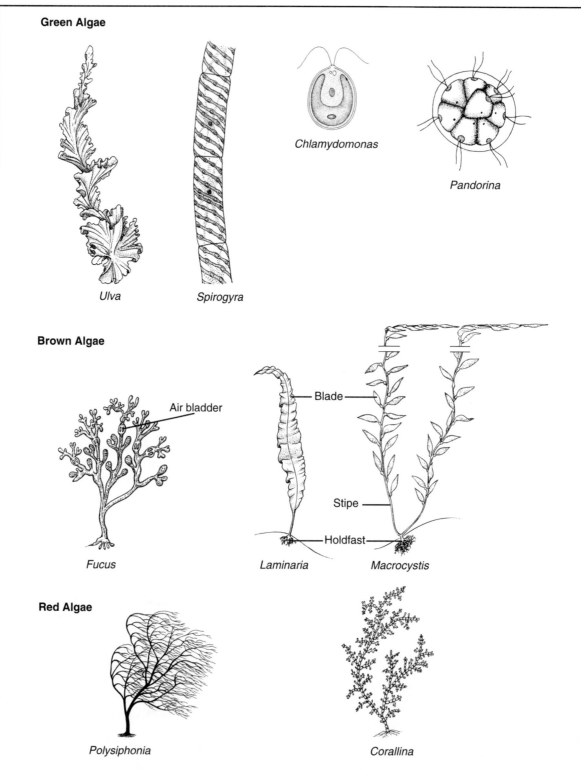

Figure 26.1. Examples of green, brown, and red algae. The drawings are not to scale. *Chlamydomonas* and *Pandorina* are microscopic in size, while *Macrocystis* is a giant kelp that may be over 100 ft in length.

chlorophylls a and **c** in the chloroplasts of these algae. The typical structure of a brown alga includes a **holdfast,** which anchors the alga to a rock, a **stipe** (stalk), and **blades,** which are leaf-like structures. See Figure 26.1. All parts of the plant carry on photosynthesis, but the blades are the most important photosynthetic organs. Blades of large brown algae often have air-filled bladders associated with them, to hold them near the surface of the water. See Figure 26.1 and Plates 8.2–8.4.

Red Algae (Division Rhodophyta)

Most red algae are marine and occur at greater depths than brown algae. They have a delicate body structure. Like brown algae, they are attached to rocks by holdfasts. See Figure 26.1 and Plate 8.5. The presence of red pigments, known as **phycobilins,** in addition to **chlorophylls a** and **d,** allows the red algae to absorb the deeper-penetrating wavelengths of light for photosynthesis.

Reproduction in Algae

Asexual reproduction is by mitotic cell division in unicellular algae, and by fragmentation in colonial and multicellular species. Sexual reproduction ranges from conjugation in colonial green algae to the more complex alternation of generations in multicellular forms of green, brown, and red algae.

Consider **conjugation** in *Spirogyra,* a filamentous, green alga, shown in Figure 26.2. When opposite mating types are in contact with each other, a tube forms between cells of the two filaments. The contents of each cell condense, forming a **gamete.** One gamete migrates through the tube from one cell to the other and fuses with the gamete in the receiving cell, forming a diploid (2n) **zygote.** This then forms a **zygospore** that can withstand unfavorable conditions and that germinates under favorable conditions. Germination is by meiosis, forming four haploid nuclei. Three of these nuclei disintegrate, leaving the cell with one haploid nucleus. Mitotic division of this cell results in the formation of a new haploid filament.

Although **alternation of generations** in a plant life cycle first appeared in advanced multicellular algae, it is most highly developed in terrestrial plants, where you will study it in some detail. Here, it is important for you to understand its basic characteristics.

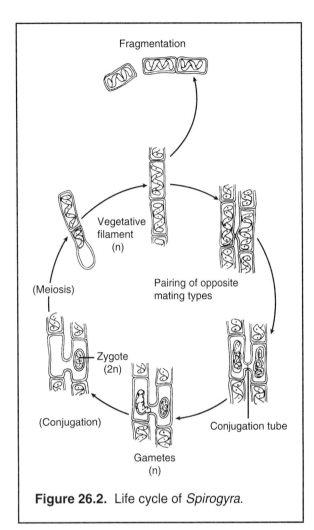

Figure 26.2. Life cycle of *Spirogyra.*

Examine Figure 26.3. Note that there are two adult generations: a gametophyte generation alternates with a sporophyte generation. The haploid (n) **gametophyte** produces gametes (eggs and sperm) by mitosis. Union of sperm and egg forms a diploid (2n) **zygote** that grows to become a diploid (2n) **sporophyte.** The sporophyte produces haploid (n) **spores** by meiotic division. The spores then germinate and grow into gametophytes of the next generation.

Materials

Per student
Colored pencils

Per lab
Medicine droppers
Microscope slides and cover glasses
Cultures of:

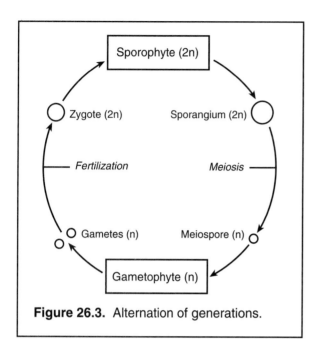

Figure 26.3. Alternation of generations.

Chlamydomonas
Spirogyra
Volvox
Demonstration microscope setups of:
    Chlamydomonas, 1,000×
    Volvox, 40×
Demonstration of the effect of light on *Chlamy-domonas* distribution in a partially shaded Petri dish
Prepared slides of *Spirogyra* conjugation showing gametes and zygospores
Representative brown, green, and red algae

Assignment 1

1. **Complete item 1a on Laboratory Report 26 that begins on page 403.**
2. Examine *Chlamydomonas* set up under a demonstration microscope. Compare your observations with Figure 26.1 and Plate 7.1.
3. Examine the distribution of *Chlamydomonas* in the partially shaded Petri dish. *Do not move the dish.* Lift the black-paper light shield to make your observations, then replace it in its prior position.
4. Examine *Volvox* set up under a demonstration microscope and note its manner of movement. *Volvox* is a colony formed of many *Chlamydomonas*-like cells. Compare your ob-

servations with Plate 7.2. **Complete items 1b-1e on the laboratory report.**
5. Make a water-mount slide of a few strands of *Spirogyra,* a nonmotile, filamentous alga, and examine them at 100× and 400×. Locate the cellular parts shown in Figure 26.1. Compare your specimen with Plate 7.3. **Complete items 1f and 1g on the laboratory report.**
6. Examine the demonstration specimens of brown, green, and red algae. Compare the specimens with Figure 26.1 and Plates 8.1–8.5. **Complete items 1h and 1i on the laboratory report.**
7. Study the life cycle of *Spirogyra* in Figure 26.2. Color the diploid stages of the cycle green. Examine a prepared slide of *Spirogyra* conjugation. Note the gametes and zygospores. **Complete item 1 on the laboratory report.**

MOSS PLANTS (DIVISION BRYOPHYTA)

Mosses, liverworts, and hornworts compose the bryophytes. These plants occur in habitats that are moist for at least part of the year, because they require surface water for their sperm to swim to the eggs. Most bryophytes are not well adapted for terrestrial life because they lack vascular tissues (xylem and phloem) that conduct water, minerals, and nutrients. They also lack true leaves, stems, and roots.

Study the moss life cycle in Figure 26.4. Locate the gametophytes and the sporophyte. In this moss, there are separate male and female gametophytes, but some moss species have a single gametophyte that contains both male and female reproductive organs.

The dominant generation is the **gametophyte,** meaning that the gametophyte is larger and lives longer than the sporophyte. A gametophyte has leaf-like photosynthetic organs that are attached to a stem-like stalk from which **rhizoids** extend, anchoring the gametophyte to the soil. Rhizoids do not absorb water and minerals from the soil like roots.

A **sporophyte** consists of a stalk (seta) attached to a gametophyte, and a terminal **sporangium.** The sporophyte is partially parasitic on the gametophyte, since it has limited photosynthetic capabilities.

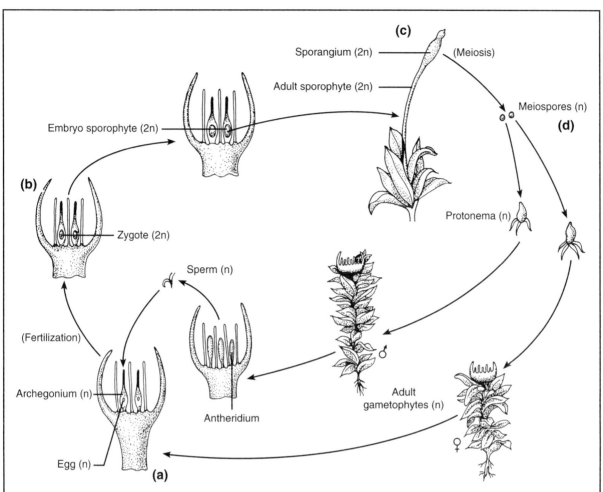

Figure 26.4. Life cycle of a moss. (a) A male gametophyte (n) develops sperm-forming antheridia at its tip, and a female gametophyte (n) develops archegonia, each forming one egg. When released, a flagellated sperm swims into an archegonium and fertilizes the egg cell. (b) Union of sperm (n) and egg (n) forms a zygote (2n) that develops into an embryo sporophyte (2n). (c) The embryo sporophyte grows to become an adult sporophyte that remains attached to the base of the archegonium. (d) Meiospores (n) are formed by meiosis within the sporangium and released. Upon germination, they grow into protonemas (n) that subsequently develop into adult male and female gametophyte (n).

Materials

Per student
Colored pencils
Compound microscope

Per lab
Moss gametophytes and sporophytes, living
Representative bryophytes, living or plasto-
  mounts
Prepared slides of *Mnium*:
  antheridia with sperm, l.s.

archegonia with eggs, l.s.
protonema
sporangium (capsule) with spores, l.s.

Assignment 2

1. Study Figure 26.4 and color all diploid phases of the life cycle green.
2. Examine a living gametophyte with sporophyte

attached. Locate the rhizoids, stalk, and leaf-like organs of the gametophyte, and locate the stalk and sporangium (capsule) of the sporophyte. **Complete item 2a on the laboratory report.**

3. Examine prepared microscope slides of antheridia with sperm, archegonia with eggs, sporangium with spores, and protonema at 40× and 100×. **Complete item 2 on the laboratory report.**

FERNS (DIVISION PTEROPHYTA)

Ferns are the most common and best known of the nonseed vascular plants. Vascular plants possess two vascular tissues: xylem and phloem. **Xylem** conducts water and dissolved minerals upward from roots to stem and leaves. **Phloem** conducts organic nutrients either upward or downward within the plant. Plants that possess vascular tissue have true **roots, stems,** and **leaves.** Without vascular tissue, plants would not be much bigger than mosses and could not live in drier habitats.

Examine the fern life cycle in Figure 26.5. A fern sporophyte is the dominant generation, and it is what you recognize as a fern plant. A sporophyte has an underground stem, a **rhizome,** that is anchored by numerous **roots. Leaves** arising from the stem are usually large and subdivided into many leaflets. Leaflets of some leaves have small brown spots, known as **sori,** on their undersurface. A single sorus, consists of many spore-forming sporangia that are usually protected by a thin cover, the indusium.

The separate gametophyte is small, flat, and roughly heart-shaped. It is anchored to the soil by rhizoids. Archegonia are clustered together on its undersurface near the notch, while antheridia are more widely scattered closer to the bases of the rhizoids.

Materials

Per student
Colored pencils
Compound microscope
Dissecting microscope

Per lab
Demonstration microscope setup of archegonium with an egg cell
Fern sporophytes with sori, living

Microscope slides and cover glasses
Prepared slides of fern:
   antheridium with sperm
   gametophyte, w.m.
   sorus, x.s.

Assignment 3

1. Study Figure 26.5. Color the diploid portions of the fern life cycle green.
2. Examine a prepared slide of a fern gametophyte. Locate the rhizoids, antheridia, and archegonia. Examine a prepared slide of an antheridium and the demonstration microscope setup of an archegonium. How many eggs are located within its swollen base? **Complete items 3a–3c on the laboratory report.**
3. Examine a living fern sporophyte. Locate the rhizome, roots, leaves, and sori.
4. Examine a prepared slide of a sorus, x.s., showing sporangia and indusium. Locate the spores within the sporangia.
5. Remove a leaflet with sori and examine a sorus with a dissecting microscope. If an indusium is present, remove it with a dissecting needle to expose the sporangia. Place the leaflet on a microscope slide without water or a cover glass, and examine the exposed sporangia at 40× with your compound microscope. Watch what happens as the sporangia dry out. **Complete item 3 on the laboratory report.**

SEED PLANTS

Seed plants are better adapted to terrestrial life than ferns, but maintain a life cycle involving alternation of generations. The dominant sporophyte—the generation that you recognize as a plant—consists of a root system and a shoot system. The **root system** usually lies below ground. It provides anchorage for the plant and it absorbs water and minerals. The **shoot system** consists of stems, leaves, and reproductive organs. It produces organic nutrients via photosynthesis.

A seed plant contains well-developed vascular tissue that provides structural support for its roots, stems, and leaves, as well as permitting

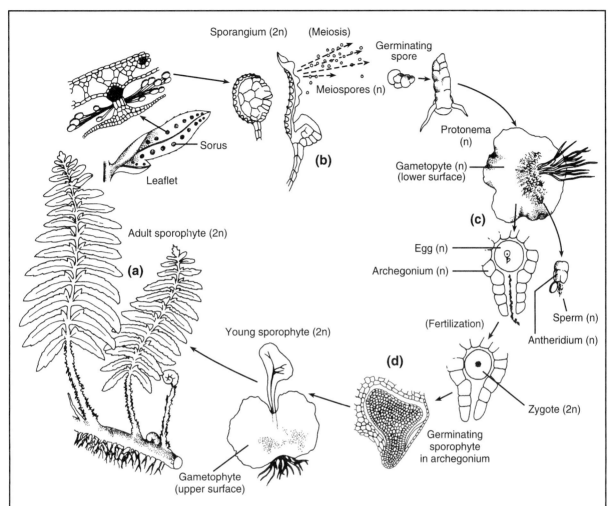

Figure 26.5. Life cycle of a fern. (a) A mature fern sporophyte (2n) forms sporangia on the undersurface of the leaflets of some leaves. Sporangia are clustered in groups called sori. (b) Meiospores (n), formed by meiosis, are released from sporangia. The spores germinate and develop into protonemas (n) that grow to become gametopytes (n). (c) Sperm-forming antheridia and egg-containing archegonia are located on the lower surface of a gametophyte. When released, a flagellated sperm swims into an archegonium to fertilize the egg. (d) Union of sperm (n) and egg (n) forms a zygote (2n) that develops into an embryo sporophyte within the archegonium. Growth of the embryo sporophyte ultimately produces an independent adult sporophyte (2n).

the transport of materials. Leaves, the main photosynthetic organs, are supported by **veins** composed of vascular tissue.

Sporophytes of seed plants form two different types of meiospores. **Microspores** are formed in **microsporangia** and mature to form **pollen grains,** which in turn develop into microscopic **male gametophytes. Megaspores** are formed by **megasporangia** and develop into **female gametophytes.**

Seed plants are not dependent on water for sperm transport. Instead, pollen grains are transferred to the female gametophyte by wind or insects, a process called **pollination.** Then the pollen grain develops into a male gametophyte with a **pollen tube** that carries **sperm nuclei** to the **egg** in the female gametophyte. Fertilization results in a zygote, the first cell of a new sporophyte generation.

The embryo sporophyte develops within the

female gametophyte. The embryo sporophyte, female gametophyte, and stored nutrients compose a **seed** that has a resistant seed coat enabling dormancy during unfavorable conditions. When conditions are favorable, a seed germinates, producing a new sporophyte generation.

Seed plants are divided into two large groups. **Gymnosperms** have their seeds exposed on the surface of modified leaves. **Angiosperms** produce flowers and have their seeds enclosed within a **fruit.**

Cone-Bearing Plants (Division Coniferophyta)

Conifers are the best known and largest group of gymnosperms. The cones of conifers contain reproductive organs, and two types of cones are formed. **Staminate** (pollen-producing) **cones** are small with paper-thin scales. **Ovulate** (seed forming) **cones** are large with woody scales. Pollen is transferred from staminate cones to ovulate cones by wind. Seeds are borne exposed on the upper surface of the scales of mature ovulate cones. The leaves are either needlelike or scalelike. Conifers may attain considerable size and may live in rather dry habitats because their vascular tissue is well developed, their leaves restrict water loss, and they do not require water for sperm transport.

Study the life cycle of the pine illustrated in Figure 26.6.

Materials

Per student
Colored pencils
Compound microscope
Dissecting instruments

Per lab
Cones and leaves of representative conifers
Demonstration microscope setups:
  pine female gametophyte with archegonium and egg
  sectioned pine seed showing embryo sporophyte
Ovulate pine cones with seeds
Staminate pine cones with pollen
Pine seeds
Prepared slides of pine microsporangium with pollen

Assignment 4

1. Study the pine life cycle in Figure 26.6 and color the diploid stages green. **Complete item 4a on the laboratory report.**
2. Examine the male and female pine cones. Compare their size and weight. Locate the microsporangia of a male cone and the seeds on the upper surfaces of the scales of a female cone.
3. Examine the cones and scale-like and needle-like leaves of representative conifers.
4. Examine a prepared slide of pine microsporangia with pollen. Note the structure of pollen. **Draw a few pollen grains in item 4b on the laboratory report.**
5. Examine a pine female gametophyte (megasporangium) with archegonium set up under a demonstration microscope. Locate the egg in an archegonium. **Draw a female gametophyte with archegonium and egg in item 4b on the laboratory report.**
6. Examine a sectioned pine seed set up under a demonstration dissecting microscope. Locate the embryo sporophyte embedded in tissue of the female gametophyte.
7. **Complete item 4 on the laboratory report.**

Flowering Plants (Division Anthophyta)

Flowering plants are the most advanced plants. Their success in colonizing the land is due to well-developed vascular tissues and **flowers** (reproductive organs) that greatly enhance reproductive success. Most flowers attract insects that bring about pollination, a process leading to the fertilization of egg cells by sperm nuclei. The seeds of flowering plants are enclosed within **fruits** that facilitate dispersal. The reproductive patterns of flowering plants show some major adaptations over gymnosperms.

1. Reproductive structures are grouped in **flowers** that usually contain both microsporangia and megasporangia.
2. Pollination is usually by wind in grasses, but it is by insects in most flowering plants. Insects are attracted to flowers by color and nectar.

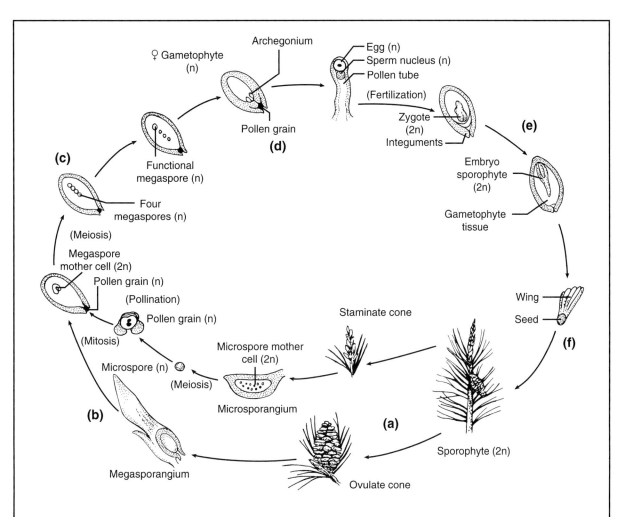

Figure 26.6. Life cycle of a pine. (a) The sporophyte (2n) produces staminate and ovulate cones. Each "leaf" of a staminate cone contains two microsporangia, and each "leaf" of an ovulate cone contains two megasporangia. (b) Microsporangia form microspores (n) by meiosis, and microspores develop into pollen grains. Pollen grains are released and carried by wind to the megasporangia, where they begin to develop into male gametophytes. (c) Megasporangia form four megaspores (n) by meiosis. Three megaspores disintegrate, and one megaspore grows to form a female gametophyte. (d) The pollen tube of a male gametophyte grows into an archegonium and discarges two sperm nuclei (n), one of which fertilizes the egg (n). (e) The resultant zygote (2n) grows to become an embryo sporophyte embedded within the female gametophyte. The wall of the megasporangium forms the seed coat of the new seed. (f) The seed is released and dispersed by wind. It germinates to grow into a new sporophyte (2n) generation.

3. Portions of the flower develop to form a fruit that encloses the seeds and enhances seed dispersal by wind in some plants, but by animals in most.

Study the life cycle of a flowering plant illustrated in Figure 26.7.

Flower Structure

The basic structure of a flower is shown in Figure 26.8, but this fundamental organization has many variations.

The **receptacle** supports the flower on the stem, and the reproductive structures are

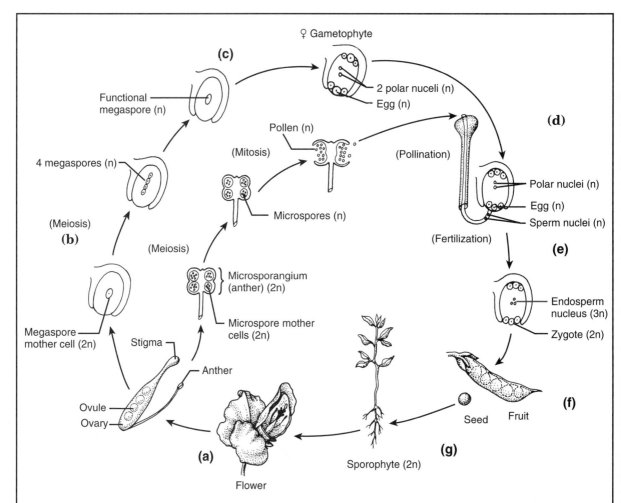

Figure 26.7. Life cycle of a pea plant. (a) The sporophyte (2n) produces flowers with anthers (microsporangia) and ovules (megasporangia). (b) Anthers produce microspores (n) by meiosis. Each microspore develops into a pollen grain. Ovules form four megaspores (n), but only one remains functional. (c) The functional megaspore grows to form a female gametophyte (n) containing an egg and two polar nuclei within the ovule. (d) Pollen grains are transferred to the stigma of a flower and develop to become male gametophytes (n) that grow pollen tubes down the style. A pollen tube carries two sperm nuclei (n) to a female gametophyte. (e) One sperm nucleus fertilizes the egg, forming a zygote (2n), and the other fuses with the two polar nuclei, forming the endosperm nucleus (3n). The zygote grows to become an embryo sporophyte, and the endosperm develops to provide stored nutrients for the embryo. (f) The ovule, endosperm, and embryo sporophyte form the seed that is contained within a fruit—which is an enlarged, ripened ovary. Fruit and/or seeds are disseminated by wind or animals. (g) Upon germination, seeds grow to form a new sporophyte (2n) generation.

enclosed within two whorls of modified leaves. The inner whorl consists of **petals** that are usually colored to attract pollinating insects. The outer whorl consists of **sepals** that are typically smaller than the petals and are usually green.

The **stamens** consist of an **anther** supported by a **filament.** Anthers contain the microsporangia that produce pollen. The **pistil** consists of three parts. The basal portion is the **ovary,** which contains **ovules** (megasporangia). The tip of the pistil is the **stigma,** which receives pollen

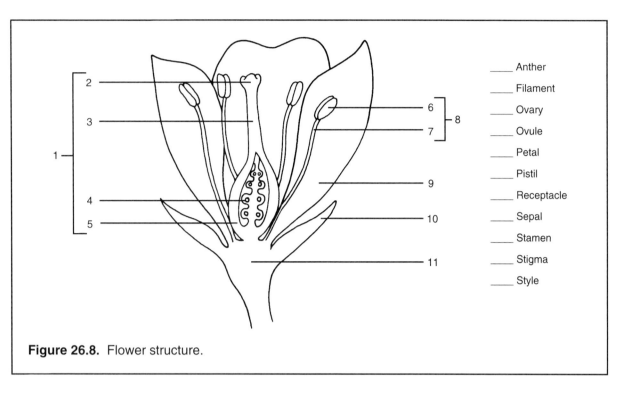

Figure 26.8. Flower structure.

_____ Anther

_____ Filament

_____ Ovary

_____ Ovule

_____ Petal

_____ Pistil

_____ Receptacle

_____ Sepal

_____ Stamen

_____ Stigma

_____ Style

and secretes enzymes promoting pollen germination. The **style** is a slender stalk that joins the stigma and ovary. Nectar is secreted near the base of the ovary.

Fruits and Seeds

As the seeds develop in the ovary, the ovary grows and ripens to form a fruit that provides protection for the seeds and facilitates seed dispersal. The three basic types of fruits are shown in Figure 26.9.

1. **Dry dehiscent fruits** split open when sufficiently dry, casting out the seeds, sometimes with considerable force. Pea and bean pods are examples.
2. **Dry indehiscent fruits** do not open, and the ovary wall tightly envelops the seed. Acorns and fruits of corn and other cereals are examples of fruits of this type.
3. **Fleshy fruits** remain moist for a considerable period, and are usually edible and colored. Animals scatter the seeds by feeding on the fruits.

A seed (Figure 26.10) consists of a protective **seed coat** that is derived from the wall of the ovule, stored nutrients, and a dormant **embryo sporophyte.** In monocots and some dicots, the stored nutrients compose the endosperm. In most dicots, the embryonic leaves, known as **cotyledons,** contain many of the stored nutrients, and the endosperm is reduced. Seeds are able to withstand unfavorable conditions, and tend to germinate only when favorable conditions exist.

Materials

Per student
Colored pencils
Compound microscope
Dissecting instruments
Dissecting microscope

Per lab
Dropping bottles of iodine solution
Microscope slides and cover glasses
Model of a flower
Bean seeds, soaked
Corn fruits, soaked
Lily or *Gladiolus* flowers
Pea pods, fresh
Representative flowers, fruits, and seeds
Prepared slides of lily anthers, x.s.

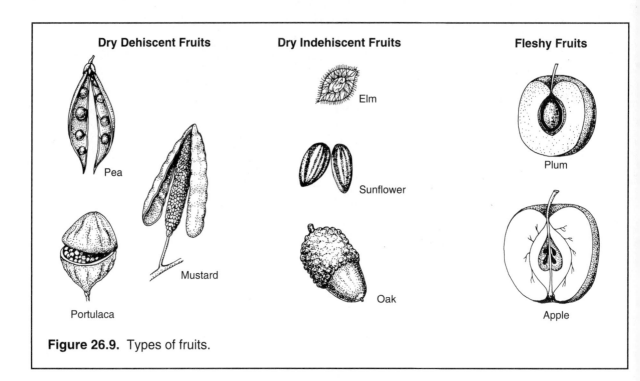

Figure 26.9. Types of fruits.

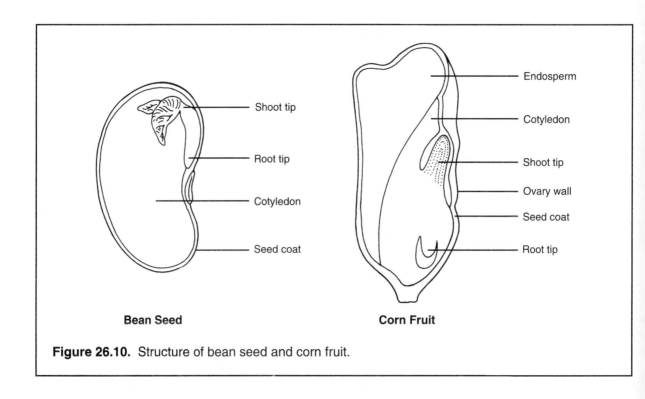

Figure 26.10. Structure of bean seed and corn fruit.

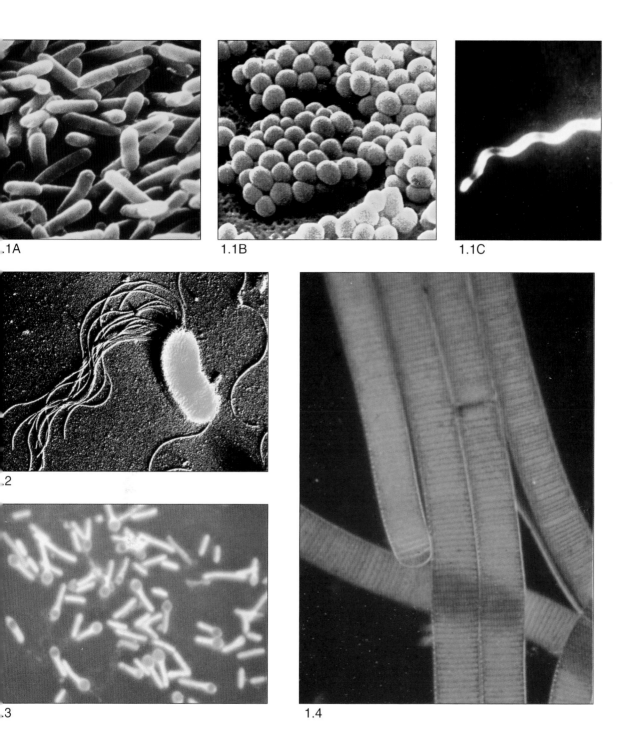

1.1A 1.1B 1.1C

.2

.3 1.4

e 1.1 The three basic shapes of bacteria. (A) -shaped (bacillus) cells are shown in this e-color scanning electron micrograph (SEM) of idomonas aeruginosa. *P. aeruginosa* is a rine-resistant species that may cause skin ctions in swimmers and users of hot tubs. (B) erical (coccus) cells are evident in this e-color SEM of *Staphylococcus aureus*, a species sing pimples and abscesses. (C) Spiral-shaped rillum) cells are illustrated in this SEM of onema pallidum, the bacterium causing hilis.

Plate 1.2 Flagella are clearly visible in this false-color scanning electron micrograph of *Pseudomonas flourescens*, a motile soil bacterium.

Plate 1.3 The rod-shaped bacterium *Clostridium tetani* causes tetanus (lockjaw). Some cells contain an endospore at one end of the cell. An endospore is resistant to unfavorable conditions and grows into a new bacterium when conditions become favorable.

Plate 1.4 Filaments of *Oscillatoria*, a photosynthetic cyanobacterium, are composed of thin, circular cells joined together like stacks of coins.

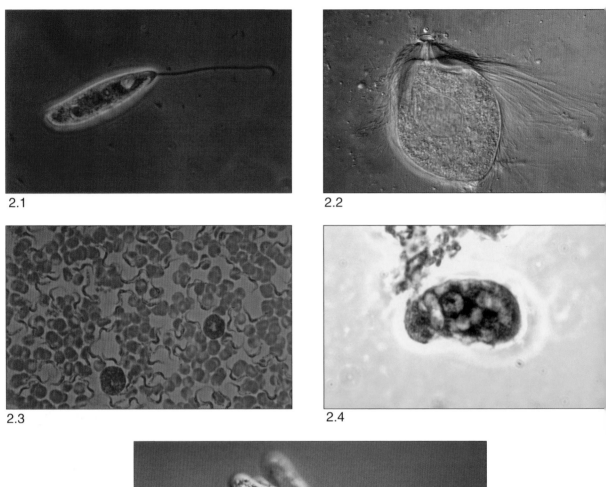

2.1

2.2

2.3

2.4

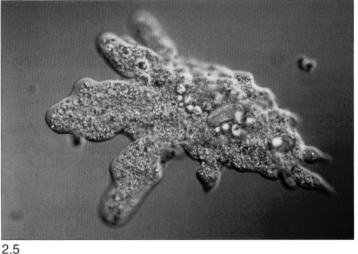

2.5

Plate 2.1 *Peranema tricophorum,* a freshwater flagellated protozoan, is propelled by a thick anterior flagellum held rigid and straight except for its tip. It engulfs prey through an opening near the base of the flagellum. A thin trailing flagellum is also present but is not visible here.

Plate 2.2 *Trichonympha* is a multiflagellated protozoan that lives in the gut of termites. It digests wood particles, forming nutrients that can be used by the termites.

Plate 2.3 *Trypanosoma brucei* among human blood cells. It causes African sleeping sickness and is transmitted by the bite of tsetse flies.

Plate 2.4 *Entamoeba histolytica* invades the lining of the intestine, causing amoebic dysentery in humans.

Plate 2.5 *Amoeba proteus* is a common freshwater amoeboid protozoan. Note the pseudopodia. The lighting used here shows the granular nature of the cytoplasm and light colored food vacuoles.

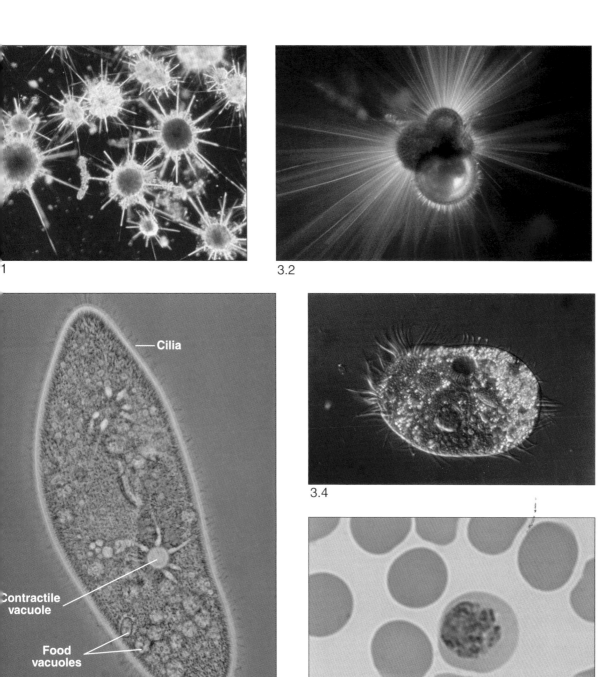

3.2

Cilia

Contractile
vacuole

Food
vacuoles

3.4

3.5

te 3.1 Radiolarians are marine amoeboid
-tozoans that secrete a siliceous shell around
 central cell body. Slender pseudopodia radiate
ough tiny holes in the shell and capture tiny
d particles that are carried back to the cell
 y by streaming cytoplasm.

te 3.2 A living foraminiferan. Foraminiferans
 marine amoeboid protozoans that secrete a
careous shell. Thin raylike pseudopodia extend
ough tiny openings in the shell.

te 3.3 *Paramecium*, a freshwater ciliate. Beat-
 cilia propel *Paramecium* rapidly through the

water. Food organisms are digested in food vacuoles. Excess water is collected and pumped out
by two contractile vacuoles.

Plate 3.4 *Euplotes* is an advanced ciliate. Tufts of
cilia unite to form cirri that function almost like
legs as the ciliate "walks" over the bottom of a
pond. Note the food vacuoles and a contractile
vacuole that appears like a cavity in this view.

Plate 3.5 *Plasmodium vivax*, a sporozoan, in a
human red blood cell. It causes malaria and is
transmitted by the bite of female anopheline
mosquitos.

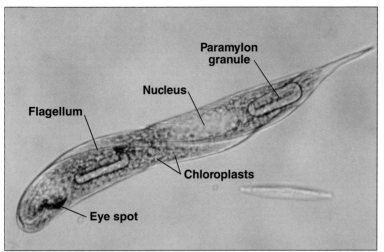

4.1

Paramylon
granule

Nucleus

Flagellum

Chloroplasts

Eye spot

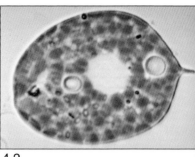

4.2

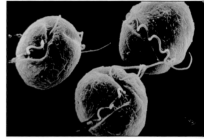

4.3

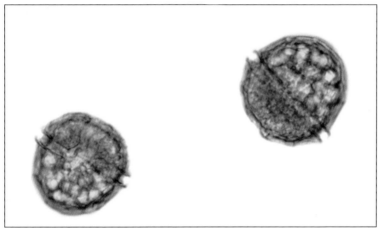

4.4

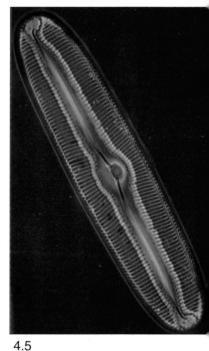

4.5

Plate 4.1 *Euglena* is a photosynthetic euglenoid. The anteriorly located flagellum provides locomotion. It is recurved along the body in this view. An eye spot, located near the base of the flagellum, detects light intensity. Food reserves are stored as paramylon granules.

Plate 4.2 *Phacus* is a freshwater euglenoid. Note the eye spot and chloroplasts. The nucleus appears as a large clear area in the center of the cell. An anterior flagellum, not visible in this photomicrograph, provides locomotion.

Plate 4.3 Like all dinoflagellates, *Gymnodiniur* marine form, possesses two flagella. One lies in equatorial groove in the cell wall, and the ot hangs from the end of the cell SEM.

Plate 4.4 *Peridinium* a freshwater dinoflagell common in acid-polluted lakes. Here, chloropla and equatorial groove are visible, but the flag are not.

Plate 4.5 The diatom *Pinnularia mobilis*. The silice walls of diatoms are formed of two parts that together like the top and bottom of a pillbox.

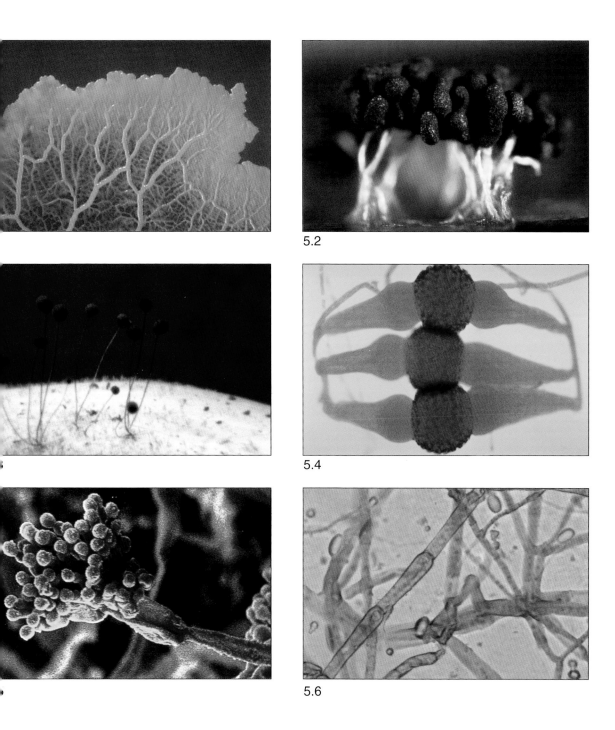

5.2

5.4

5.6

Plate 5.1 The plasmodium of the slime mold *Physarum* lives like a giant, multinucleate amoeba that creeps along, absorbing nutrients and engulfing food organisms.

Plate 5.2 When either food or moisture becomes inadequate, a *Physarum* plasmodium transforms into spore-forming sporangia supported by short stalks.

Plate 5.3 In *Rhizopus stolonifer*, black bread mold, sporangiophore hyphae support sporangia that contain asexual spores. Each sporangiophore is attached to the substrate by rhizoid hyphae.

Plate 5.4 The fusion of (+) and (−) gametes in *Rhizopus stolonifer* results in the formation of a zygospore sandwiched between a pair of gametangia. Later, each zygospore (2n) undergoes meiosis, producing a haploid mycelium

Plate 5.5 This false-color SEM of the asexual conidiospores of *Penicillium* shows their spherical shape and attachment to a sporangiophore. 5,750X.

Plate 5.6 The hyphae of *Nectria* are composed of separate cells joined end to end. Such hyphae are said to be septate because cross walls separate the cells.

6.1A

6.1B

6.2

6.3

6.4

Plate 6.1 (A) Cup-shaped ascocarps (fruiting bodies) of *Peziza aurantia* a sac fungus. (B) Sac-like asci containing ascospores lines the ascocarps of *P. aurantia*. When the asci rupture, ascospores are released and dispersed by wind.

Plate 6.2 Basidiocarps (fruiting bodies) of shelf fungi, a club fungus, contain spore-forming basidia.

The basidiospores are released through tiny p on the lower surface of the basidiocarps.

Plate 6.3 Each basidium of a club fungus duces four basidiospores, as shown in this SE

Plate 6.4 A discharge of millions of basidiosp from a puffball, a club fungus.

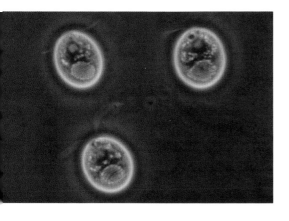

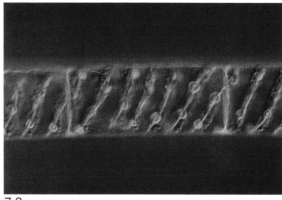

7.3

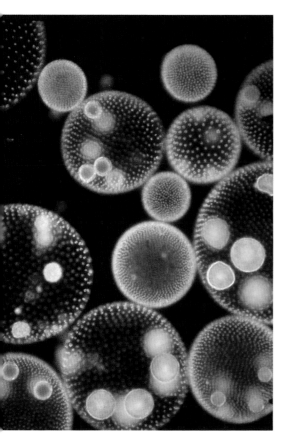

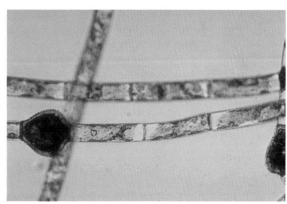

7.4

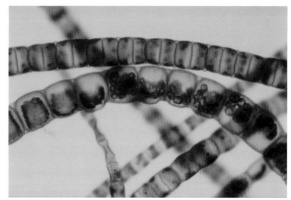

7.5

Plate **7.1** *Chlamydomonas* is a tiny unicellular green alga that moves by means of two flagella. Only the basal portion of the cup-shaped chloroplast is visible here. An eyespot, appearing as a blue dot, enables detection of light intensity.

Plate **7.2** *Volvox* colonies may consist of up to 50,000 *Chlamydomonas*-like cells. *Volvox* may reproduce asexually by forming daughter colonies within the sphere. Some forms also have a few cells specialized for sexual reproduction.

Plate **7.3** The spiral chloroplast in a cell of *Spirogyra*, a filamentous green alga. The enlarged regions are pyrenoids, sites of starch storage.

Plate **7.4** Cellular specialization for sexual reproduction is evident in *Oedogonium*, a filamentous green alga. The enlarged, dark cells are oogonia that contain an oospore formed after union of sperm and egg cells.

Plate **7.5** Cellular specialization for reproduction also occurs in *Ulothrix*, another filamentous green alga. The cells in the lower curved filament are sporangia containing asexual zoospores.

8.1

8.2

8.3

8.4

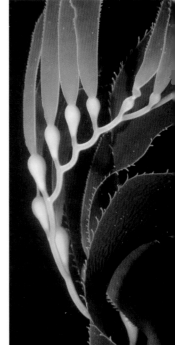

8.5

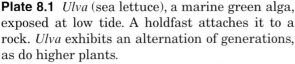

Plate 8.1 *Ulva* (sea lettuce), a marine green alga, exposed at low tide. A holdfast attaches it to a rock. *Ulva* exhibits an alternation of generations, as do higher plants.

Plate 8.2 Brown marine algae are characterized by a robust body that can withstand strong wave action. Note the holdfast, stipe, and divided blade in *Durvillaea*, a brown alga from the Australian coast.

Plate 8.3 A forest of giant kelp, *Macrocystis*, along the California coast provides an important habitat for marine animals. The long stipes may be (100 ft. in length.

Plate 8.4 The delicate, feathery body structur red algae sharply contrasts with the robust br algae. Red algae, like *Callithamnion*, live at oc depths where wave action is minimal.

Plate 8.5 Gas-filled floats of *Macrocystis* lift blades and stipe toward the water surface wh more sunlight is available.

Assignment 5

1. Study the life cycle of a flowering plant in Figure 26.7. Color the diploid stages green. **Complete item 5a on the laboratory report.**
2. Label Figure 26.8. Color the male portion of the flower yellow and the female portion green. **Complete items 5b and 5c.**
3. Obtain a lily flower and compare it with Figure 26.8 and the flower model to identify its parts.
4. Examine a prepared slide of an anther, x.s. Note the microsporangia and enclosed pollen. Remove an anther from your flower and observe it with a dissecting microscope. Note the pollen. Place a few pollen grains on a microscope slide without water and observe them at 40× and 100× with a compound microscope. **Complete item 5d on the laboratory report.**
5. Use a scalpel to make a cross section of the ovary and examine it with a dissecting microscope. Locate the ovules. **Complete item 5e on the laboratory report.**
6. Examine the representative flowers and note the variations in their structure. Locate the male and female portions of each.
7. Examine the representative seeds and fruits. Use Figure 26.9 to classify the fruits according to type. Obtain and examine a pea pod. **Complete items 5f and 5g on the laboratory report.**
8. In Figure 26.10, color the embryos green and the cotyledons blue. Obtain a soaked bean seed and corn kernel. Open the "halves" of the bean seed as in Figure 26.10 to expose the embryo. Use a scalpel to section the corn kernel and examine its cut surface. Examine both with a dissecting microscope to locate their components shown in Figure 26.10.
9. Add a drop of iodine solution to the bean cotyledons and embryo, and to the cut surface of the corn kernel. After 3 min, rinse, blot dry, and examine.
10. **Complete item 5 on the laboratory report.**

Assignment 6

Your instructor has set up numbered "unknown" plants for you to identify as a mini-practicum. **Record your responses in item 6 on the laboratory report.**

27 Structure of Flowering Plants

OBJECTIVES

After completion of the laboratory session, you should be able to:
1. Identify the external structure of a flowering plant.
2. Identify monocots and dicots.
3. Identify types of root systems and leaves.
4. Identify tissues composing roots, stems, and leaves.
5. Define all terms in bold print.

Flowering plants are the most advanced vascular plants. They have well-developed vascular tissues that provide support as well as material transport, and, as you learned in Exercise 26, their life cycle involves flowers, pollination, and seeds enclosed in fruits. In this exercise, you will study the external and internal structure of roots, stems, and leaves of flowering plants.

Flowering plants are subdivided into two major classes: **monocotyledonous** (monocots) **plants** and **dicotyledonous** (dicots) **plants.** Their distinguishing characteristics are shown in Figure 27.1.

Monocots include grasses, palms, and lilies. Most flowering plants are dicots, and they fall into one of two major categories: herbaceous or woody. **Herbaceous dicots** are usually annuals or biennials. Annuals complete their life cycle in a single growing season, as in the case of most wildflowers, beans, and tomatoes. Biennials require two growing seasons; flowers are produced only in the second season. **Woody dicots** are usually perennials that live several years and produce flowers each year, such as oaks and roses.

GENERAL EXTERNAL STRUCTURE

Figure 27.2 shows the basic parts of a flowering plant. The shoot system consists of a **stem** that supports the **leaves, flowers,** and **fruits.** Leaves branch from the stem at sites called **nodes.** A section of stem between nodes is an **internode.** Leaves are the primary photosynthetic organs of the plant, and they exhibit two types of venation: net venation and parallel venation. Leaves with **net venation** have a central vascular bundle (vein) called a midrib, from which smaller lateral veins branch. Such leaves consist of a thin, expanded portion called a **blade** and a leaf stalk called a **petiole.** Leaves with **parallel venation** lack a midrib and have veins running parallel to each other for the length of the blade. Such leaves usually lack a petiole.

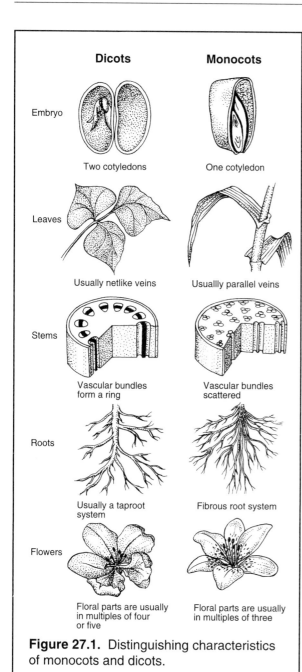

Figure 27.1. Distinguishing characteristics of monocots and dicots.

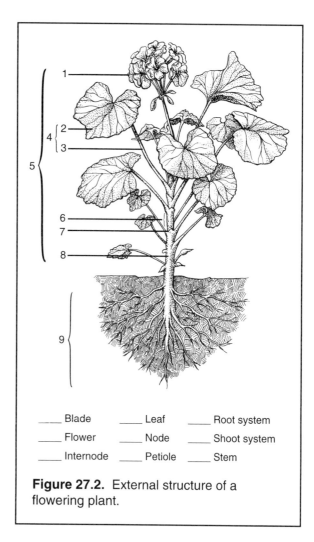

____ Blade	____ Leaf	____ Root system
____ Flower	____ Node	____ Shoot system
____ Internode	____ Petiole	____ Stem

Figure 27.2. External structure of a flowering plant.

The root system is located in the ground, and it may be even more highly branched than the shoot system. Roots not only anchor the plant, but also absorb water and nutrients.

Materials

Per student
Dissecting microscope

Per lab
Coleus seedlings
Coleus and corn stems in solution of red food coloring
Corn seedlings
Representative dicots and monocots
Razor blades, single-edged

Assignment 1

1. Label Figure 27.2. **Complete items 1a and 1b on Laboratory Report 27 that begins on page 407.**
2. Obtain corn and *Coleus* seedlings. Gently wash the soil from their roots and lay them on a paper towel for observation. Compare the roots, stems, and leaves with Figures 27.1 and 27.2.

3. Stems of several *Coleus* and corn seedlings have been placed in a water-soluble dye that stains vascular tissue as it is carried up the stem. Use a razor blade to cut thin cross sections from the *Coleus* and corn stems and examine them with a dissecting microscope. Observe the arrangement of the vascular bundles and compare them with Figure 27.1.

4. Identify the "unknown" plants as dicots or monocots. **Complete item 1 on the laboratory report.**

ROOTS

Roots perform three important functions: (1) anchorage and support, (2) absorption and transport of water and minerals, and (3) storage and transport of organic nutrients.

The Root Tip

The basic structure of a root tip is shown in Figure 26.3. Cells formed in the **region of cell division** become either part of the root proper or form the root cap. The **root cap** is composed of rather large cells that protect the region of cell division. It also provides a sort of lubrication as its cells are eroded by the root growing through the soil. Cells of the root are enlarged in the **region of elongation,** and this accounts for the greatest increase in the linear growth of the root. As the cells become older, they develop their specialized characteristics in the **region of differentiation,** where the **primary root tissues** are formed. Note that the cells of the root are arranged in columns. The column in which a cell is located determines the type of cell it will become. For example, cells in the outermost columns become epidermal cells, and those in the innermost columns become xylem cells. **Root hairs** are extensions of epidermal cells in the region of differentiation. They greatly increase the surface area of the root tip and are the primary sites of water and mineral absorption.

Growth in length of both roots and stems results from the formation of new cells in the region of cell division and the subsequent enlargement of these cells in the region of elongation. Thus, growth of both roots and stems occurs at their tips, and this growth is continuous throughout the life of the plant.

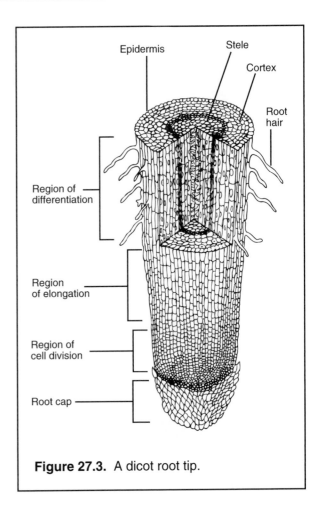

Figure 27.3. A dicot root tip.

Root Tissues

Figure 27.4 shows the tissues found in a root of a herbaceous dicot (*Ranunculus*) as viewed in cross section. Note the three major structural divisions of the root: epidermis, cortex, and stele.

The **epidermis** is the outermost layer of cells that provides protection for the underlying tissues and reduces water loss.

The **cortex** composes the bulk of the root and consists mostly of large, thin-walled cells used for food storage. The **endodermis,** the innermost layer of the cortex, is composed of thick-walled, water-impermeable cells and a few water-permeable cells with thinner walls. This ring of cells controls the movement of water and minerals into and out of the xylem.

The **stele** is that portion of the root within the endodermis. It is sometimes called the central cylinder. The outer layer of the stele is composed of thin-walled cells, the **pericycle,** from which branch roots originate. The thick-walled cells in

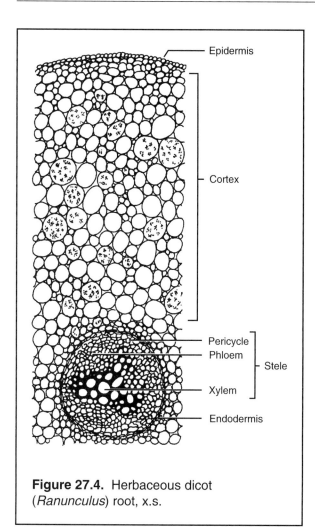

Figure 27.4. Herbaceous dicot (*Ranunculus*) root, x.s.

Materials

Per student
Colored pencils
Compound microscope
Dissecting microscope
Dissecting instruments

Per lab
Dropping bottles of methylene blue, 0.01%
Microscope slides and cover glasses
Examples of adventitious, fibrous, and tap roots
Germinated grass (or radish) seeds
Prepared slides of:
     Allium (onion) root tip, l.s.
     Ranunculus (buttercup) root, x.s.

Assignment 2

1. Color-code the root cap and regions of cell division, cell elongation, and cell differentiation in Figure 27.3 and the xylem and phloem in Figure 27.4.
2. Obtain a germinated grass seed. Place it on a microscope slide in a drop of water. Examine the young root with a dissecting microscope and at 40× with a compound microscope. Note the root hairs. Locate the oldest (longest) and youngest root hairs. Use a scalpel to cut the seed from the root. Add a drop of methylene blue and a cover glass. Examine the root at 100×. Note the attachment of a root hair to an epidermal cell. Try to locate a cell nucleus in a root hair. **Complete items 2a-2c on the laboratory report.**
3. Examine a prepared slide of *Allium* root tip, l.s., with a compound microscope at 40×. Locate the root cap and the regions shown in Figure 27.3, except for the region of differentiation, which is not shown on your slide. Compare the size of the cells in each region. **Complete item 2d on the laboratory report.**
4. Examine a prepared slide of *Ranunculus* root, x.s., with a compound microscope at 40×. Locate the tissues shown in Figure 27.4. Note the starch granules in the cells of the cortex and the arrangement of xylem and phloem.
5. Examine the examples of root types. Note their characteristics. **Complete item 2 on the laboratory report.**

the center of the stele are the **vessel cells** of the **xylem.** Between the rays of the xylem are the **sieve tubes** and **companion cells** of the **phloem**.

Root Types

There are three types of roots. **Tap root systems**, have a single dominant root from which branch roots arise. **Fibrous root systems** consist of a number of similar-sized roots that branch repeatedly. Fibrous roots are characteristic of monocots. Both types of root systems may be shallow or deep, depending on the species of plant, but tap roots are capable of the deepest penetration. **Adventitious roots** are unique in that, unlike other roots, they do not grow from the primary root of the embryo. They originate from stems or leaves and are typically fibrous.

STEMS

The stem serves as a connecting link between the roots and the leaves and reproductive organs of flowering plants. It also may serve as a site for food storage. The arrangement of vascular tissue in stems varies among the subgroups of vascular plants.

Monocot Stems

Figure 27.5 illustrates the cross-sectional structure of part of a corn stem. Note the *scattered* **vascular bundles** surrounded by large thin-walled cells, a characteristic of monocots. Each vascular bundle has thick-walled, fibrous cells around its edges, surrounding the large xylem vessels and the smaller sieve tubes and companion cells of phloem. Most of the support for the stem is provided by the xylem and fibrous cells.

Herbaceous Dicot Stems

Figure 27.6 shows the structure of a portion of an alfalfa stem in cross section. Note that the vascular bundles are arranged in a *broken ring* just interior to the **cortex.** This pattern is characteristic of herbaceous dicots. Each vascular bundle is composed of three tissues (from outside): **phloem, vascular cambium,** and **xylem**; the central portion of the stem, the **pith,** is composed of large, thin-walled cells.

Woody Dicot Stems

The structure of a young woody stem in cross section is shown, in part, in Figure 27.7. Note the characteristic *continuous ring* of primary vascular tissue and the arrangement of the tissues within the stem. The **wood** of woody plants is actually xylem tissue.

In each growing season, the vascular cambium forms new (secondary) xylem and phloem. Each growing season is identifiable by an **annual ring** of xylem, which is composed of large-celled **spring wood** (lighter color) and small-celled **summer wood** (darker color). No cells are formed in fall or winter in temperate climates. The stem grows in diameter by the formation and enlargement of new xylem and phloem. Since the growth is actually from the inside of the stem, and since the epidermis and

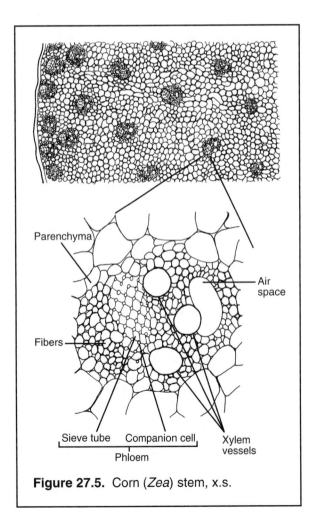

Figure 27.5. Corn (*Zea*) stem, x.s.

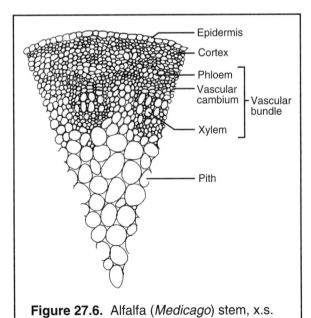

Figure 27.6. Alfalfa (*Medicago*) stem, x.s.

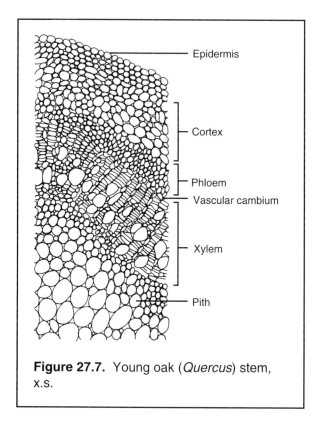

Figure 27.7. Young oak (*Quercus*) stem, x.s.

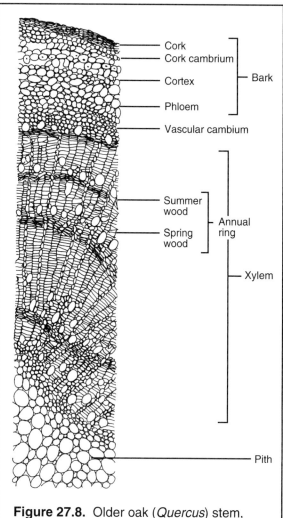

Figure 27.8. Older oak (*Quercus*) stem, x.s.

cortex cannot grow, they tend to fracture and slough off. The **cork cambium** forms in the cortex, and it produces cork cells that assume the function of protecting the underlying tissues and preventing water loss. At this stage, the stem is subdivided into (1) bark, (2) vascular cambium, and (3) wood. The **bark** is everything exterior to the vascular cambium: phloem, cortex, cork cambium, and cork. Figure 27.8 illustrates the structure of an older woody stem.

Materials

Per student
Colored pencils
Compound microscope

Per lab
Tree stem, x.s.
Prepared slides of:
    Quercus (oak) stem, 1 yr, x.s.
    Quercus stem, 4 yr, x.s.
    Medicago (alfalfa) stem, x.s.
    Zea (corn) stem, x.s.

Assignment 3

1. Color-code the xylem and phloem in Figures 27.5, 27.6, and 27.7. In Figure 27.8, color-code the spring wood, summer wood, vascular cambium, phloem, cork, and cork cambium.
2. Examine a prepared slide of *Zea* stem, x.s., and locate the parts shown in Figure 27.5. Note the arrangement of the vascular bundles in the stem. **Complete items 3a and 3b on the laboratory report.**
3. Examine a prepared slide of *Medicago* stem, x.s., and locate the parts shown in Figure 27.6. Note the arrangement of the vascular bundles. **Complete item 3c on the laboratory report.**

4. Examine prepared slides of *Quercus* stems, 1 yr and 4 yr, x.s. Locate the parts shown in Figures 27.7 and 27.8.
5. Determine the age of the demonstration section of tree stem. **Complete item 3 on the laboratory report.**

LEAVES

Leaves are the primary organs of photosynthesis, and they are sites of gas exchange, including water loss by evaporation. The major part of a leaf is the broad, thin **blade** that is attached to a stem by a leaf stalk, the **petiole,** in most dicots. Some leaves, especially in monocots, lack a petiole, so the blade is attached directly to a stem.

Venation

The blade of a leaf is supported by many **veins** that are composed of vascular tissue plus a fibrous vascular bundle sheath. Leaves of monocots have **parallel venation** in which veins are arranged in parallel rows running the length of the leaf. See Figure 27.9e. Leaves of dicots have **net venation** in which a large central vein, the **midrib,** extends the length of the leaf. Veins branching from the midrib branch repeatedly into smaller and smaller veins that form a network throughout the leaf. See Figure 27.9a.

The pattern of veins in net-veined leaves follows one of two major types. In **pinnate** leaves, there is a single midrib with several lateral branches, forming a pattern resembling the structure of a feather. In **palmate** leaves, there are several major veins branching from the petiole, resembling the fingers extending from the hand. If a leaf is divided into a number of leaflets (parts), it is said to be **compound.** If not, it is said to be **simple.** The combination of these patterns determines the type of venation, and it is used in classifying plants. See Figure 27.9.

Internal Structure

The internal structure of a lilac (*Syringa*) leaf is shown in Figure 27.10. An upper and lower **epidermis** form the surfaces of the leaf, and each is coated with a waxy **cuticle** that retards evaporative water loss through the epidermis. Epidermal cells do not contain chlorophyll, except for the scattered **guard cells** that surround tiny openings called **stomata** (stoma is singular). Gas exchange between leaf tissues and the atmosphere occurs through the stomata.

Stomata are often found only on the lower epidermis. This location prevents direct exposure of

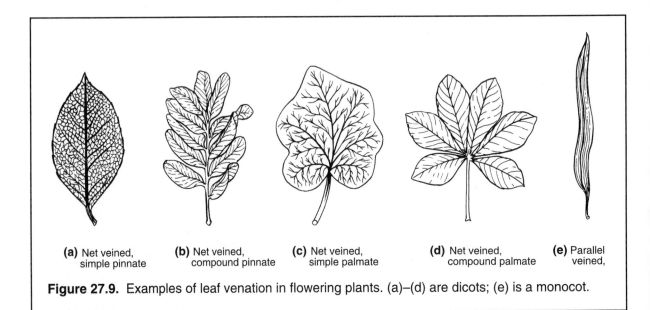

(a) Net veined, simple pinnate   **(b)** Net veined, compound pinnate   **(c)** Net veined, simple palmate   **(d)** Net veined, compound palmate   **(e)** Parallel veined,

Figure 27.9. Examples of leaf venation in flowering plants. (a)–(d) are dicots; (e) is a monocot.

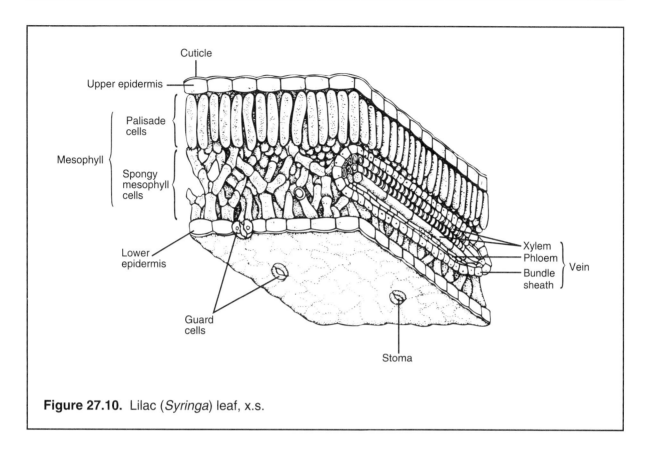

Figure 27.10. Lilac (*Syringa*) leaf, x.s.

stomata to sunlight and helps to reduce evaporative water loss. When guard cells are photosynthesizing, the accumulation of sugars causes them to swell with water, which enters them by osmosis. The swelling bends the guard cells, opening the sto mata, which increases gas exchange. At night, water leaves the guard cells, causing them to straighten out and close the stomata.

The **mesophyll** is the main photosynthesizing tissue in a leaf. It consists of two types. **Palisade mesophyll** consists of closely packed, upright cells located just under the upper epidermis, and it performs most of the photosynthesis. **Spongy meosphyll** consists of a meshwork of cells with many spaces between them, and it occurs in the lower half of a leaf. This arrangement aids the movement of gases within the leaf.

The leaf veins not only provide support but also movement of materials. Xylem brings water and minerals from roots to the leaf cells, and phloem carries away sugar to stem and root cells for use or storage.

Materials

Per student
Colored pencils
Compound microscope

Per lab
Leaves with different types of venation
Leaves with "unknown" venation
Prepared slides of:
    Syringa (lilac) leaf, x.s.
    Syringa leaf epidermis with stomata
    Zea (corn) leaf, x.s.

Assignment 4

1. Examine the leaves with different types of venation. Identify the types of venation for the

"unknowns." **Complete items 4a and 4b on the laboratory report.**

2. Study Figure 27.10. Color the mesophyll and guard cells green.

3. Examine a prepared slide of *Syringa* leaf, x.s. Locate the cells and tissues shown in Figure 27.10. **Complete items 4c–4e on the laboratory report.**

4. Examine a prepared slide of *Syringa* epidermis with stomata. Note the density of the stomata and the arrangement of the guard cells. **Complete items 4f–4g on the laboratory report.**

5. Examine a prepared slide of *Zea* (corn) leaf, x.s. Compare it to the *Syringa* leaf, noting differences and similarities. **Complete item 4h on the laboratory report.**

Assignment 5

Your instructor has set up several slides and plant parts as a mini-practicum. Your task is to identify the structures indicated by numbered pins, tags, or pointers of numbered microscopes. **Complete item 5 on the laboratory report.**

28

Simple Animals

OBJECTIVES

After completion of the laboratory session, you should be able to:
1. Compare the animal groups studied as to their distinguishing characteristics.
2. Identify and classify to phylum members of the groups studied.
3. Identify the major structural components of organisms studied and describe their functions.
4. Define all terms in bold print.

Members of the kingdom **Animalia** are characterized by (1) a **multicellular body** formed of different types of **eukaryotic cells** that lack a cell wall and plastids, (2) **heterotrophic nutrition,** and (3) movement by means of **contractile fibers.** Animals are diploid (2n), and they reproduce sexually by haploid (n) **gametes**—large nonmotile eggs and small motile sperm that are formed by meiotic division. A few simple animals also reproduce asexually.

Figure 28.1 shows the presumed evolutionary relationships of the major phyla of animals. These relationships are based on fundamental similarities and differences that will become clear as you work through the next three exercises.

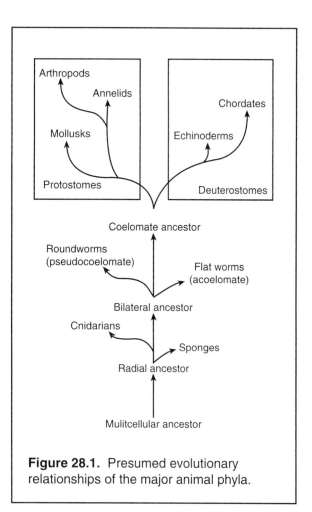

Figure 28.1. Presumed evolutionary relationships of the major animal phyla.

TABLE 28.1
Distinctive Characteristics of Sponges, Cnidarians, Flatworms, and Roundworms

	Sponges	*Cnidarians*	*Flatworms*	*Roundworms*
Symmetry	Asymmetrical or radial symmetry	Radial symmetry	Bilateral symmetry	Bilateral symmetry
Level of organization	Cellular-tissue level	Tissue level	Organ level	Organ system level
Germ layers	None	Ectoderm and endoderm	Ectoderm, mesoderm, and endoderm	Ectoderm, mesoderm, and endoderm
Coelom	N/A	N/A	Acoelomate	Pseudocoelomate
Body plan	N/A	Sac-like	Sac-like	Tube within a tube

N/A = Not applicable.

You will study the major phyla from simplest to most complex. Your primary goal is to learn the recognition characteristics of each group and to understand the major evolutionary advances exhibited by each group.

In this exercise, you will study the simpler phyla of animals. Table 28.1 indicates a few basic characteristics that distinguish sponges, coelenterates, flatworms, and roundworms from one another. Refer to it as you watch for the gradual increase in complexity among these groups.

SPONGES (PHYLUM PORIFERA)

Sponges are the simplest animals. They probably evolved from colonial flagellated protozoans. Most sponges are marine; only a few live in fresh water. The nonmotile adults live attached to rocks or other objects, but the larvae are ciliated and motile. The cells of an adult are so loosely organized that the level of organization is intermediate between the cellular and tissue levels, a **cellular-tissue level.** Sponges are categorized according to which one of three types of **spicules,** or skeletal elements, they possess: chalk (calcium carbonate), glass (silica), or fibrous (protein). Note their basic characteristics in Table 28.1.

Figure 28.2 illustrates the basic structure of a simple sponge. Note the numerous pores in the body wall. The beating flagella of **collar cells** create a steady current of water flowing through the **incurrent pores** into **incurrent canals,** through **pore cells** into **radial canals,** on into the **spongocoel** (central cavity), and out the **os-**

culum. Sponges are filter feeders. Tiny food particles in the water are engulfed by collar cells and digested intracellularly within food vacuoles. Nutrients are then distributed to other cells by diffusion that is aided by wandering amoebocytes. Gas exchange and removal of metabolic wastes occur by diffusion.

Sponges release their gametes into the water, where fertilization occurs. The zygote develops into a motile, ciliated larva that ultimately settles to the bottom, attaches, and grows into an adult sponge.

Materials

Per student
Colored pencils
Dissecting microscope

Per lab
Syracuse dishes
Demonstration microscope setups of:
    Grantia, l.s., at 40×
    Grantia, l.s., showing collar cells at 400×
Preserved *Grantia*

Assignment 1

1. Color-code the collar cells, epidermis, spicules, and amoebocytes in the enlargement of the body wall in Figure 28.2.
2. Examine the demonstration skeletons of representative sponges, noting the type of each one.
3 Place a preserved *Grantia* in a Syracuse dish

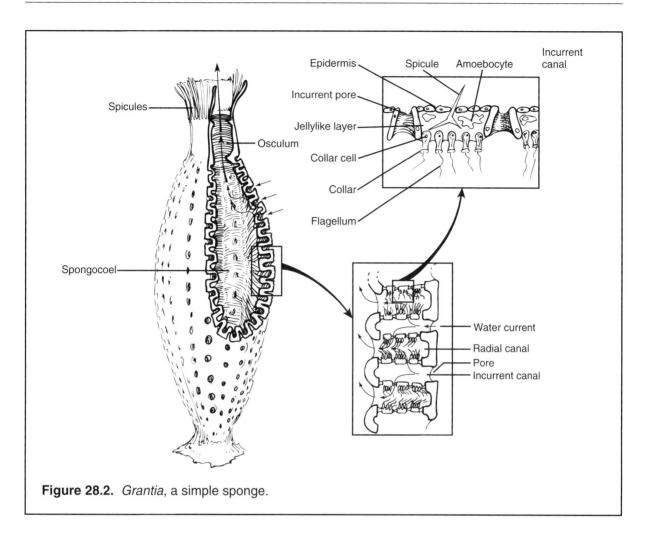

Figure 28.2. *Grantia*, a simple sponge.

and add sufficient water to cover it. Examine it with a dissecting microscope and locate the osculum, spicules around the osculum, and incurrent pores.

4. Examine a slide of sectioned *Grantia* set up under a demonstration microscope. At 40×, locate the incurrent pores, radial canals, spongocoel, and osculum. At 400x, observe the collar cells. **Complete item 1 on Laboratory Report 28 that begins on page 411.**

COELENTERATES (PHYLUM CNIDARIA)

Cnidarians include jellyfish, sea anemones, corals, and hydroids. Most forms are marine, but a few forms occur in fresh water. Note their basic characteristics in Table 28.1.

Cnidarians exhibit two distinct body forms, as shown in Figure 28.3. The tubular body of a **polyp** is attached to the substrate at one end and has a mouth surrounded by tentacles at the other. A **medusa** is a free-swimming form. Medusae have bell-shaped bodies with tentacles hanging from the edge of the bell and a mouth located at its center. Pulsating contractions of the bell produce a swimming motion. Most species of cnidarians occur as either polyps or medusae, but some species exhibit both forms. In such species, a sexually reproducing medusa alternates with an asexually reproducing polyp, as shown in Figure 28.4.

The structure of *Hydra*, a freshwater form, illustrates the basic coelenterate structure. See Figure 28.5. The body wall consists of two cell layers, an **epidermis** (ectoderm) and a **gastrodermis** (endoderm), separated by a noncellular

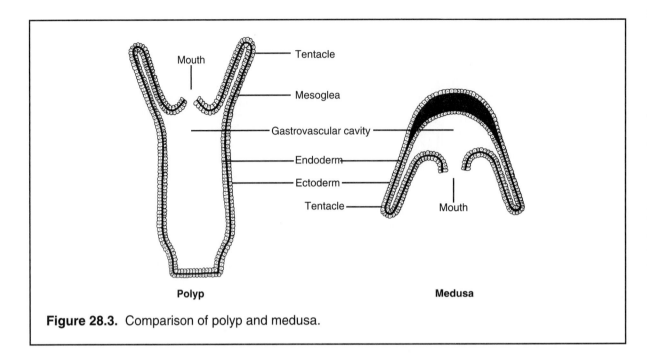

Figure 28.3. Comparison of polyp and medusa.

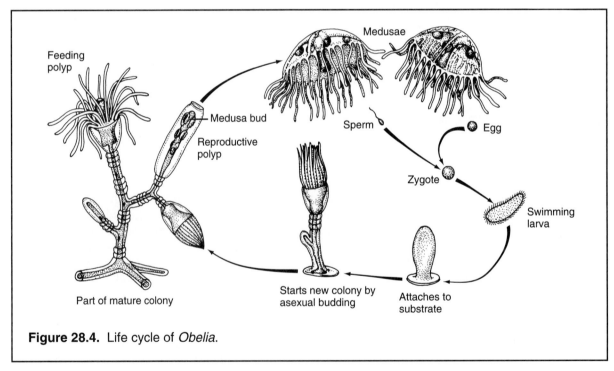

Figure 28.4. Life cycle of *Obelia*.

mesoglea. Contraction of **myofibrils** in cells of each layer enables *Hydra* to move. A **nerve net**, formed of interconnected neurons, lies within the mesoglea and coordinates body functions.

Tentacles of all cnidarians contain **stinging cells** (cnidocytes) that eject dart-like projectiles (nematocysts) for offense and defense. Prey organisms are paralyzed by the stinging darts and engulfed through the mouth into the **gastrovascular cavity,** where partial digestion occurs. Then the partially digested particles are engulfed by gastrodermal cells, and digestion is

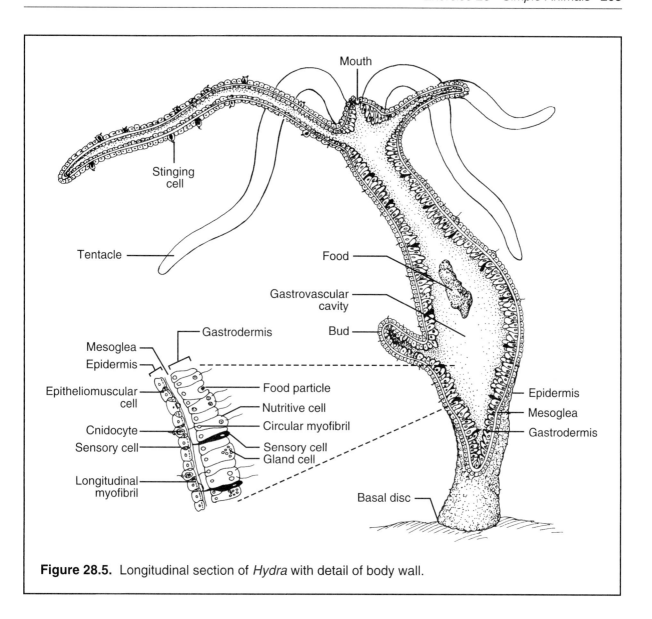

Figure 28.5. Longitudinal section of *Hydra* with detail of body wall.

completed within food vacuoles. The distribution of nutrients, gas exchange, and removal of metabolic wastes occur by diffusion.

Hydra reproduces asexually by budding as well as sexually by gametic reproduction. A *Hydra* develops an egg within an ovary and sperm within a testis. The sperm use their single flagellum to swim to the egg through the water. After fertilization, the zygote develops into a new individual.

Materials

Per student
Colored pencils
Compound microscope

Per lab
Cultures of:
   Daphnia, small
   Hydra
Demonstration setups of:
   Hydra in small aquarium feeding on small *Daphnia*
   Representative cnidarians
Demonstration microscope setup of a tentacle of a living *Hydra,* showing stinging cells
Prepared slides of:
   Hydra, l.s.
   Obelia medusae, w.m.
   Obelia polyps, w.m.

Assignment 2

1. Color the epidermis green and the gastrodermis yellow in the enlargement of the *Hydra* body wall in Figure 28.5.
2. Examine a prepared slide of *Hydra* l.s., and locate as many of the structures labeled in Figure 28.5 as you can.
3. Examine prepared slides of *Obelia* polyps and medusae. Locate the tentacles and mouth in each body form. Note the buds forming on the asexually reproducing polyps.
4. Observe the living *Hydra* in the small aquarium (or beaker) as they feed on *Daphnia*. Note their feeding strategy and behavior. Do some *Hydra* have buds?
5. Examine the representative coelenterates. Note which are polyps and which are medusae.
6. Examine a tentacle of a living *Hydra* set up under a demonstration microscope. Locate a stinging cell and its hairlike trigger. **Complete item 2 on the laboratory report.**

FLATWORMS (PHYLUM PLATYHELMINTHES)

Flatworms have a sac-like body plan like coelenterates, but their dorsoventrally flattened body is **bilaterally symmetrical** with anterior and posterior ends. All three embryonic tissues are present, and the embryo develops into an adult with an **organ level of development.** See Table 28.1.

All flatworms are **hermaphroditic.** Sperm is exchanged with another flatworm during copulation, but self-fertilization may also occur, especially in parasitic species. Internal fertilization produces fertile eggs laid by the female. Many flatworms are free-living, and they feed on nonliving organic material. Some are important parasites of humans and other vertebrates.

Free-Living Flatworms

A common freshwater planarian, *Dugesia*, is typical of nonparasitic forms of flatworms. See Figures 28.6 and 28.7. When feeding, the **pharynx** is extended through a midventral opening, the **mouth,** to engulf food into a branched **gastrovascular cavity.** Digestion begins within the gastrovascular cavity and is completed within gastrodermal cells, as in coelenterates. Nutrients are distributed by diffusion.

Neurons in flatworms are more specialized than those in coelenterates, and are organized into specific structures that are concentrated at the anterior end. The **eyespots** of flatworms are photoreceptors, and chemical and tactile receptors are located on the auricles. Body movements are coordinated by the anterior **cerebral ganglia** ("brain") and two **ventral nerve cords** that extend posteriorly. Movement in *Dugesia* results from the beating cilia of epidermal cells on the ventral body surface.

Specialized excretory organs first appear in flatworms. Metabolic wastes are collected via diffusion into **excretory canals.** The beating of long cilia of numerous flame cells creates a current propelling the liquid wastes out of the body through **exretory pores** scattered over the body surface.

Parasitic Flatworms

Tapeworms and **flukes** are flatworms that parasitize humans and other vertebrates. Their life cycles include one or two intermediate hosts and a final host in which the adults live.

Tapeworm adults live in the small intestine of their hosts, where they attach to the intestinal wall by hooks and/or suckers of a **scolex.** Nutrients are absorbed through the body wall; a gastrovascular cavity is absent.

A tapeworm body consists of a series of repeating units, called **proglottids,** that are continuously produced by the scolex. When mature, proglottids are self-fertilizing, reproductive factories. When proglottids become filled with fertile eggs, they separate from the posterior end of the tapeworm and pass from the host in feces.

Intermediate hosts (herbivores) are infected by accidentally eating the eggs. Larvae hatch from the eggs in the intermediate host's intestine, burrow into blood vessels, and are carried by blood to skeletal muscles, where they encyst. When a carnivore or omnivore eats muscles of the intermediate host infested with encysted tapeworm larvae (bladder worms), the larvae are activated and attach to the new final host's intestine. See Figure 28.8.

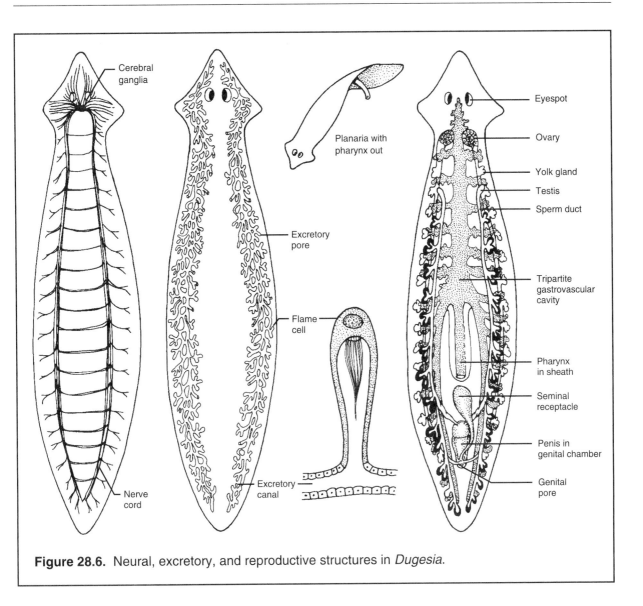

Figure 28.6. Neural, excretory, and reproductive structures in *Dugesia*.

Adult flukes infest the lungs, liver, and blood of vertebrates, including humans. Intermediate hosts in the life cycle often include snails and fish. For example, the human liver fluke is prevalent in portions of Asia where sanitation is poor and the eating of raw or undercooked freshwater fish is practiced.

Materials

Per student
Compound microscope
Dissecting microscope

Per lab
Camel's hair brushes

Watch glasses
Culture of *Dugesia*
Representative flatworms
Demonstration setup of planaria's response to light
Preserved specimens:
   Ascaris
   Tapeworms
Prepared slides of:
   Dugesia, w.m., injected gastrovascular cavity
   Taenia sp. scolex, w.m.
   Taenia sp. larvae (cysticerci) encysted in
      muscle
   Taenia sp. gravid proglottid

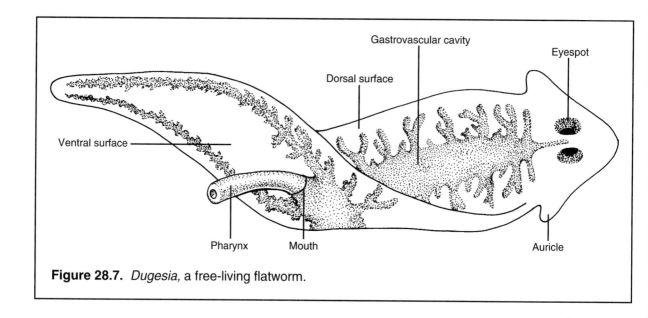

Figure 28.7. *Dugesia,* a free-living flatworm.

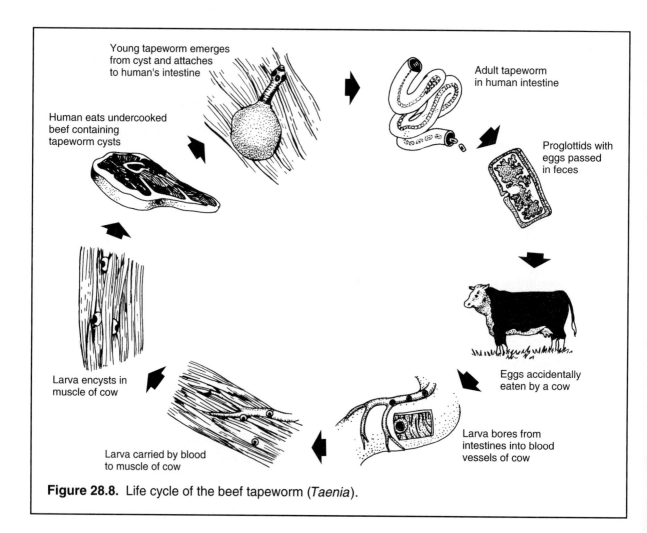

Figure 28.8. Life cycle of the beef tapeworm (*Taenia*).

Assignment 3

1. Use a camel's hair brush to place a living planarian in a watch glass with a small amount of water. Examine it with a dissecting microscope. Note the eyespots, auricles, and manner of movement. Touch the posterior end gently with a pencil and note the response. Touch an auricle with a pencil and note the response.
2. Examine a prepared slide of *Dugesia* and note the highly branched gastrovascular cavity.
3. Your instructor has placed a few *Dugesia* in water in a partially shaded Petri dish. Lift the light shield to observe their distribution. Do they prefer shade or light? **Complete items 3a-3d on the laboratory report.**
4. Examine prepared slides of *Taenia* encysted larvae, scolex, and gravid proglottids. Note the hooks and suckers on the scolex and the dark eggs in the expanded uterus of the gravid proglottid. Study the life cycle of the beef tapeworm in Figure 28.8. **Complete items 3e and 3f on the laboratory report.**
5. Observe the demonstration flatworms. Estimate the lengths of the tapeworms. Note the infestation sites of the adult flukes. **Complete item 3 on the laboratory report.**

ROUNDWORMS (PHYLUM NEMATODA)

Roundworms are advanced over flatworms by having an **organ system level** of development, a **tube-within-a-tube body plan,** and a **false coelom.** Animals with a tube-within-a-tube body plan have a complete digestive tract that has two openings: a **mouth** at one end and an **anus** at the other. The space between the digestive tract and the body wall is the **coelom** or body cavity. It is a **false coelom** (pseudocoelom) in roundworms, because it is not completely lined with mesodermal tissue. See Figure 28.9. The sexes of roundworms are separate. Most roundworms are freeliving, but some are important parasites of plants and animals.

Ascaris is a large roundworm parasite of swine and humans. Adults live in the small intestine, where they feed on partially digested

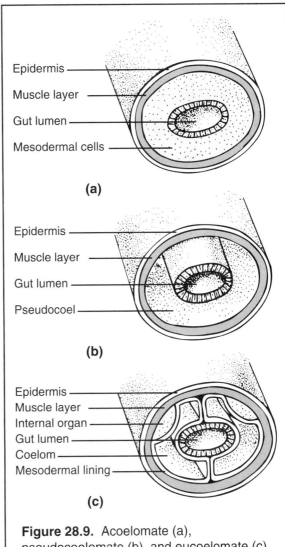

Figure 28.9. Acoelomate (a), pseudocoelomate (b), and eucoelomate (c), conditions in bilaterally symmetrical animals.

food. An external cuticle protects *Ascaris* from digestive juices of the host. Dorsal and ventral nerve cords coordinate body movements. Metabolic wastes are collected in laterally located excretory canals and excreted through an excretory pore near the anterior end of the body. See the internal structure of a female in Figure 28.10. Note that the reproductive system is the dominant feature, a common trait of endoparasites. Each tubular half of the bipartite uterus is continuous with an oviduct and ovary.

A female *Ascaris* is about 30 cm long. A male is smaller and has a curved posterior end.

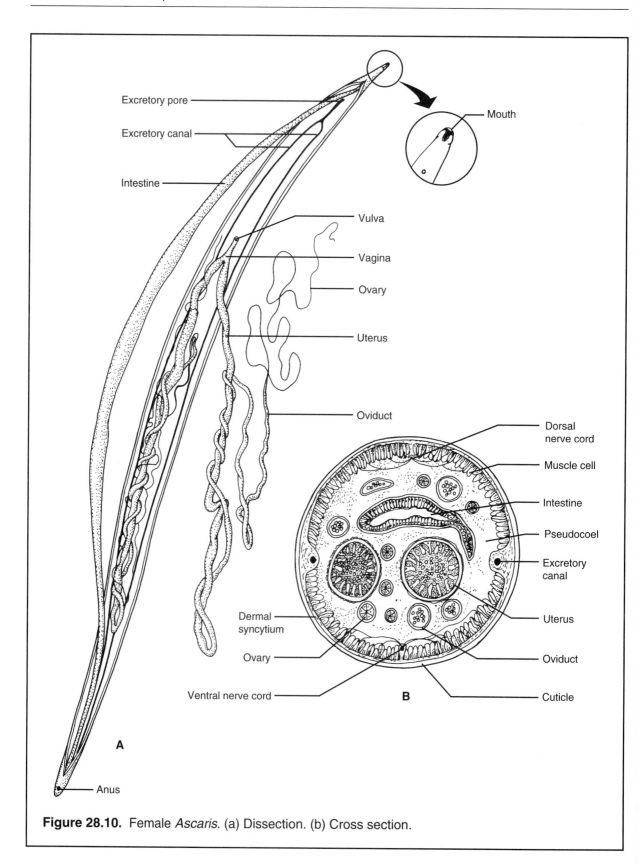

Figure 28.10. Female *Ascaris*. (a) Dissection. (b) Cross section.

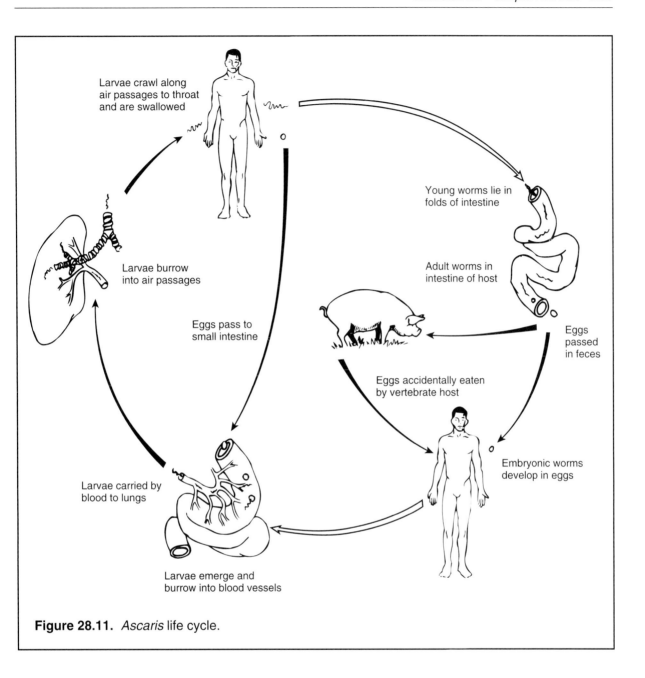

Figure 28.11. *Ascaris* life cycle.

Labels in figure:
- Larvae crawl along air passages to throat and are swallowed
- Young worms lie in folds of intestine
- Larvae burrow into air passages
- Adult worms in intestine of host
- Eggs pass to small intestine
- Eggs passed in feces
- Eggs accidentally eaten by vertebrate host
- Larvae carried by blood to lungs
- Embryonic worms develop in eggs
- Larvae emerge and burrow into blood vessels

After mating in the small intestine, the female releases eggs that are passed in the host's feces. New hosts are infected by accidentally eating the eggs. Note the life cycle in Figure 28.11.

A number of important roundworm parasites infest humans. Infestation by one of these, *Trichinella spiralis,* occurs by eating undercooked pork or wild game in which microscopic encysted larvae are present. The larvae emerge in the intestines, attach to the intestinal wall, and grow into adults that feed by sucking blood from the intestinal wall. After copulation, females release larvae that burrow into blood vessels of the intestine. The larvae are carried by the blood to skeletal muscles, where they burrow through the vessel walls and encyst within muscle tissue. Note that each host is infested with both adult and larval forms.

Materials

Per student
Compound microscope
Dissecting instruments

Per lab
Dissecting pans, wax bottomed
Dissecting pins
Medicine droppers
Microscope slides and cover glasses
Culture of *Turbatrix*
Demonstration setups of:
  Ascaris, dissected female
  representative roundworms
Preserved *Ascaris*
Prepared slides of:
  Ascaris, x.s.
  Trichinella, encysted larvae
Representative roundworms

Assignment 4

1. Make a slide of a drop of *Turbatrix* culture and observe these free-living roundworms that can live in vinegar. Note their swimming motion. **Complete item 4a on the laboratory report.**
2. Examine the preserved male and female *Ascaris*. Note their sizes and distinctive shapes.
3. Examine the demonstration dissected female *Ascaris*. Locate the parts shown in Figure 28.10a.
4. Examine a prepared slide of *Ascaris* female, x.s., and locate the structures labeled in Figure 28.10b. **Complete items 4b-4d on the laboratory report.**
5. Examine a prepared slide of *Trichinella* larvae encysted in muscle tissue. Can you see the larvae with the naked eye? **Complete item 4 on the laboratory report.**

Assignment 5

Review a few of the characteristics of the phyla studied by completing item 5 on the laboratory report.

Assignment 6

Your instructor has set up several specimens for you to identify as a mini-practicum. Your task is to identify the phylum to which each specimen belongs. **Complete item 6 on the laboratory report.**

29

Mollusks, Annelids, and Arthropods

OBJECTIVES

After completion of the laboratory session, you should be able to:
1. Compare the animal groups studied as to their distinguishing characteristics.
2. Identify and classify to phylum and class the members of the groups studied.
3. Identify the major structural components of organisms studied and describe their functions.
4. Define all terms in bold print.

Mollusks, annelids, and arthropods compose the **protostomates,** which are so named because the blastopore of the embryo becomes the mouth of the adult. Additionally, all protostomates share these characteristics:

1. Organ system level of development
2. Bilateral symmetry
3. Tube-within-a-tube body plan
4. True coelom (lined with mesodermal tissue)
5. Extracellular digestion in the digestive tract.
6. Transport of materials by a circulatory system

In this exercise, your primary goal is to learn to recognize members of each phylum and to understand the major characteristics of each group.

MOLLUSKS (PHYLUM MOLLUSCA)

The molluscan body plan is characterized by (1) a **ventral muscular foot,** (2) a **dorsal visceral mass,** and (3) a **mantle** that envelops the visceral mass and usually secretes a **shell.** In addition, most mollusks have a **radula,** a file-like mouthpart unique to mollusks, that is used to scrape off bits of food. Table 29.1 lists the distinguishing characteristics of the molluscan classes.

Most molluscan characteristics are shown in the structure of a clam in Figure 29.1. The dorsal visceral mass is enclosed within a mantle, and the muscular foot extends ventrally. The mantle secretes an exterior shell composed of two valves (halves) that are hinged dorsally. The valves of the shell may be tightly closed by the adductor muscles. The foot is used to dig into mud or sand and bury the clam, often with only the siphons extending above the substrate. Muscle contractions are controlled by the nervous system.

Unlike other molluscan classes, pelecypods—the class to which the clam belongs—are filter feeders. Water is brought into the mantle cavity

TABLE 29.1
Distinguishing Characteristics of Molluscan Classes

Class	Characteristics
Monoplacophora (*Neopilina*)	Primitive; remnants of segmentation; single shell; dorsoventrally flattened foot; radula; marine
Polyplacophora (chitons)	Shell of eight overlapping plates; dorsoventrally flattened foot; radula; marine
Scaphapoda (tooth shells)	Conical shell open at each end; digging foot; radula; marine
Gastropoda (snails and slugs)	Shell absent or single, often coiled; dorsoventrally flattened foot; radula; marine, freshwater, and terrestrial
Bivalvia (clams, oysters)	Shell of two valves, hinged dorsally; laterally flattened food; no radula; filter feeder; marine or freshwater
Cephalopoda (octopi, squids)	Single shell or none; foot modified to form tentacles; image-producing eyes; horny beak and radula; predaceous; water-jet propulsion; marine

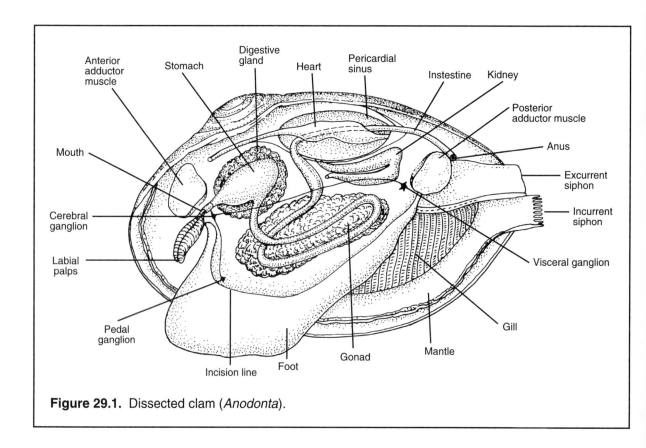

Figure 29.1. Dissected clam (*Anodonta*).

via the **incurrent siphon,** and it exits via the **excurrent siphon.** As water flows over the gills, gas exchange occurs between the water and blood in the **gills,** and tiny food particles are trapped in mucus on the gill surfaces. Both mucus and en-

trapped food are carried by beating cilia to the **labial palps** and on into the **mouth.**

Digestion is completed in the digestive tract, and nutrients are absorbed into the blood for transport. The **open circulatory system** empties blood into tissue spaces from which the blood slowly flows back to the heart.

Materials

Per student
Colored pencils

Per lab
Demonstration dissections of a freshwater clam
Representative mollusks that are numbered but
  not labeled
Squid, fresh or frozen

Assignment 1

1. Color-code the digestive tract, heart and blood vessels, and kidney of the clam in Figure 29.1.
2. Examine the demonstration dissected clam and locate as many of the structures labeled in Figure 29.1 as possible.
3. Examine a squid and note how it is adapted for a predaceous mode of life. The foot is modified into arms and tentacles with suction cups. The body is streamlined, and swimming is powered by water-jet propulsion as water is ejected forcefully from the siphon. The shell is reduced to a thin supporting membrane, and the image-forming eyes resemble vertebrate eyes but are developed differently.
4. Examine the representative mollusks and use Table 29.1 to determine the class of each. **Complete item 1 on Laboratory Report 29 that begins on page 413.**

ANNELIDS (PHYLUM ANNELIDA)

The major characteristic of the phylum annelida is a **segmented body** formed of repeating units called **somites.** Most of the somites are basically similar to one another, including the arrangement of many internal structures. Most annelids are marine, but some occur in freshwater and terrestrial habitats. Table 29.2 lists the characteristics of the major classes of annelids.

The most familiar annelids are earthworms (class Oligochaeta), which literally eat their way through the soil, digesting whatever organic material is ingested. Each somite of the earthworm has four pairs of **setae** (bristles) that provide traction for a crawling worm. The smooth band around the body is the **clitellum.** It secretes a mucus band around copulating worms and later secretes an egg case for the fertile eggs.

Figures 29.2, 29.3, and 29.4 show the internal structures of an earthworm. Note the specializations of the anterior portion of the digestive tract. For example, the **pharynx** aids the ingestion of food, the crop temporarily holds food, and the **gizzard** grinds food into tiny particles. Digestion is completed in the intestine, and nutrients are absorbed into the blood for transport. Annelids have a **closed circulatory system,** and their blood is therefore always contained within blood vessels.

Longitudinal and circular muscles occur in the body wall and in the wall of the digestive tract. Contractions are coordinated by the nervous system. Metabolic wastes are removed by a pair of **nephridia** in each somite. Gas exchange occurs by diffusion through the moist epidermis.

TABLE 29.2
Distinguishing Characteristics of Annelid Classes

Class	Characteristics
Polychaeta	Head with simple eyes and tentacles; segments with lateral extensions (parapodia) and many bristles (setae); sexes usually separate; predaceous; marine
Oligochaeta	No head; segments without extensions and with few, small bristles; hermaphroditic; terrestrial or freshwater
Hirudinea (leeches)	No head; segments with superficial rings and without lateral extensions or bristles; anterior and posterior suckers; hermaphroditic; parasitic, feeding on blood; terrestrial or freshwater

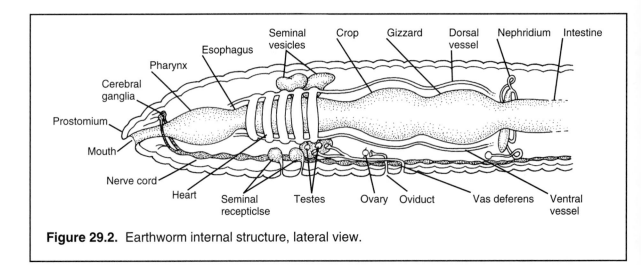

Figure 29.2. Earthworm internal structure, lateral view.

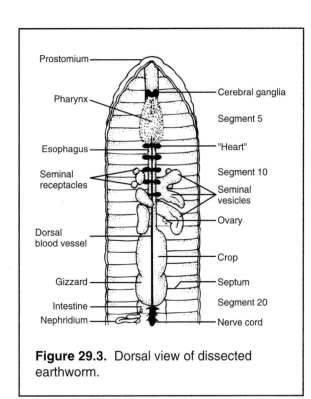

Figure 29.3. Dorsal view of dissected earthworm.

Earthworms are hermaphroditic, and they exchange sperm during copulation. **Seminal vesicles** hold sperm formed by testes until they are released during copulation. **Seminal receptacles** receive sperm during copulation and later release sperm to fertilize eggs as they are released from the oviducts.

The nervous system includes a "brain" (supraesophageal ganglia) and a double **ventral nerve cord** with **segmental ganglia** and lateral nerves.

Materials

Per student
Colored pencils
Dissecting instruments

Per lab
Dissecting pans, wax bottomed
Dissecting pins
Representative annelids that are numbered but not labeled
Preserved earthworms

Assignment 2

1. Study the internal structure of an earthworm in Figures 29.2–29.4. Color-code the digestive tract, blood vessels, and nephridia in Figures 29.2 and 29.4.
2. **Complete items 2a-2c on the laboratory report.**
3. Obtain an earthworm and place it in your dissecting pan. Locate the somites, mouth, anus, setae, and clitellum.
4. Dissect the earthworm as follows. Locate the structures shown in Figure 29.2.
   a. Pin the earthworm dorsal (convex) side up to the bottom of the dissecting pan by placing a pin through the prostomium, a small projection over the mouth, and another pin through a somite posterior to the clitellum.

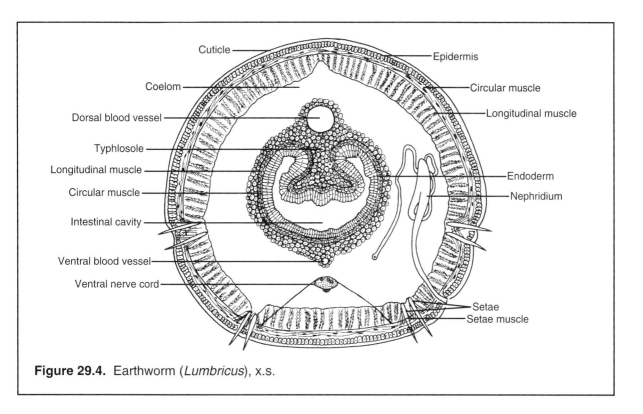

Figure 29.4. Earthworm (*Lumbricus*), x.s.

b. Use a sharp scalpel to cut through the body wall *only*, along the dorsal midline from the prostomium to the clitellum. *Be careful not to damage underlying organs.*

c. Use a dissecting needle to break the septa separating the somites as you spread out and pin down the body wall to expose the internal organs.

d. Add water to cover the worm in order to prevent it from drying out.

e. Locate the structures labeled in Figures 29.2–29.4. The ovaries and testes and their associated ducts are tiny and will be difficult to see.

f. Carefully use a dissecting needle and forceps to remove tissue in order to expose the brain.

g. Remove the crop, gizzard, and part of the intestine to expose the ventral nerve cord with its segmental ganglia.

h. Cut open the crop and gizzard and compare the thickness of their walls.

i. Dispose of the worm as directed by your instructor.

5. Examine the representative annelids and use Table 29.2 to determine the class of each. **Complete item 2 on the laboratory report.**

ARTHROPODS (PHYLUM ARTHROPODA)

There are more species of arthropods than all other animals combined. The characteristics enabling such success include a well-developed nervous system, complex sense organs, and a fusion of body segments to form specialized body regions. In addition, all arthropods exhibit these distinguishing characteristics: (1) **jointed appendages,** (2) an **exoskeleton** of chitin, (3) **compound** or **simple eyes,** and (4) an **open circulatory system.** Table 29.3 lists the characteristics of the major classes of arthropods.

Crayfish (Crustacea)

The freshwater crayfish exhibits the basic arthropod and crustacean characteristics. Note its external structure in Figure 29.5. The segmented body is divided into a **cephalothorax,** covered by the carapace, and an **abdomen.** Each body segment has a pair of jointed appendages, and most are modified for special functions. For example, **antennae** contain chemical and tactile sensory receptors, **compound eyes** are photoreceptors, **mouthparts**

TABLE 29.3
Distinguishing Characteristics of Arthropod Classes

Class	Characteristic
Arachnida (spiders, scorpions)	Cephalothorax and abdomen; four pairs of legs; simple eyes; no antennae; mostly terrestrial
Crustacea (crabs, shrimp)	Cephalothorax and abdomen; compound eyes; two pairs of antennae; five pairs of legs; mostly marine or freshwater
Chilopoda (centipedes)	Elongate, dorsoventrally flattened body; each body segment with a pair of legs; one pair of antennae; simple eyes; terrestrial
Diplopoda (millipedes)	Elongate, dorsally convex body; segments fused in pairs, giving *appearance* of two pairs of legs per segment; one pair of antennae; simple eyes; terrestrial
Insecta (insects)	Head, thorax, and abdomen; three pairs of legs on thorax; wings, if present, on thorax, simple and compound eyes; one pair of antennae; freshwater or terrestrial

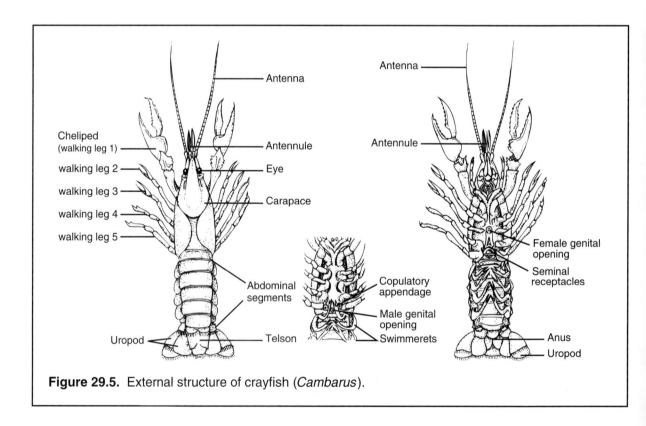

Figure 29.5. External structure of crayfish (*Cambarus*).

are used in feeding, **walking legs** enable movement and grasping, and **uropods** are used in swimming. The first pair of walking legs, the **chelipeds,** is modified for capturing food. The **swimmerets** present on abdominal appendages are relatively unmodified, but the first two pairs are modified for sperm transfer in males.

The exoskeleton is hardened for protection, but it remains flexible at joints, enabling movement. It is shed periodically to allow growth. The carapace covers the **gills** (gas exchange organs) that are located laterally and attached to either the body wall or the bases of the walking legs.

Figure 29.6 shows the internal structure of a crayfish. Note the **heart** and **arteries** of the

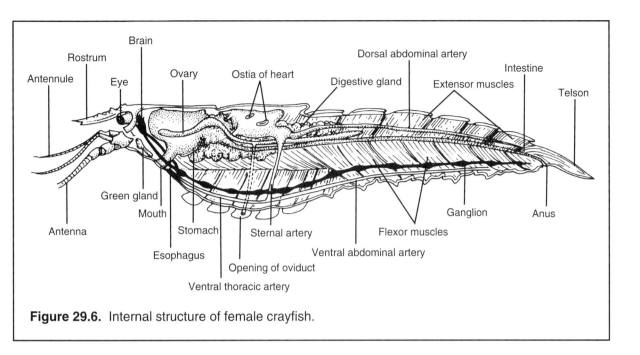

Figure 29.6. Internal structure of female crayfish.

open circulatory system. Blood is emptied into the coelom and slowly returns to reenter the heart through tiny valved openings, the **ostia.** A pair of elongated **gonads,** either ovaries or testes, lies just ventral to the heart. A large **digestive gland** surrounds the gonads and **stomach,** filling much of the cephalothorax. Metabolic wastes are removed by **green glands** located near the bases of the antennae. The highly developed nervous system includes a **"brain"** (supraesophageal ganglia) and a double ventral nerve cord with segmental ganglia and lateral nerves.

Grasshopper (Insecta)

A grasshopper illustrates typical insect characteristics. Note in Figure 29.7 that the body is divided into three distinct parts: a **head, thorax,** and **abdomen.** The head contains both simple and compound eyes, only one pair of antennae, and the **mouthparts,** which are modified appendages. The three pairs of legs and two pairs of wings are attached to the thorax. (Not all insects have wings.) Visible on the abdomen are the **tympanum,** a sound receptor comparable to your eardrum, and **spiracles,** which are openings into the **tracheal system,** a complex of tubules that carry air directly to body tissues.

Internally, the digestive tract is divided into crop, stomach, and intestine. **Gastric caeca** are glands that aid digestion. Metabolic wastes are removed from the hemolymph by **Malpighian tubules** into the intestine. Hemolymph is pumped by a series of "hearts" anteriorly through the "aorta" and emptied into the coelom. It then slowly returns through the coelom to the "hearts." Note the double **ventral nerve cord** with **segmental ganglia** and the **"brain"** (supraesophageal ganglia).

Most insects undergo **metamorphosis,** a hormonally controlled process that changes the body form of the insect at certain stages of its life cycle. In **complete metamorphosis,** as in butterflies, there are four life stages: egg; larva, a wormlike feeding stage; pupa, a nonfeeding stage; and adult. In **incomplete metamorphosis,** as in grasshoppers, the pupal stage is absent. The egg hatches into a nymph (in terrestrial forms) or naiad (in aquatic forms) that grows through several molts, with the adult emerging from the last molt.

Materials

Per student
Colored pencils

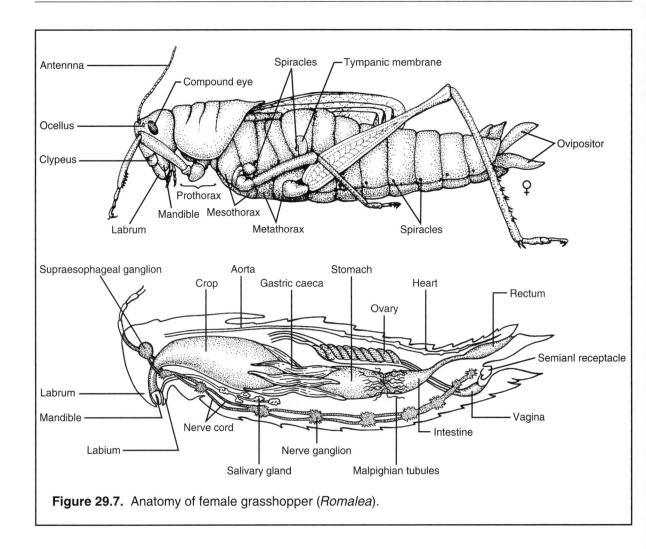

Figure 29.7. Anatomy of female grasshopper (*Romalea*).

Per lab

Representative arthropods that are labeled

Representative arthropods that are numbered but not labeled

Preserved crayfish and grasshoppers

Assignment 3

1. **Complete item 3a on the laboratory report**
2. Examine a preserved crayfish and locate the external structures shown in Figure 29.5. Examine an eye with a dissecting microscope. Note the individual facets forming this compound eye. Note the differences of the walking legs. **Complete item 3b on the laboratory report.**
3. Color-code the digestive tract, heart and blood vessels, and green gland of the crayfish in Figure 29.6, and study its internal structure.
4. Dissect a crayfish as described below and locate the parts labeled in Figure 29.6.
   a. Hold the crayfish dorsal side up with its head facing away from you. Use scissors to make incision 1 as shown in Figure 29.8. This removes the left side of the carapace, exposing the gills.
   b. Note the gills. How are they arranged? To what are they attached? Note the thin, transparent body wall medial to the gills.
   c. Make incision 2 by extending incision 1 from the cervical groove to the eye and

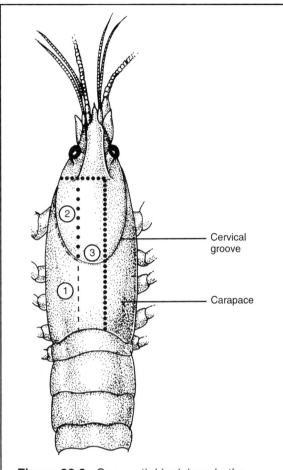

Figure 29.8. Sequential incisions in the crayfish dissection.

downward to remove the left anterior lateral part of the carapace. This exposes the lateral wall of the stomach and a large mandibular muscle.

d. Make incision 3 as shown, and *gently* lift off the median strip of carapace to expose the heart. Note the arteries extending from the heart and the ostia.

e. Carefully remove organs as you work your way from dorsal to ventral, identifying each organ encountered. When the nerve cord is exposed, locate the segmental ganglia.

f. Make a transverse cut through the abdomen and examine the cut surface. Locate the muscles, intestine, and nerve cord.

g. Dispose of your crayfish as directed by your instructor.

5. Obtain a preserved grasshopper. Locate the external structures shown in Figure 29.7. **Complete item 3c on the laboratory report.**

6. Examine the labeled representative arthropods. Compare them with Table 29.3.

7. Use Table 29.3 to determine the class of the "unknowns." **Complete item 3 on the laboratory report.**

30

Echinoderms and Chordates

OBJECTIVES

After completion of the laboratory session, you should be able to:
1. Compare the animal groups studied as to their distinguishing characteristics.
2. Classify to phylum and class members of the groups studied.
3. Identify the major structural components of organisms studied and describe their functions .
4. Define all terms in bold print.

Echinoderms and chordates are called **deuterostomates** because the embryonic blastopore becomes the anus, and the mouth is formed from the second embryonic opening to appear. This embryonic characteristic suggests that echinoderms are more closely related to chordates than to protostomates.

ECHINODERMS (PHYLUM ECHINODERMATA)

These "spiny-skinned" animals are named after the spines that protrude from their calcareous **endoskeleton,** which lies just under a thin epi-dermis. All echinoderms are marine. A unique characteristic is their **water vascular system.** Although larval stages are bilaterally symmetrical, adults are radially symmetrical. Adults have an **oral surface,** where the mouth is located, and an **aboral surface,** where the anus is located. There are no anterior or posterior ends nor dorsal or ventral surfaces. The characteristics of the classes of echinoderms are listed in Table 30.1.

Note the echinoderm characteristics in the dissected sea star in Figure 30.1. The typical pentamerous (five-part) structure is obvious by the five **arms** that radiate from a **central disc.** The water vascular system consists of interconnecting tubes filled with seawater. Seawater is filtered as it enters through the **madreporite. Radial canals** extend into each arm from the **ring canal,** and the **lateral canals** end in **tube feet** that have suction cups at their ends. Contraction and relaxation of an **ampulla** extends and retracts a tube foot. The oral surface of each arm has an **ambulacral groove** that is bordered by tube feet.

A very short digestive tract of mouth, stomach, intestine, and anus is housed in the central disc, but a pair of digestive glands fills most of each arm. A pair of gonads is also located in each arm. Sexes are separate. Nervous and circulatory systems are greatly reduced.

TABLE 30.1
Classes of Echinoderms

Class	Characteristics
Asteroidea (sea stars)	Star-shaped forms with broad-based arms; ambulacral grooves with tube feet that are used for locomotion
Ophiuroidea (brittle stars)	Star-shaped forms with narrow-based arms that are used for locomotion; ambulacral grooves covered with ossicles or absent
Echinoidea (sea urchins)	Globular forms without arms; movable spines and tube feet for locomotion; mouth with five teeth
Holothuroidea (sea cucumbers)	Cucumber-shaped forms without arms or spines; tentacles around mouth
Crinoidea (sea lilies)	Body attached by stalk from aboral side; five arms with ciliated ambulacral grooves and tentacle-like tube feet for food collection; spines absent

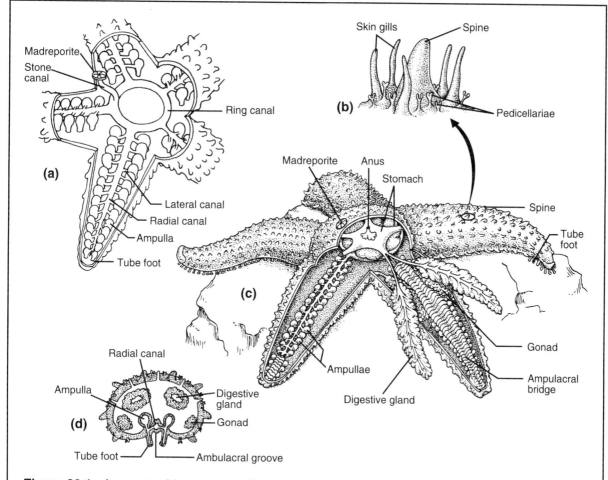

Figure 30.1. A sea star (*Asterias*). (a) Water vascular system. (b) Skin gills and pedicellariae. (c) Dissection. (d) Arm cross section.

Materials

Per student
Colored pencils
Dissecting instruments
Dissecting pans
Preserved sea star

Per lab
Representative echinoderms that are numbered but not labeled

Assignment 1

1. **Complete items 1a and 1b on Laboratory Report 30 that begins on page 415.**
2. Study the structure of a sea star in Figure 30.1 and color-code the water vascular system.
3. Obtain a sea star and note its external structure. Perform the dissection as follows to locate the structures labeled in Figure 30.1.
   a. Use scissors to cut off one of the arms. Examine the cut end and locate the structures labeled in Figure 30.1d.
   b. Locate the anus and madreporite on the aboral surface of the central disc, and the mouth on the oral surface.
   c. Use scissors to remove the aboral surface of the central disc. Locate the short intestine, stomach, and attachment of the digestive glands.
   d. Use scissors to remove the aboral surface of one of the arms, and locate the digestive glands and gonads. Then locate the ambulacral bridge and the ampulla of the tube feet.
   e. Remove the stomach to locate the ring canal.
   f. Dispose of the sea star as indicated by your instructor.
4. Examine the representative echinoderms and use Table 30.1 to identify the classes to which they belong. **Complete item 1 on the laboratory report.**

CHORDATES (PHYLUM CHORDATA)

The four distinctive characteristics of chordates are present in the embryo but may not persist in the adult. They are: (1) a **dorsal tubular nerve cord,** (2) **pharyngeal gill slits,** (3) a **noto-chord,** and (4) a **post-anal tail**. There are three major groups of chordates: tunicates, lancelets, and vertebrates.

Tunicates (Subphylum Urochordata)

All chordate characteristics are present in the motile larvae but some are absent in the non-motile adults. See Figure 30.2. Both larval and adult tunicates (sea squirts) are filter feeders with pharyngeal gills providing both gas exchange and food collection. Collected food particles are passed by cilia into the stomach.

Lancelets (Subphylum Cephalochordata)

In contrast to the tunicates, all chordate characteristics are present in adult lancelets, such as amphioxus, shown in Figure 30.3. The notochord, an evolutionary forerunner of a vertebral column, provides support for the body, and the segmentally arranged muscles enable a weak swimming motion. Note the simple dorsal and caudal fins.

Amphioxus lives buried in mud with its anterior end sticking out. Water is brought in through the buccal cavity, passes over the gills into the atrium, and flows out the atriopore. Like tunicates, lancelets are filter feeders. The pharyngeal gills serve as both gas exchange organs and food collectors. Food particles are moved by cilia into the intestine.

Vertebrates (Subphylum Vertebrata)

You are most familiar with this group of animals. Adult vertebrates may not exhibit all of the four chordate characteristics, although all four are present in their embryos. A vertebrate body is usually divided into **head, trunk,** and **tail,** but a tail is not present in all adults. Most forms have **pectoral** and **pelvic appendages.** An **endoskeleton** of cartilage or bone provides support, including a **vertebral column** that usually completely replaces the notochord.

The **brain** is encased in a cranium, and the **nerve cord** is surrounded by the vertebral column. Metabolic wastes are removed by **kidneys,** and gas exchange occurs by either **pharyngeal gills** or **lungs.** The **closed circulatory system** includes a ventral heart of 2–4

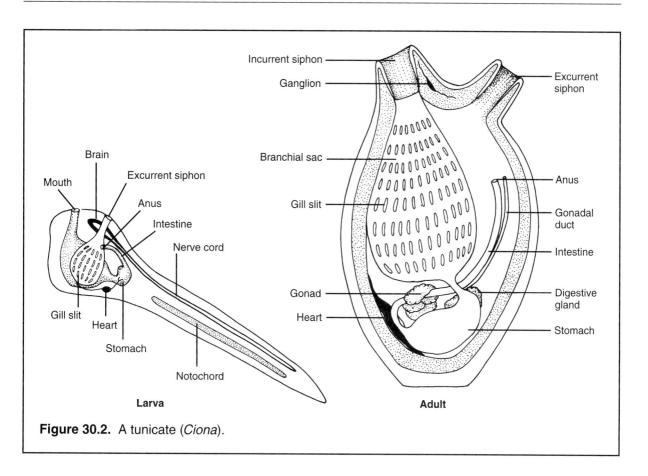

Figure 30.2. A tunicate (*Ciona*).

chambers. The distinctive characteristics of vertebrate classes are listed in Table 30.2.

Jawless Fish (Class Agnatha). Present day agnaths (lampreys and hagfish) are degenerate forms living as scavengers and parasites on other fish, which probably explains the absence of scales and paired fins and the presence of a cartilaginous skeleton. External features of a lamprey are shown in Figure 30.4.

Cartilaginous Fish (Class Chondrichthyes). A side branch of jawed bony fish that appeared about 425 million years ago, sharks and rays have an endoskeleton of cartilage. Thick **pectoral** and **pelvic fins** provide stabilization when swimming. **Dermal scales,** a basic fish characteristic, are formed from the dermis of the skin. They are thick and heavy and provide protection without hindering body flexibility.

Bony Fish (Class Osteichthyes). Modern bony fish are better adapted for a swimming lifestyle. They have **lightweight dermal scales** that provide protection while reducing body weight, and a **swim bladder** that provides controllable buoyancy. The **bony endoskeleton** is also usually lightweight. As in all fish, a **lateral line system** detects underwater vibrations.

Amphibians (Class Amphibia). Amphibians (frogs, toads, and salamanders) usually live in moist environments near ponds and streams. They must return to an aquatic habitat to reproduce, since fertilization is external, and embryonic and larval development requires free water. The larvae have gills as gas exchange organs, and a fish-like tail for swimming. These features are absorbed, and legs and lungs develop as amphibians mature into air-breathing adults. Because adult amphibians lack scales, excessive water loss through the skin is avoided by living in a moist habitat.

Reptiles (Class Reptilia). Reptiles were the first truly land animals because they do not have to return to water to reproduce. Fertilization is

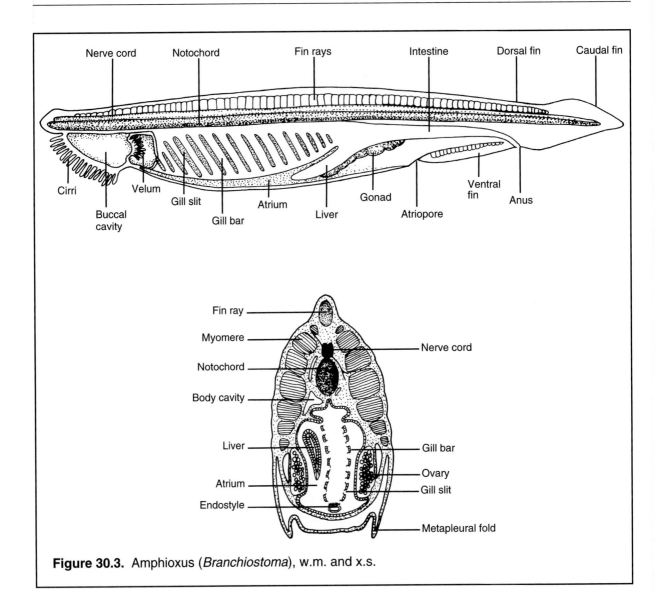

Figure 30.3. Amphioxus (*Branchiostoma*), w.m. and x.s.

internal, and **leathery shelled eggs** (amniote eggs) contain nutrients and a "private pond" for the developing embryo. Further, reptiles have **epidermal scales** that retard water loss through the skin, so they can live in dry habitats. **Claws** on their toes are modified epidermal scales. Ribs join to the sternum to form a **rib cage** that provides protection for the heart and lungs. Like fish, amphibians, and all lower animals, reptiles are **poikilothermic** (their body temperature fluctuates with the ambient temperature) and **ectothermic** (heat for body temperature comes mostly from external sources).

Birds (Class Aves). Birds possess many adaptations for flight, including (1) **feathers** (derived from epidermal scales) providing a lightweight airfoil, (2) hollow, lightweight bones, (3) air sacs associated with the lungs, enhancing buoyancy, (4) a fused skeleton except for the neck, tail, and appendages, and (5) pectoral appendages modified into **wings.** Jaws are modified into a **horny beak** lacking teeth. Fertilization is internal, and hard-shelled eggs contain nutrients and an aqueous habitat for embryonic development. Parental care is greatly increased.

Unlike lower forms, birds are **homeothermic**

TABLE 30.2
Classes of Vertebrates

Class	Characteristics
Agnatha (jawless fish)	No jaws; scaleless skin; median fins only; cartilaginous endoskeleton; persistent notochord; two-chambered heart
Chondrichthyes (cartilaginous fish)	Jaws; subterminal mouth; placoid dermal scales; median and paired fins; cartilaginous endoskeleton; two-chambered heart
Osteichthyes (bony fish)	Jaws; terminal mouth; membranous median and paired fins; bony endoskeleton; lightweight dermal scales; two-chambered heart; operculum covers gills; swim bladder
Amphibia (amphibians)	Aquatic larvae with gills; air-breathing adults with lungs; scaleless skin; paired limbs; bony endoskeleton; three-chambered heart
Reptilia (reptiles)	Bony endoskeleton with rib cage; well-developed lungs; epidermal scales; claws; paired limbs in most; leathery-shelled eggs; three-chambered heart
Aves (birds)	Anterior limbs modified into wings; feathers, epidermal scales; claws; endoskeleton of lightweight bones; rib cage; horny beak without teeth; hard-shelled eggs; four-chambered heart
Mammalia (mammals)	Hair; reduced or absent epidermal scales; claws or nails; mammary glands; diaphragm; bony endoskeleton with rib cage; young develop in uterus and are nourished via placenta; four-chambered heart

(body temperature is relatively constant) and **endothermic** (heat from cellular respiration maintains body temperature).

Mammals (Class Mammalia). The key characteristics of mammals are **hair, mammary glands,** and a **diaphragm** that separates the abdominal cavity from the thoracic cavity. Like birds, mammals are homeothermic and endothermic.

Fertilization is internal and, except for primitive forms, offspring develop in the uterus of the female and receive nutrients from the mother via a placenta. Female mammals nurse their young for variable lengths of time after birth.

Materials

Per student
Colored pencils
Dissecting microscope

Per lab
Demonstration specimens of lamprey, shark, perch, and frog life cycle
Demonstration dissections
    lamprey, cross section through the tail
    shark, cross section through the tail

Demonstration microscope setups of:
    tunicate (ammocoete) larva
    amphioxus, w.m.
Preserved amphioxus and lamprey
Representative chordates that are numbered but not labeled

Assignment 2

1. Color-code the nerve cord green, the notochord blue, and pharyngeal slits red in the tunicates in Figure 30.2 and in the amphioxus in Figure 30.3.
2. Examine the slide of a tunicate larva set up under a demonstration microscope. Compare it with Figure 30.2 and locate the four chordate characteristics.
3. Obtain a preserved amphioxus and examine it with a dissecting microscope. Locate the opening to the buccal cavity, the dorsal and caudal fins, and the segmentally arranged body muscles.
4. Examine a slide of amphioxus, w.m., set up under a demonstration microscope. Compare it with Figure 30.3 and locate the four chordate

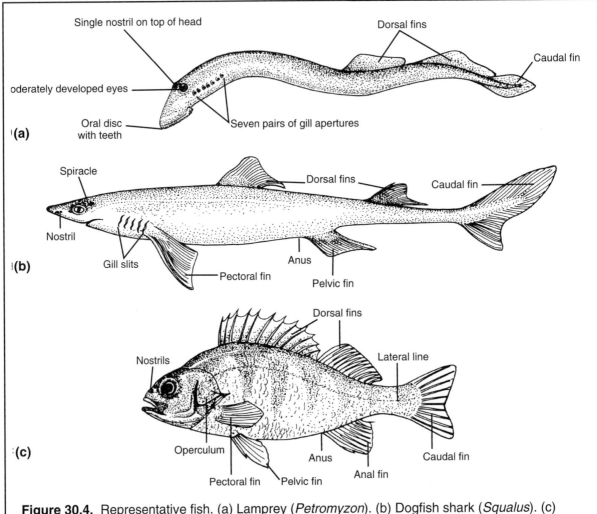

Figure 30.4. Representative fish. (a) Lamprey (*Petromyzon*). (b) Dogfish shark (*Squalus*). (c) Perch (*Perca*).

characteristics. **Complete items 2a and 2b on the laboratory report.**

5. Compare a lamprey, shark, and perch as to: (a) type of fins, (b) number of gill slits, (c) type of dermal scales, and (d) location of jaws, if present. Locate the lateral line. **Complete items 2c-2f on the laboratory report.**

6. Examine cross sections through the tail of a lamprey and a shark. Locate the cartilaginous vertebrae, nerve cord, and notochord, if pres-

ent. **Complete item 2g on the laboratory report.**

7. Examine the demonstration of the frog life cycle. Why are amphibians considered to be poorly adapted to terrestrial life? **Complete items 2h-2m on the laboratory report.**

8. Examine the representative chordates and identify the subphylum or class to which each belongs. **Complete the laboratory report.**

Part 5   Evolution and Ecology

31

Human Evolution

OBJECTIVES

After completion of the laboratory session, you should be able to:
1. Contrast the skulls of modern apes and modern humans.
2. Distinguish skulls of modern apes and fossil hominids.
3. Identify and contrast the hominid skulls studied.
4. Define all terms in bold print.

The origin of modern humans is a topic of great interest to all of us. Who were our ancestors? What did they look like? The precise answers to these questions are not known, but little by little more information about human evolution is being discovered.

The fossil record of human evolution is sparse, and considerable disagreement exists among paleontologists about the relationships among the known fossils. However, most agree on the following points.

1. Our closest living relatives are chimpanzees and gorillas.
2. The great apes and hominids (human and human-like forms) evolved from a common ancestor living in Africa.

3. Hominid evolution involves adaptive radiation and a branching evolutionary tree.

Figure 31.1 illustrates the structural differences between the skulls of a gorilla and a modern human. Although this comparison is between two modern species, it suggests skull features likely to be present in intermediate forms if a common ancestry is assumed.

Assignment 1

1. Study Figure 31.1 to become familiar with the identified differences between gorilla and modern human skulls.
2. **Complete item 1 on Laboratory Report 31 that begins on page 417.**

HOMINIDS

Hominids are distinguished from the great apes by their erect posture and bipedal walking. The separation of great apes and hominids probably took place about 5 million years ago, and their common ancestor is not known.

The oldest known hominid lived about 3.7 million years ago, while the earliest known human appeared about 2 million years ago. Modern hu-

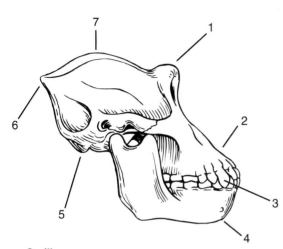
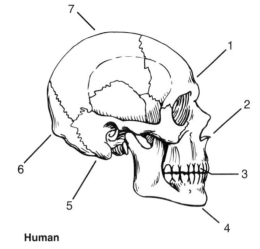

Gorilla

1. Heavy brow ridges; low-crowned, sloping cranium
2. Muzzlelike faces
3. Large canines
4. No chin
5. Posterior skull attachment
6. Occipital crest for attachment of heavy neck muscles
7. Sagittal crest for attachment of heavy jaw muscles

Human

1. Almost no brow ridges; high forehead and high-crowned cranium
2. Flattened face
3. Small canines
4. Well-developed chin
5. Central skull attachment
6. No occipital crest, neck muscles not as heavy and attached lower
7. No sagittal crest, lighter jaw muscles attached lower

Figure 31.1. Comparison of gorilla and human skulls.

mans appeared about 40,000 years ago, and their anatomy has not changed significantly since that time.

As you read the following descriptions of known hominids, locate their positions on Figure 31.2, which shows one of several possible patterns of hominid evolution.

Australopithecus

The earliest known hominid is *Australopithecus ramidus,* based on a few bone fragments found in East Africa and dated at 4.4 million years ago (mya). The second oldest known australopithecine is *Australopithecus afarensis,* which lived from 3.7 to 2.7 million years ago in Africa. Some believe that *A. afarensis* is ancestral to the later appearing australopithecines—*A. africanus* (3 to 2 million years ago), *A. boisei* (2.5 to 1.4 million years ago), and *A. robustus* (2.3 to 1.3 million years ago)—and to the earliest member of the genus *Homo.* Others believe humans had branched from an unknown ancestor before the appearance of *A. afarensis.*

Australopithecines were small, 4 to 4.5 ft tall, erect, bipedal, and had a brain a bit larger than a chimpanzee's. Both plants and animals composed their diet.

Homo habilis

Homo habilis was the first human. It lived 2 million to 1.4 million years ago, and it coexisted with *A. boisei* and *A. robustus. Homo habilis* is thought to be ancestral to *H. erectus.* Some authorities believe that *H. habilis* is only an advanced form of *A. africanus.* The skeletal features of *H. habilis,* especially the teeth and

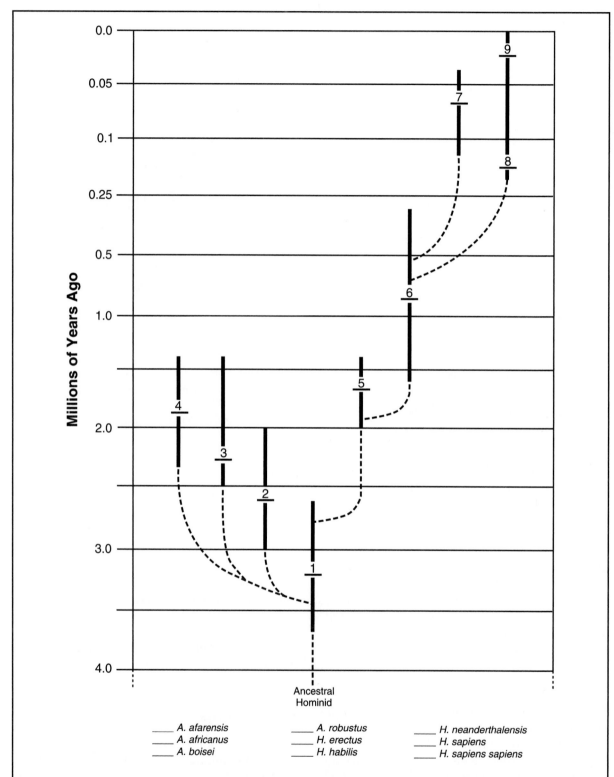

Figure 31.2. Possible hominid evolutionary relationships. Heavy lines indicate when each species is known to have lived, based on fossil remains. Dotted lines suggest possible evolutionary relationships.

cranial size, are more human-like than are those of *Australopithecus*. *H. habilis* apparently used crude stone tools.

Homo erectus

Homo erectus lived from 1.6 million to 300,000 years ago. Numerous fossils of this species have been found in Africa, Asia, and Europe. *H. erectus* was taller (5 ft), larger brained, and more efficiently bipedal than was *Australopithecus*. Evidence indicates the species engaged in construction and use of stone tools, use of fire, and a hunting-and-gathering life-style. The skull is characterized by (1) heavy brow ridges, (2) sloping forehead, (3) long, low-crowned cranium, (4) no chin, and (5) protruding face.

Homo neanderthalensis

Neanderthals lived from 130,000 to 35,000 years ago. They are named for the Neander Valley in Germany where early fossils were found. Some paleoanthropologists think Neanderthals are an evolutionary dead end that rose from *H. erectus* and were replaced by *H. sapiens sapiens*. Others believe that they are a subspecies of *H. sapiens* (*H. sapiens neanderthalensis*) that became extinct through interbreeding with *H. sapiens sapiens*.

Neanderthals were stocky with heavy muscles, and about 5 ft tall. Skulls exhibit (1) heavy brow ridges, (2) large protruding faces, (3) no chin, and (4) a sloping, low-crowned but large cranium. They constructed stone, bone, and stick tools, used fire, wore clothing, were skilled hunters, and were able to cope with the frigid climate of the glacial periods. They also buried flowers with their dead. They coexisted briefly with modern humans before their extinction. The cause of their extinction is not understood.

Homo sapiens

Some authorities believe early *H. sapiens* evolved about 200,000 years ago from *H. erectus* in Africa and migrated to other parts of the world replacing other *Homo* species. Others believe different populations of *H. sapiens* evolved independently in various parts of the world from different populations of *H. erectus*. Most paleoanthropologists agree that modern humans, *H. sapiens sapiens*, evolved from early *H. sapiens*.

TABLE 31.1
Average Cranial Capacity (cc) of Hominids

Australopithecus afarensis	414
Australopithecus africanus	441
Australopithecus robustus	530
Homo habilis	640
Homo erectus	990
Homo neanderthalensis	1465
Homo sapiens sapiens	1350

Modern humans appeared about 40,000 years ago. The fossil record shows that they ranged throughout western and central Europe, although they may have originated elsewhere. The early modern humans represented by **Cro-Magnon man** were (1) excellent tool and weapon makers, using stone, sticks, bone, and antlers as raw materials; (2) skilled hunters; and (3) fine artists, as evidenced by cave paintings and sculptures. The radiation of these early modern humans ultimately produced the various races observed today.

Assignment 2

1. Examine Table 31.1.
2. Label Figure 31.2.
3. **Complete item 2 on the laboratory report.**

SKULL ANALYSIS

An understanding of hominid evolution is based on the study of fossil bones, teeth, and artifacts, such as tools, associated with specimens. Analysis of the fossils involves detailed measurements and careful comparisons of the fossils. In this section, you will analyze certain skeletal materials and fossil replicas by making measurements and calculating indexes as well as conducting qualitative comparisons of the available specimens.

Indexes are useful for comparative purposes because they overcome the problems caused by differences in the size of specimens. An index is a ratio calculated by dividing one measurement

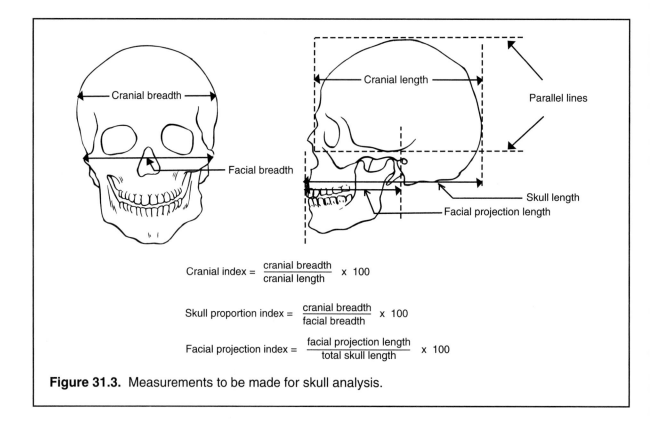

Figure 31.3 labels: Cranial breadth, Facial breadth, Cranial length, Parallel lines, Skull length, Facial projection length

$$\text{Cranial index} = \frac{\text{cranial breadth}}{\text{cranial length}} \times 100$$

$$\text{Skull proportion index} = \frac{\text{cranial breadth}}{\text{facial breadth}} \times 100$$

$$\text{Facial projection index} = \frac{\text{facial projection length}}{\text{total skull length}} \times 100$$

Figure 31.3. Measurements to be made for skull analysis.

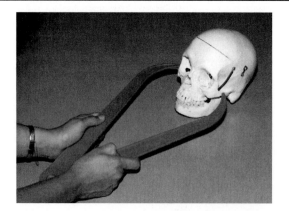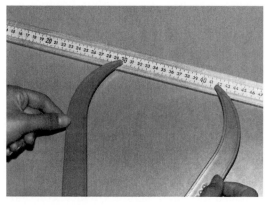

Figure 31.4. Procedures for making measurements for skeletal analysis.

by another, and it indicates the proportional relationship of the two measurements. Calipers and meter sticks are to be used in making the measurements, and all measurements are to be in millimeters. Figures 31.3 and 31.4 show how to make the measurements. The indexes are calculated by dividing one measurement by another (carried to two decimal places) and multiplying by 100.

Handle the bones and fossil replicas carefully since they are easily damaged. Avoid making marks or scratches on the specimens. The measurements are best made by working in groups of two to four students.

Materials

Calipers
Metersticks
Skull replicas of:
   Australopithecus africanus
   Homo erectus
   Homo neanderthalensis
   Homo sapiens sapiens (Cro-Magnon)
Skull of modern human

Cranial (Cephalic) Index

When this index is determined for living forms, it is called a **cephalic index**. The term **cranial index** is used in reference to nonliving specimens. An index of less than 75 indicates a long-headed condition. An index of 80 or more identifies a round-headed individual.

$$\text{Cranial index} = \frac{\text{cranial breadth}}{\text{cranial length}} \times 100$$

Cranial breadth: maximum width of the cranium
Cranial length: maximum distance between the posterior surface and the small prominence (glabella) between the brow ridges

Skull Proportion Index

The skull is composed of the face and the cranium, and the skull proportion index identifies the proportional relationship between these two components. The greater the value of the index, the larger the cranium is in relation to the face.

$$\text{Skull proportion index} = \frac{\text{cranial breadth}}{\text{facial breadth}} \times 100$$

Cranial breadth: maximum breadth of the cranium
Facial breadth: maximum distance between the lateral surfaces of the cheekbones (zygomatic arches)

Facial Projection Index

A projecting, muzzlelike face is a primitive condition among primates. This index indentifies the degree of facial projection in a specimen. The greater the value of the index, the greater the degree of facial projection.

$$\text{Facial projection index} = \frac{\text{facial projection length}}{\text{total skull length}} \times 100$$

Facial projection length: distance between the anterior margins of the auditory canal and upper jaw (maxilla)
Total skull length: maximum distance between the posterior surface of the cranium and the anterior margin of the upper jaw (maxilla)

Assignment 3

1. Determine the cephalic index for each member of your laboratory group. Exchange data with all groups in the class. Determine the range and average index value for the class.
2. **Complete items 3a–3d on the laboratory report.**
3. Determine the (1) cranial index, (2) skull proportion index, and (3) facial projection index for each of the skulls available. **Record your data in item 3e on the laboratory report.**
4. Compare the available skulls as to the following characteristics and rate each on a 1–5 scale. **Record your responses in item 3e on the laboratory report.**

Skull and vertebra attachment
   1 = most posterior
   5 = most central

Brow ridges
   1 = most pronounced
   5 = least pronounced

Length of canines
   1 = longest
   5 = no longer than incisors

Forehead
   1 = most sloping
   5 = best developed

Chin
   1 = no chin
   5 = best developed

5. **Complete the laboratory report.**

32

Ecological Relationships

OBJECTIVES

After completion of the laboratory session, you should be able to:
1. Describe the components of an ecosystem.
2. Describe the roles of members of a community in the flow of energy.
3. Contrast the flow of energy and the cycling of materials in an ecosystem.
4. Explain how to use indicator organisms in bioassays to determine the effect of pollutants.
5. Describe the resistance of *Daphnia* to the pollutants tested.
6. Explain how sowbugs orient to their habitat.
7. Define all terms in bold print.

Life on earth is maintained by the delicate balance between organisms and their environment. All organisms, including humans, are subject to the ecological "laws" that govern these interrelationships.

Ecology is the branch of biology that investigates the environmental relationships of organisms, and it is primarily concerned with populations, communities, and ecosystems. A **population** is a group of interbreeding members of the same species living in a defined area.

A **community** is composed of all populations of organisms in a defined area. An **ecosystem** consists of both the community and the nonliving portion (e.g., water, air, and soil) of the environment in a defined area.

ENERGY FLOW

One of the most important interactions of organisms involves the flow of energy through a community. In the flow of energy, members of a community perform one of three ecological roles: producer, consumer, or decomposer.

Producers are photosynthetic autotrophs that capture radiant energy and convert it into the chemical energy of organic nutrient molecules, such as carbohydrates, fats, and proteins, which compose bodies. Plants and plant-like protists are producers.

Consumers are heterotrophic organisms that must obtain energy by consuming (feeding on) producers or other consumers. Animals are consumers. There are three major types of consumers. **Herbivores** feed directly on plants or plant-like protists, **carnivores** feed only on other animals, and **omnivores** feed on both plants and other animals.

Decomposers are heterotrophs that obtain energy by consuming organic wastes of organ-

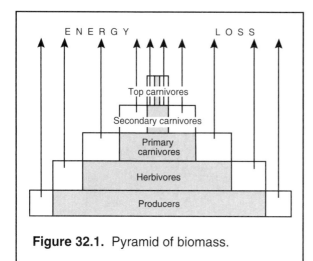

Figure 32.1. Pyramid of biomass.

ENERGY LOSS
Top carnivores
Secondary carnivores
Primary carnivores
Herbivores
Producers

isms and dead organisms. Most fungi and bacteria are decomposers.

Energy flows from producer to consumer to decomposer. It is not recycled because life processes ultimately convert the usable energy in organic molecules into nonusable heat energy. If you could follow a given amount of energy captured by producers as it flows through the community, you would find that the amount of available energy rapidly decreases in the transfer from producer to consumer to decomposer. In fact, decomposers use the last bit of available energy. Thus, a continuous supply of energy is needed for life on earth to continue, and anything that interferes with photosynthesis or energy flow through a community jeopardizes one or more populations of a community.

The amount of energy available for consumers determines the number of each of the three types of consumers that can be supported in the community. Figure 32.1 shows this relationship in a pyramid of biomass, the dry weight that is proportional to its contained energy. Each layer of the pyramid is a trophic (feeding) level. Note that as you move up the pyramid, the biomass of each trophic level decreases. In fact, only 10% of the energy at each trophic level is transferred to the next trophic level. Although there may be a very large biomass of producers, the biomass of the top carnivore is quite small.

Assignment 1

1. **Complete item 1a on Laboratory Report 32 that begins on page 421.**
2. Examine Figure 32.1, and **complete items 1b–1d on the laboratory report.**
3. Examine Figure 32.2, which shows a food web of a simplified community. The arrows indicate the feeding relationships of the consumers. Determine the roles of each type of organism by labeling the figure. **Complete item 1 on the laboratory report.**

CYCLING OF MATERIALS

Although energy flows once through the community, chemical substances composing organisms repeatedly cycle between the community and the nonliving environment. When energy-containing organic molecules pass from producer to consumer, some molecules become part of the body structure of the consumer, and some molecules are broken down by cellular respiration to release energy to power life processes. When organisms die, decomposers break down the organic molecules in their bodies to release energy and convert them into inorganic molecules that can be reused by producers. This is the ultimate recycling program.

The carbon cycle in Figure 32.3 is an example of how materials cycle in an ecosystem. Recall that carbon atoms form the skeleton of the organic molecules that compose organisms.

Assignment 2

1. Follow the cycling of carbon through the ecosystem and label the processes that occur at each step in Figure 32.3.
2. **Complete item 2 on the laboratory report.**

ENVIRONMENTAL POLLUTION

Pollution results when the by-products of human activity accumulate in soil, air, or water at levels that are harmful to living organisms. Pollution may result from a wide variety of sources,

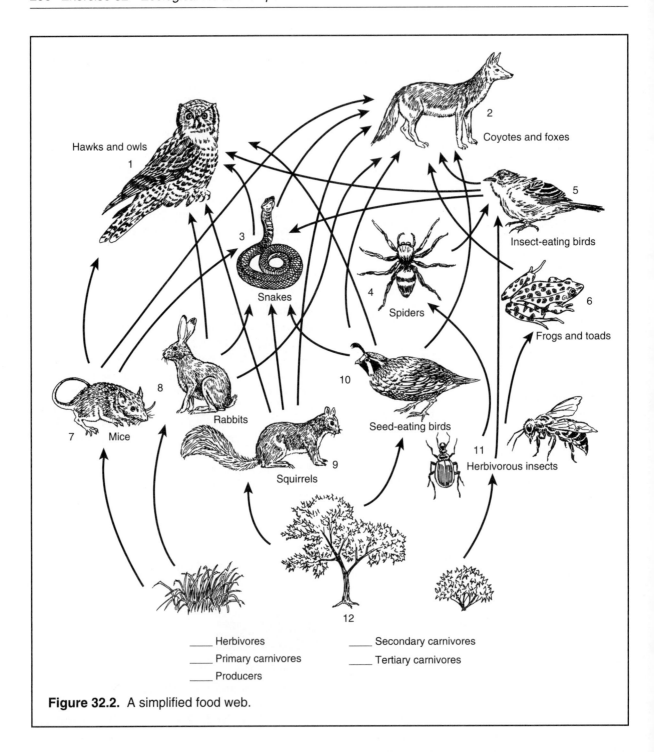

Hawks and owls
1

Coyotes and foxes
2

3
Snakes

4
Spiders

5
Insect-eating birds

6
Frogs and toads

8
Rabbits

7   Mice

9
Squirrels

10
Seed-eating birds

11
Herbivorous insects

12

_____ Herbivores

_____ Primary carnivores

_____ Producers

_____ Secondary carnivores

_____ Tertiary carnivores

Figure 32.2. A simplified food web.

such as agricultural practices, factories, the combustion of fossil fuels, thermonuclear plants, mining, landfills, and so on.

One way ecologists assess the hazard that a pollutant poses to natural ecosystems is to determine the effect of the pollutant on **indicator organisms.** The sensitivity of an indicator or-

ganism to pollutants provides a **bioassay** of the impact of a given pollutant. The indicator organism is exposed to a variety of concentrations of pollutant in order to determine the concentration at which a harmful effect is clear. One method is to determine the LC_{50} (lethal concentration$_{50}$), the concentration that kills 50% of the

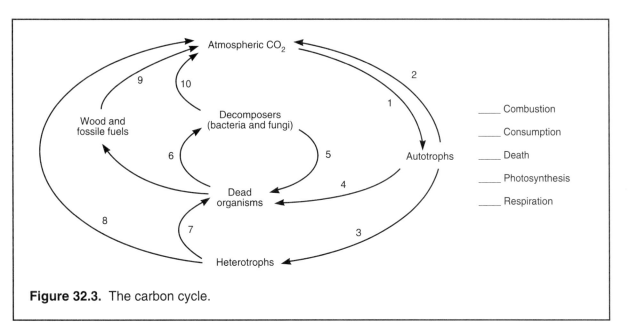

Figure 32.3. The carbon cycle.

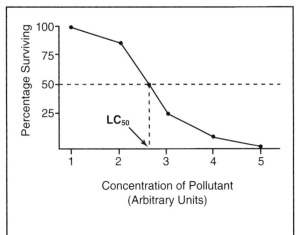

Concentration of Pollutant
(Arbitrary Units)

Figure 32.4. Determination of the LC_{50}.
1. Plot the percentage surviving at each concentration of pollutant. 2. Connect the dots with straight lines to form a survivor curve. 3. Draw a perpendicular line from the point where the survivor curve intersects the 50% surviving line. 4. The point where the perpendicular line intersects the X axis indicates the LC_{50}.

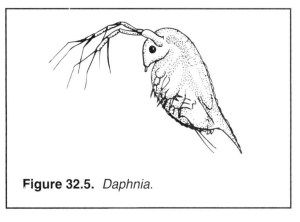

Figure 32.5. *Daphnia.*

indicator organisms within a specific time of exposure. Figure 32.4 illustrates how to determine the LC_{50}.

In this section, you will use *Daphnia magna* (Figure 32.5), a small freshwater crustacean, as an indicator organism to investigate the effect of pollution. *Daphnia* feed on microscopic algae and protists, and in turn serve as a prime food source for small fish. They have a jerky swimming motion powered by rapid strokes of paired, feathery antennae, and while swimming they sweep tiny food organisms into their mouths with their thoracic legs. *Daphnia* normally live in water that has a pH of 5.5–7.0, summer temperatures of 15–25 °C, and no pesticide content.

The following experiments are best done by groups of 3 to 4 students. Work out a division of labor for your lab group. **Caution.** Handle the acid and pesticide solutions with care. If you spill any on your hands, quickly wash with soap and water. Notify your instructor of any spills.

When transferring *Daphnia*, be sure to immerse the tip of the pipette below the surface of the receiving pond water before ejecting the

Daphnia. This prevents getting air under the carapace which will cause them to float on the surface and die. *Be sure to rinse the pipette in tap water between transfers to prevent contamination.*

Materials

Per student group
Beakers, 250 ml, 2
Glass-marking pen
Glass vials, 1" × 3", 5 per experiment
Thermometers, Celsius, 1
Vial holders, 3

Per lab
Cultures of *Daphnia*
Flasks of pond water
Flasks of pond water adjusted to: pH 7; pH 6; pH5; pH 4; and pH 3
Flasks of pond water containing these percentages of Diazinon Plus: 0.001%; 0.002%; 0.003%; and 0.004%
Ice, crushed
Pipettes modified to transfer *Daphnia*
Water baths at 30° C and 35° C

Assignment 3

1. Perform one or more of the following experiments as assigned by your instructor and pool the results of the entire class.
2. **Record the results in item 3 on the laboratory report.**
3. After completing the experiements, **complete item 3 on the laboratory report.**

Experiment 1: Acid Pollution

1. Label five vials, 1 to 5, and fill each vial about 2/3rds full with pond water adjusted to these pH values.
Vial 1 = pH7	Vial = pH 4
Vial 2 = pH 6	Vial = pH 3
Vial 3 = pH 5	
   Place the vials in a rack or holder at your work station so they will not spill.
2. Starting with Vial 1, transfer four healthy *Daphnia* to each vial. Discard any floating *Daphnia* and replace them with healthy *Daphnia*. Rinse the transfer pipette in tap water between each transfer. Why? Record the time.
3. **Record the number of surviving Daph-**

nia after 1 hour on the laboratory report.

Experiment 2: Pesticide Pollution

1. Label five full vials, 1 to 5, and fill each vial about 2/3rds full with pond water containing these percentages of Diazinon Plus.
Vial 1 = 0.000%	Vial 4 = 0.003%
Vial 2 = 0.001%	Vial 5 = 0.004%
Vial 3 = 0.002%	
   Place the vials in a rack or holder at your work station so they will not spill.
2. Starting with Vial 1, transfer four healthy *Daphnia* to each vial. Discard any floating *Daphnia* and replace them with healthy *Daphnia*. Rinse the transfer pipette in tap water between each transfer. Why? Record the time.
3. **Record the number of surviving Daphnia after 1 hour on the laboratory report.**

Experiment 3: Thermal Pollution

1. Fill two 250-ml beakers half-full with tap water to make small water baths. Use ice to adjust the temperature of one to 20° C and leave the other one at room temperature. Large water baths are available in the lab at 30° C and 35° C. Record the temperature of each water bath.
2. Label four vials, 1 through 4, and fill each about 2/3rds full with pond water.
3. Place one vial of pond water in each water bath.
   Vial 1 = 20° C
   Vial 2 = room temperature
   Vial 3 = 30° C
   Vial 4 = 35° C
   Allow five minutes for temperature equilibration.
4. Transfer four healthy *Daphnia* to each vial. Discard any floating *Daphnia* and replace them with healthy *Daphnia*. Record the time.
5. **Record the number of surviving Daphnia after 1 hour on the laboratory report.**

BEHAVIORAL ECOLOGY

If you have looked carefully at organisms in an ecosystem, you have noticed that they tend to occupy specific **habitats,** particular places where they live. For example, some plants live in sunlit areas, others thrive in shade, and some prefer partial sunlight. Animals also exhibit habitat preferences, and some of their preferences result from responses to specific stimuli. One task of biologists is to determine an organism's habitat and how the organism is adapted to it.

In this section, you will investigate the responses to certain stimuli that may explain why sowbugs (wood lice) tend to be found under logs, rocks, and fallen leaves. You will test the hypothesis that light and moisture affect the orientation of sowbugs. The experiments will offer alternative locations for sowbugs to congregate: (1) light or shade, and (2) moisture or dryness.

Materials

Per student group
Sowbugs
Black construction paper
Dropping bottle of tap water
Laboratory lamp, preferably fluorescent
Masking tape
Petri dish, plastic, 14 cm in diameter
Filter paper, 14.5 cm in diameter

Assignment 4

1. Obtain a large Petri dish and fit a circle of filter paper into the bottom of it. This provides a dry habitat for testing the sowbugs' response to light.

2. Use black construction paper to make a roof-like light shield that will shade half of the Petri dish. The roof must be low enough to provide heavy shade, but high enough so you can see under it to observe the sowbugs. You may need to anchor the light shield to the lab table with masking tape.

3. Position the lamp so that it shines on the Petri dish. If it has an incandescent light source, keep it at least 3 ft from the Petri dish to prevent heat from affecting your experiment.

4. Place 5 sowbugs in the Petri dish and replace the lid. Before you place it under the light shield, observe the movement of the sowbugs.

5. **Complete items 4a and 4b on the laboratory report.**

6. Place the dish under the light shield so that half of the dish is heavily shaded, so as to provide a choice for the sowbugs of light or shade. Observe the sowbugs and **record their positions (i.e., light or shade) at 3-min intervals for 15 min in item 4c, and complete item 4d on the laboratory report.**

7. Remove the light shield and turn off the laboratory lamp. Remove the lid of the Petri dish and add 5 drops of water to one spot at the edge of the filter paper. Replace the lid. The sowbugs now have a choice between moist and dry areas. Observe the sowbugs and **record their positions (i.e., dry or moist) at 3-min intervals for 15 min in item 4e, and complete the laboratory report.**

33 Population Growth

After completion of the laboratory session, you should be able to:
1. Distinguish between population growth and population growth rate.
2. Compare theoretical and realized population growth curves.
3. Describe the role of the following in population growth:
   a. Biotic potential
   b. Environmental resistance
   c. Density-dependent factors
   d. Density-independent factors
   e. Environmental carrying capacity
4. Contrast the growth of human and natural animal populations.
5. Compare the growth of human populations in major regions of the world.
6. Define all terms in bold print.

A **population** is a group of interbreeding members of the same species within a defined area. Populations possess several unique characteristics that are not found in individuals: **density, birth rate, death rate, age distribution, biotic potential, dispersion,** and **growth form**. One of the central concerns in ecology is the study of population growth and factors that control it.

Natural populations are maintained in a state of dynamic equilibrium with the environment by two opposing factors: biotic potential and environmental resistance. **Biotic potential** is the maximum reproductive capacity of a population that is theoretically possible in an unlimiting environment. It is never realized except for brief periods. **Environmental resistance** includes all **limiting factors** that prevent the biotic potential from being attained.

The limiting factors may be categorized as **density-dependent factors,** the effects of which increase as the population increases. Space, food, water, waste accumulation, and disease are examples of such factors. In contrast, **density-independent factors** exert the same effect regardless of the population size. Climatic factors and the kill of individual predators generally function in this manner.

Assignment 1

Complete item 1 on Laboratory Report 33 that begins on page 425.

GROWTH CURVES

There are two basic types of population growth patterns. **Theoretical population growth** is the growth that would occur in a population if the biotic potential of the species were realized, that is, if all limiting factors were eliminated. This type of growth does not occur in nature except for very short periods of time. **Realized population growth** is the growth of a population that actually occurs in nature. Let's consider both types of population growth in bacteria.

Theoretical Growth Curves

In the absence of limiting factors, bacteria exhibit tremendous theoretical population growth potential. Bacterial population size is measured as the number of bacteria in a milliliter (ml) of nutrient broth, a culture medium. Assume that bacterial cells divide at half-hour intervals. If the initial population density were 10,000 bacteria per milliliter of nutrient broth (10×10^3 bacteria/ml), a half hour later the population would be 20,000 bacteria per milliliter (20×10^3 bacteria/ml), and the population would have doubled. See Table 33.1. In another half hour, the population would double again to 40×10^3 bacteria/ml. In an unlimiting environment, this population would double every half hour. If the population size is calculated for each half-hour interval and plotted on a graph, a line joining the points on the graph yields a **theoretical growth curve**.

Realized Growth Curves

Natural populations exhibit realized growth that results from the opposing effects of biotic potential and environmental resistance. Table 33.2 indicates the growth of a bacterial population in a tube of nutrient broth—a limited environment which results in the ultimate death of the entire population. The population was sampled at half-hour intervals, and no additional nutrients were added.

Assignment 2

1. Determine and record the theoretical growth of the bacterial population which doubles at half-hour intervals and record the data in Table 33.1.
2. **Plot the theoretical growth in item 2a on the laboratory report.**
3. **Plot the realized growth curve in item 2a on the laboratory report.** Use a different symbol or color for this curve to distinguish it from the theoretical curve.
4. Compare the theoretical and realized population growth curves. The theoretical curve is a J-shaped curve. The realized growth curve starts off in the same manner, but it is soon changed by environmental resistance. As the

TABLE 33.1
Theoretical Growth of a Bacterial Population

Time (hr)	Population Density (10^3 bacteria/ml)
0.0	10
0.5	20
1.0	_____
1.5	_____
2.0	_____
2.5	_____
3.0	_____
3.5	_____
4.0	_____

TABLE 33.2
Realized Growth of a Bacterial Population

Time (hr)	Population Density (10^3 bacteria/ml)
0.0	10
0.5	20
1.0	40
1.5	80
2.0	150
2.5	290
3.0	450
3.5	520
4.0	520
4.5	515
5.0	260
5.5	80
6.0	0

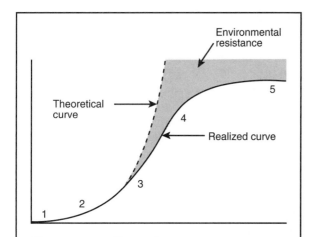

Figure 33.1. Theoretical and realized growth curves. 1. lag phase, 2. exponential growth phase, 3. inflection point, 4. decreasing growth phase, 5. carrying capacity.

limiting factors take effect, the curve becomes S shaped. In the artificial environment of a test tube, the bacterial population dies out. In natural populations where materials can cycle and energy is continuously available, the population size levels off where it is at equilibrium with the environment.

5. Study Figure 33.1, which compares generalized theoretical and realized population growth curves. The theoretical growth curve exhibits exponential growth, with the slope of the curve becoming ever steeper. In contrast, the effect of environmental resistance causes the realized growth curve to become S shaped.

Several parts of the realized growth curve are recognized. Growth is slow, initially, in the **lag phase,** due to the small number of organisms present. This is followed by the **exponential growth phase,** when the periodic doubling of the population yields explosive growth. As environmental resistance takes effect, the growth gradually slows, leading to the **inflection point** that indicates the start of the **decreasing growth phase** as the curve turns to the right. Continued environmental resistance causes the curve to level off at the **carrying capacity** of the environment, where the population size is in balance with the environment.

The carrying capacity is the maximum pop-

ulation that can be supported by the environment—a balance between biotic potential and environmental resistance. The population remains stable because births equal deaths.

6. **Complete item 2 on the laboratory report.**

HUMAN POPULATION GROWTH

One of the major concerns of modern society is the rapid growth of the human population in recent decades. Table 33.3 shows the estimated size of the world population at various times in history and projects a rather conservative growth in the near future.

Growth Rate

Ultimately, a population is limited by a decrease in **birth rate,** an increase in **death rate,** or both. In natural nonhuman populations, an increase in death rate is the usual method. This allows the environment to select for those better-adapted individuals who will contribute the greater share of the genes to the next generation. While this mode of selection occurred in preindustrialized human societies, it has been highly modified, but not stopped, by technology in modern societies.

TABLE 33.3
Estimated World Human Population

Year	Population Size (millions)
8000 B.C.	5
4000	86
1 A.D.	133
1650	454
1750	728
1800	906
1850	1,130
1900	1,610
1950	2,515
1960	3,019
1970	3,698
1980	4,450
1990	5,292
2000	6,400*

* Conservative projection.

Concern over the growth rate of the human population gained worldwide attention in the 1960s and persists today. The growth rate is the difference between the number of persons born (birth rate) and the number who die (death rate) per year. For humans, the growth rate usually is expressed per 1,000 persons. For example, if a human population shows 25 births and 10 deaths per 1,000 persons in a year, the growth rate may be determined as follows:

$$\frac{25 - 10}{1,000} = \frac{15}{1,000} = \frac{1.5}{100} = 1.5\% \text{ growth rate}$$

The growth rate reached a peak of 2% in 1965 and declined to 1.7% in 1984, but this slight decline does not mean that the population growth is no longer a concern.

The growth rate is often expressed as the **doubling time,** that is, the time required for the population to double in size. In 1850, the population required 135 yr to double, but this was reduced to only 35 yr in 1965. All of the efforts to decrease population growth only extended the doubling time to 46.7 yr by 1997.

The doubling time of a population is determined by dividing 70 yr (a demographic constant) by the growth rate. For example, if the world population growth rate is 1.7%, the doubling time (d) would be determined as follows:

$$d = \frac{70 \text{ yr}}{\text{growth rate}} = \frac{70 \text{ yr}}{1.5} = 46.7 \text{ yr}$$

This means that the entire world population will double in only 46.7 years. If the present standard of living is to be maintained, the available resources must also double in 46.7 yr. Is this likely? Are the earth's resources infinite?

Obviously, the standard of living varies throughout the world. Developed (industrialized) countries have a relatively high standard of living, and developing countries have a relatively low standard of living. The standard of living is usually expressed as the resources per person and is determined by dividing the gross domestic product (GDP) by the population size. The contrast in the standard of living in developed and developing countries is evident in the GDP per capita for the United States and India in 1995.

United States = $26,037
India = $365

Assignment 3

1 Use the data in Table 33.3 to **plot the human population growth curve in item 3a on the laboratory report.**
2. **Complete items 3b–3i on the laboratory report.**
3. Calculate the growth rate and doubling time for the human population in the countries or regions shown in Table 33.4.
4. **Complete the laboratory report.**

TABLE 33.4
1997 Human Population Data

Country or Region	Population Size (10⁶)	Birth Rate (per 10³)	Death Rate (per 10³)	Growth Rate (% inc./yr)	Doubling Time (yr)
World	5,849	24	9	1.5	46.7
Africa	758	41	14	_____	_____
Asia	3,538	24	8	_____	_____
Europe	729	12	11	_____	_____
Latin America	492	25	7	_____	_____
North America	302	15	9	_____	_____
Oceania	29	19	8	_____	_____

Source: 1997 Demographic Yearbook. Published by the United Nations, 1999.

Part 6 Laboratory Reports

LABORATORY REPORT 1

Orientation

1. Laboratory Procedures

a. Place an X by those activities that are to be completed before coming to the laboratory.

_____ Perform the experiments.   _____ Color-code diagrams.

_____ Learn the meaning and   _____ Tear lab report out of
         spelling of new terms.   lab manual.

_____ Label diagrams.   _____ Answer lab report questions
_____ Understand the objectives.   on background information.

b. Place an X by those events in the laboratory that should be brought to the attention of your instructor.

_____ Problems with equipment   _____ Minor injury

_____ Spillage of a liquid   _____ Directions not understood

_____ Breakage of glassware   _____ Problems with supplies

2. Biological Terms

a. Use Appendix A to determine the literal meaning of these terms.

Biology _____

Cardiac _____

Hypodermic _____

Erythrocyte _____

Dermatitis _____

b. Use Appendix A to construct terms with these literal meanings.

Pertaining to something within a cell _____

Study of the heart _____

Nerve disease _____

Cancerous tumor _____

Pertaining to a single cell _____

3. Measurements: Length

a. Indicate the name and value of these metric symbols.

Symbol	Name of Unit	Value of Unit
ml	_____	_____
cm	_____	_____
mm	_____	_____
kg	_____	_____

b. Measure the diameter of a penny in millimeters.   _____ mm

Convert your measurement to centimeters and meters.   _____ cm;   _____ m

c. Measure the length of your little finger. _____ mm; _____ cm

d. Complete the chart showing the number of students with each finger length.

Length in mm	No. of Students	Length in mm	No. of Students	Length in mm	No. of Students	Length in mm	No. of Students
51	_____	59	_____	67	_____	75	_____
52	_____	60	_____	68	_____	76	_____
53	_____	61	_____	69	_____	77	_____
54	_____	62	_____	70	_____	78	_____
55	_____	63	_____	71	_____	79	_____
56	_____	64	_____	72	_____	80	_____
57	_____	65	_____	73	_____	81	_____
58	_____	66	_____	74	_____	82	_____

e. Using the data from the chart in item 3d, indicate:
The range of little finger lengths. _____ mm to _____ mm
The average of little finger lengths. _____ mm
(The average is calculated by dividing the sum of the finger lengths by the number of fingers measured.)

f. A graph presents data in a visual way. Plot the data from the chart in item 3d on the graph below by placing a dot where an imaginary vertical line from a particular finger length on the horizontal axis intersects with an imaginary horizontal line from the number of students with that finger length on the vertical axis. Draw a heavy vertical line from each dot to the horizontal axis, producing a bar graph. Another way of showing data in a graph is by drawing a curve that fits the dots. Try it.

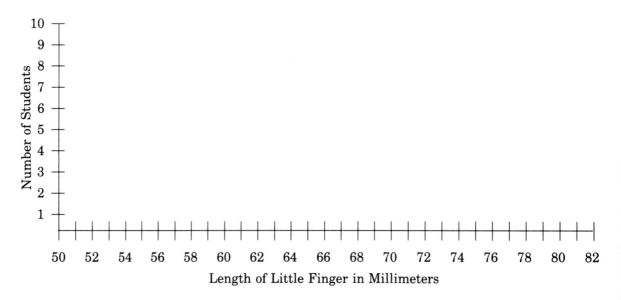

g. Referring to the graph, what is the most frequent finger length? _____ mm
   How many students have little finger lengths within 1 mm of the
   average finger length?   _____
   These students compose what percentage of the class?   _____ %
   (Percentage is calculated by dividing the number of students in this category by the number of
   students in the class and multiplying by 100.)

4. Measurements: Mass

a. Record the mass of the wood blocks.   #1 _____ g;   #2 _____ g;   #3 _____ g
b. A physician directs a patient to take 1 g of vitamin C each day. How many 250-mg tablets must
   the patient take each day? _____

c. It is recommended that adults eat 0.8 g of protein per kilogram of body mass each day. What
   should be the minimum daily protein intake of a man with a weight of 185 pounds? (Hint: First
   convert 185 pounds to kilograms.)   _____ g

5. Measurements: Volume

a. Record the dimensions of the wood blocks in centimeters.
   #1 L ____ ; W ____ ; D ____   #2 L ____ ; W ____ ; D ____   #3 L ____ ; W ____ ; D ____
b. Calculate the volume of the three wood blocks in cubic centimeters.
   #1 _____ cm^3 #2 _____ cm^3 #3 _____ cm^3

c. Calculate the density of each wood block.
   #1 _____   #2 _____   #3 _____

   d. What is the volume of the test tube?   _____ ml

   e. How many drops are in 1 ml of water?   _____

   f. Measure the mass of the 50-ml graduated cylinder.   _____ g

      Measure the mass of graduated cylinder +30 ml of water.   _____ g

      Calculate the mass of the water only.   _____ g

6. Measurements: Temperature

   a. Using a Celsius thermometer, measure the temperature of the following and convert the temperature to °F.

      Air in the room.   _____ °C;   _____ °F

      Cold tap water.   _____ °C;   _____ °F

   b. Normal body temperature is 37 °C (98.6 °F). If a patient's temperature is 38 °C, what is the temperature in °F? _____ °F

7. Scientific Method

   a. Why is a control group important in an experiment? _____

      _____

   b. Why is a prediction important? _____

      _____

   c. State a new key question raised by the results and conclusion of the hypothetical example of the scientific method.

      _____

      _____

   d. Using your key question, state a new hypothesis. _____

      _____

      State a new prediction. _____

      _____

      _____

   e. Describe a controlled experiment to test your new hypothesis. _____

      _____

      _____

      _____

      _____

   f. What results would support your hypothesis? _____

      _____

      _____

LABORATORY REPORT 2

The Microscope

1. The Compound Microscope

a. List the labels for Figure 2.1.

1. _____   9. _____
2. _____   10. _____
3. _____   11. _____
4. _____   12. _____
5. _____   13. _____
6. _____   14. _____
7. _____   15. _____
8. _____

b. Write the term that matches each meaning.

1. Used as a handle to carry microscope _____
2. Lenses attached to the nosepiece _____
3. Concentrates light on the object _____
4. Lens you look through _____
5. Platform on which slides are placed _____
6. Rotates to change objectives _____
7. The shortest objective _____
8. The longest objective _____
9. Control knob used for fine focusing _____
10. Control knob used for rough focusing _____
11. Controls amount of light entering condenser _____

c. List the power of the ocular and objective lenses on *your* microscope, and calculate the total magnification for the combinations noted.

Magnification		Total
Ocular	Objective	Magnification
_____	_____	_____
	_____	_____
	_____	_____
	_____	_____

d. Write the term that matches each meaning.

1. Tissue used to clean lenses _____
2. Objective with the least working distance _____
3. Slide with an attached cover glass _____
4. Objective with the largest field _____

2. Initial Observations

a. Indicate the direction the image moves when the slide is moved:

To the left _____   Away from you _____

b. Indicate the steps to be used in focusing on an object with the high-power objective. _____

_____

_____

c. Describe the best way to relocate an object that is "lost" while viewing with the high-power objective. _____

_____

_____

d. Draw the letter *i* as it appears when observed with each of the following:

Unaided Eye	**4× Objective**	**10× Objective**	**40× Objective**

3. Depth of Field

Based on your observations, draw a side view of a large spine from a fly wing at these *total magnifications*.

40×   **100×**

4. Diameter of Field

a. Indicate the estimated diameter of field at each *total magnification* for your microscope.

Magnification	**Diameter of Field**
_____ ×	_____ mm
_____ ×	_____ mm
_____ ×	_____ mm

b. Indicate the estimated lengths of objects that extend across:

1. Two-thirds of the field at 40×   _____ mm   _____ μm
2. 25% of the field at 100×   _____ mm   _____ μm
3. Half of the field at 400×   _____ mm   _____ μm
4. 80% of the field at 40×   _____ mm   _____ μm

c. Determine and record these measurements:

The length of the letter *i*, including the dot   _____ mm   _____ μm

The diameter of the dot   _____ mm   _____ μm

5. Applications

a. When viewing the crossed hairs, with the top hair in sharp focus, is the other hair visible or in sharp focus at the following total magnifications?

Magnification	Visible	Sharp Focus
40×	_____	_____
100×	_____	_____
400×	_____	_____

b. Estimate the diameter of the blond hair. _____ μm

c. Describe the shape of the blond hair. _____

d. On the basis of your observations, are the following characteristics increased, decreased, or unchanged when magnification is increased?

Illumination _____   Depth of field _____

Working distance _____   Diameter of field _____

e. Draw five or six organisms observed in the pond water slides. Indicate the approximate size, color, and speed of motion of each. Make your drawings large enough to show details.

6. The Dissecting Microscope

a. When using the dissecting microscope, indicate the direction the image moves when the coin is moved

To the left _____ Toward you _____

b. Measure and record the diameter of field at each magnification of the dissecting microscope.

Magnification	Diameter of Field
20×	_____ mm
40×	_____ mm

7. Review

a. Contrast a prepared slide and a wet-mount slide.

Prepared slide _____

_____

Wet-mount slide _____

_____

b. How should prepared slides be handled? _____

_____

c. Describe how to determine the total magnification when using any objective of your microscope. _____

_____

d. Describe how to determine the length of an object observed with your microscope. _____

_____

_____

e. How should light intensity be adjusted when viewing nearly transparent specimens? _____

_____

_____

LABORATORY REPORT 3

The Cell

1. Prokaryotic Cells

Sketch the appearance of the prokaryotic cells of bacteria and cyanobacteria at 400× as observed from your slides.

Bacteria   **Cyanobacteria**

2. Eukaryotic Cell Structure

a. Write the term that matches each statement.
1. Sites of photosynthesis _____
2. Filaments of DNA and protein _____
3. Control center of the cell _____
4. Sites of aerobic cellular respiration _____
5. Supports plant cells _____
6. Separates the nucleus and cytoplasm _____
7. Sites of protein synthesis _____
8. Sacs of strong digestive enzymes _____
9. Packages materials in vesicles for export from the cell _____
10. Controls passage of materials into and out of the cell _____
11. Semiliquid substance in which cellular organelles are embedded _____
12. Large fluid-filled organelle in mature plant cells _____
13. Channels for the movement of materials within the cell _____
14. Assembles precursors of ribosomes within the nucleus _____
15. Nuclear components containing the genetic code controlling cellular processes _____

b. List the labels for Figures 3.2 and 3.3
Figure 3.2

1. _____   7. _____   13. _____
2. _____   8. _____   14. _____
3. _____   9. _____   15. _____
4. _____   10. _____   16. _____
5. _____   11. _____   17. _____
6. _____   12. _____   18. _____

Figure 3.3

1. _____   7. _____   13. _____
2. _____   8. _____   14. _____
3. _____   9. _____   15. _____
4. _____   10. _____   16. _____
5. _____   11. _____   17. _____
6. _____   12. _____

3. Onion Epidermal Cells

a. List the labels for Figure 3.4.

1. _____   3. _____   5. _____
2. _____   4. _____   6. _____

b. Diagram three adjacent onion epidermal cells as observed on your slide, and label the parts that you see.

c. Determine and record the average length and width of three adjacent cells.
   Width _____ mm   Length _____ mm

d. Examination of your slide will show that a few nuclei do not appear next to a cell wall. Explain that observation. _____
_____
_____

4. *Elodea* Leaf Cells

a. What structure gives an *Elodea* cell its shape? _____

b. Describe the shape of a cell. _____

c. What organelles are green? _____

d. What structure fills the greatest volume in a cell? _____

e. Draw a spine cell at the edge of the leaf at 100×. Label the observed parts.

f. Are the cells in each layer of the leaf about the same size? _____

g. Determine and record the average width and length of three cells in the upper cell layer.
   Width _____ mm   Length _____ mm

h. List the labels for Figure 3.6.

1. _____   4. _____   6. _____
2. _____   5. _____   7. _____
3. _____

5. Human Epithelial Cells

a. What structures observed in onion epidermal cells are *not* present in human epithelial cells.

_____

b. Indicate the description that best describes the human epithelial cells.

Thin and plate-like _____   Thick and box-like _____

c. Are human epithelial cells thinner than *Elodea* cells? _____

d. Diagram a few cells from your slides and label the parts observed.

e. List the labels for Figure 3.7.

1. _____   2. _____   3. _____

6. Amoeba

a. Diagram the amoeba on your slide and label the parts observed.

b. Is the plasma membrane rigid or flexible? _____

c. Is the amoeba alive? _____   Why do you think so? _____

_____

_____

d. Would you use this same evidence to determine if cells in your body are alive? _____

Explain. _____

_____

7. Review

a. Name the two characteristics of living organisms that distinguish them from nonliving things.

_____   _____

b. Indicate the structural and functional unit of life. _____

c. Based on your study and observations, indicate the presence of the following structures in animal and plant cells by placing an X in the appropriate spaces.

Structure	Animal	Green Plant	Nongreen Plant
Cell wall	_____	_____	_____
Central vacuole	_____	_____	_____
Centrioles	_____	_____	_____
Chloroplasts	_____	_____	_____
Chromatin granules	_____	_____	_____
Cytoplasm	_____	_____	_____
Endoplasmic reticulum	_____	_____	_____
Lysosomes	_____	_____	_____
Mitochondria	_____	_____	_____
Microtubules	_____	_____	_____
Nucleolus	_____	_____	_____
Nucleus	_____	_____	_____
Plasma membrane	_____	_____	_____
Ribosomes	_____	_____	_____

8. Mini-Practicum

Your instructor has set up "unknown" types of cells under several microscopes. Your task is to identify (a) the kind of cell (prokaryote, animal, or plant) and (b) the cellular structure indicated by the pointer. Record your answerers in the table below.

Microscope No.	Mitotic Stage	Cell Structure
1		
2		
3		
4		
5		

LABORATORY REPORT 4

Chemistry of Cells

1. Fundamentals

a. Write in the space provided the term that matches the definition.

_____ A substance composed of two or more elements chemically combined

_____ A substance that cannot be broken down into a simpler substance by chemical means

_____ The smallest particle of an element that exhibits the characteristics of that element

_____ The smallest particle of a compound that exhibits the characteristics of that compound

b. Consult Table 4.2 and indicate the (a) symbol and (b) number of protons, electrons, and neutrons in atoms of these elements.

Element	Symbol	Number of Protons	Number of Electrons	Number of Neutrons
Hydrogen	_____	_____	_____	_____
Oxygen	_____	_____	_____	_____
Nitrogen	_____	_____	_____	_____
Carbon	_____	_____	_____	_____
Chlorine	_____	_____	_____	_____
Calcium	_____	_____	_____	_____

2. Reactions Between Atoms

a. Using Table 4.2 and Figure 4.1, draw shell models of an atom of these elements.

Calcium   **Carbon**

b. Define:

Ionic bond _____

Covalent bond _____

c. Indicate the number of electrons in the outer shell of these neutral atoms and ions:

Na _____ K^+ _____   Mg _____ H^+ _____   Ca _____   Cl _____

d. A molecule formed by an ionic bond is shown below. Using Table 4.2, indicate the atoms involved, the chemical formula of the molecule, and the number of electrons transfered. Draw arrows to show the direction of electron transfer.

Atoms involved ————————————————

————————————————

————————————————

Chemical formula ————————————————

Number of electrons transfered ————————————————

++

17p
18n

20p
20n

–

–

17p
18n

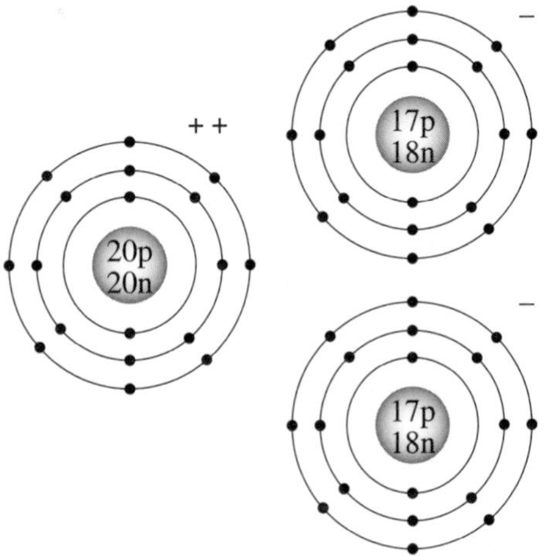

e. Using Table 4.3 as a guide, draw the structural formula of these molecules:

Methane (CH$_4$)   **Ammonia (NH$_3$)**

Carbon Dioxide (CO$_2$)   **Molecular Oxygen (O$_2$)**

3. pH

a. Define:

Acid ——

Base ——

b. Identify these pH values as either acid or base and rank them in order (left to right) of decreasing strength.

a. 10   b. 3   c. 2   d. 6   e. 8   f. 12   g. 1   h. 7.5

Acid: _____   _____   _____   _____   Base: _____   _____   _____   _____

c. Record the pH of these test solutions.

After-shave lotion _____ Vinegar _____ Household ammonia _____
Alka-Seltzer® _____ Detergent solution _____ Lemon juice _____
Unknown #1 _____ Unknown #2 _____ Unknown #3 _____

d. Indicate the number of drops of 1.0% HCl required to decrease the pH of these solutions from 7 to 6.

Distilled water _____   Buffer solution _____
Explain your results _____
_____
_____
_____
_____

4. Biological Molecules

a. Prepare a set of standards for the chemical tests, and record your results in the chart below.

Standards

Test	Organic Compound Testing for	Substance Tested	Results
Iodine	Starch	Starch	
		Water	
Benedict's	Reducing sugars	Glucose	
		Water	
Paper spot	Lipids	Corn oil	
		Water	
Sudan IV	Lipids	Corn oil	
		Water	
Biuret	Protein	Albumin	
		Water	

b. Why is the preparation of a set of standards important? _____
_____
_____

c. Perform the chemical tests on the substances provided. Compare your results with the standards and record your results in the chart below.

Indicate your test results as:

Strongly positive   +++
Moderately positive   ++
Slightly positive   +
Negative   0

Unknowns

		Test	Results			
Substance Tested	*Iodine*	*Benedict's*	*Paper spot*	*Sudan IV*	*Biuret*	Organic Compounds Present
Onion						
Potato						
Apple						
Egg white						

LABORATORY REPORT 5

Enzymes

1. Fundamentals

   a. What are the chemical subunits composing enzymes? _____

   b. What determines the specificity of an enzyme? _____

   c. What determines the three-dimensional shape of an enzyme? _____
   _____

   d. What is the role of enzymes in cells? _____
   _____

   e. Write the equation for the chemical reaction that catalase catalyzes.
   Circle the substrate and *underline* the products.

2. Catalase Activity

   a. Record the thickness (mm) of the foam layer after 1 minute.

Tube	Thickness of Foam Layer	Tube	Thickness of Foam Layer
1. Water		4. Apple	
2. Liver		5. Onion	
3. Ground beef		6. Potato	

   b. What is the function of Tube 1 containing water and peroxide? _____

   c. Do your results support the hypothesis? _____
   State a conclusion from your results. _____
   _____
   _____
   _____

   d. Among those tested, do animal or plant tissues exhibit greater catalyase activity? _____
   _____

   e. How do you explain this? _____
   _____
   _____
   _____

3. Liver Catalase and Temperature

a. Record the thickness (mm) of the foam layer after 1 minute.

Tube	Temperature	Thickness of Foam Layer
1		
2		
3	70° C	
4	100° C	

b. Do your results support the hypothesis? _____

State a conclusion from your results. _____

_____

_____

4. Liver Catalase and pH

a. Record the thickness (mm) of the foam layer after 1 minute.

Tube	pH	Thickness of Foam Layer
1	4	
2	6	
3	8	
4	10	

b. Do your results support the hypothsis? _____

State a conclusion from your results. _____

_____

_____

5. Summary

a. Is catalase active in a narrow or broad range of:

temperatures? _____   pH values? _____

b. At the molecular level, what causes the inactivation of catalase?

_____

_____

c. Do you think that there is a correlation between the widespread presence of catalase among organisms and the ranges of temperature and pH within which this enzyme is active? _____

_____   Explain. _____

_____

_____

_____

_____

LABORATORY REPORT 6

Diffusion and Osmosis

1. Brownian Movement

    a. Is Brownian movement a living process? _____

    b. Explain why the dye particles are moving. _____

    _____

2. Diffusion and Temperature

    a. What natural phenomenon enables diffusion to occur? _____

    _____

    b. Define diffusion. _____

    _____

    c. In a 10% glucose solution, what is the percentage of water? _____

    d. What is the independent variable in the experiment? _____

    e. What is the dependent variable in the experiment? _____

    f. Record the water temperature for each beaker.   A _____ °C   B _____ °C

    g. After 15 min, compare the rate of diffusion in beakers A and B.

          A _____   B _____

    h. Do the results support the hypothesis? _____

    i. What can you conclude about the relationship between temperature and the rate of diffusion?

    _____

3. Diffusion and Molecular Mass

    a. What is the independent variable in the experiment? _____

    b. What is the dependent variable? _____

    c. Record the diameter (mm) of the colored circles after 1 hr.

          Potassium permanganate _____ mm   Methylene blue _____ mm

    d. Did the results of the experiment support the hypothesis? _____

    e. What can you conclude about the relationship between molecular mass and the rate of diffusion? _____

    _____

4. Diffusion and Molecular Size

a. After 20 min, describe any color change in your setup. _____

_____

b. Explain any color change. _____

_____

_____

c. Did starch diffuse from the sac? _____ How do you know? _____

_____

d. Did glucose diffuse from the sac? _____ How do you know? _____

_____

e. Considering starch, glucose, iodine, and water, which substances were able to pass through the sac. _____

5. Osmosis

a. Define osmosis. _____

_____

_____

b. If 10% and 5% salt solutions are separated by a membrane permeable to both salt and water, indicate by an arrow the direction in which:

salt will diffuse.

10% sol'n	5% sol'n

water will diffuse.

10% sol'n	5% sol'n

c. Consider 100 ml of 10% sucrose and 10 ml of 50% sucrose. Which has the:

greater concentration of sucrose? _____

greater quantity of sucrose? _____

d. What is the independent variable? _____

e. What is the dependent variable? _____

f. What is the purpose of the sac of water? _____

g. Record the mass of the sacs in the chart below.

	20% Sucrose	10% Sucrose	100% Water
After 1 hr	_____ g	_____ g	_____ g
Start	_____ g	_____ g	_____ g
Increase	_____ g	_____ g	_____ g
% increase*	_____	_____	_____

$$* \text{ \% increase} = \frac{\text{mass increase}}{\text{mass at start}}.$$

h. Do the results support the hypothesis? _____

i. What can you conclude from the results? _____

_____

_____

_____

6. Osmosis and Living Cells

a. Considering hypotonic and hypertonic solutions, which solution has the:

greater concentration of water? _____

greater concentration of solutes? _____

b. Indicate with an arrow the direction that water moves between hypotonic and hypertonic solutions separated by a selectively permeable membrane.

Hypotonic solution	Hypertonic solution

c. Consider a 5% sucrose solution separated from a 10% sucrose solution by a semipermeable membrane.

Which solution is hypotonic? _____

Which solution is hypertonic? _____

d. Indicate the surrounding fluid that caused the celery sticks to be:

more flexible. _____ more crisp. _____

Explain what produced this result. _____

_____

_____

e. Draw the distribution of chloroplasts in an *Elodea* cell mounted in:

Water   **Salt Solution**

f. Explain what happened to *Elodea* cells mounted in salt solution. _____

_____

_____

g. What happened to the cells when the Elodea leaf was remounted in water? _____

_____

Explain why this occured. _____

_____

_____

h. Explain the basis of preserving foods with salt or sugar. _____

_____

_____

_____

LABORATORY REPORT 7

Photosynthesis

1. Carbon Dioxide and Photosynthesis

a. What is the independent variable? _____

Indicate the presence or absence of the independent variable.

Plant A _____   Plant B _____

b. What is the dependent variable? _____

c. Record your results: the presence or absence of starch.

Plant A _____   Plant B _____

d. Do the results support the hypothesis? _____

State a conclusion from your results. _____

_____

_____

_____

_____

2. Light and Photosynthesis

a. What is the independent variable? _____

b. What is the dependent variable? _____

c. Trace or sketch your leaf, showing the position of the leaf shield.

Indicate your results by showing on the diagram of your leaf where starch is present and absent.

d. Do the results support the hypothesis? _____

State a conclusion from your results. _____

_____

_____

3. Chlorophyll and Photosynthesis

   a. What is the independent variable? _____

   b. What is the dependent variable? _____

   c. Trace or sketch half a leaf, showing the distribution of chlorophyll.

   Indicate your results by showing on the diagram of your leaf where starch is present and absent.

   d. Do the results support the hypothesis? _____

   State a conclusion from your results. _____

   _____

   e. Indicate the presence or absence of sugar.

       Green leaf tissue _____   Nongreen leaf tissue _____

   Explain your results. _____

   _____

   _____

   _____

4. Summary

   a. Write the term that matches the phrase.

   1. Organelle in which photosynthesis occurs.   _____

   2. Form of energy captured by photosynthesis   _____

   3. Form of energy formed by photosynthesis   _____

   4. End product of photosynthesis   _____

   5. Storage form of carbohydrates in plants   _____

   6. Source of O_2 formed by photosynthesis   _____

   7. Source of oxygen atoms in glucose formed by
      photosynthesis   _____

   8. Source of hydrogen atoms in glucose formed by
      photosynthesis   _____

   9. Source of carbon atoms in glucose formed by
      photosynthesis   _____

   b. List the key events in each reaction:

   Light reaction _____

   _____

   Dark reaction _____

   _____

5. Chloroplast Pigments

a. Attach your chromatogram and label the original pigment line and the bands of carotene, chlorophyll a, chlorophyll b, and xanthophyll.

b. Indicate the chlorophyll type that was:

More soluble _____   Less soluble _____

c. Explain why all leaf pigments are not visible in a healthy leaf.

_____

_____

d. How do you explain the appearance of yellow pigments in many plant leaves in autumn?

_____

_____

e. Examine Figure 7.5 and determine:

1. The three parts of the spectrum responsible for most of photosynthesis.

_____   _____   _____

2. Colors of light absorbed primarily by chlorophyll a.

_____   _____

3. Colors of light absorbed primarily by chlorophyll b.

_____   _____

6. Light Quantity and the Rate of Photosynthesis

a. What is the independent variable? _____

b. What is used to indicate the rate of photosynthesis? _____

_____

Why is this a good indicator of photosynthetic rate? _____

_____

c. Record your data in the chart below.

Distance from Light Source (cm)	Readings (ml)		ml O₂/3 min	ml O₂/min
	Start	Stop		
25	_____	_____	_____	_____
	_____	_____	_____	_____
	_____	_____	_____	_____
	Average		_____	_____
50	_____	_____	_____	_____
	_____	_____	_____	_____
	_____	_____	_____	_____
	Average		_____	_____
75	_____	_____	_____	_____
	_____	_____	_____	_____
	_____	_____	_____	_____
	Average		_____	_____
100	_____	_____	_____	_____
	_____	_____	_____	_____
	_____	_____	_____	_____
	Average		_____	_____

d. Do the results support the hypothesis? _____ State a conclusion from your results.

_____

_____

_____

_____

e. Explain why your conclusion should be modified to "within the range of values tested"?

_____

_____

_____

_____

Student _____

LABORATORY REPORT 8

Lab Instructor _____

Cellular Respiration

1. **Introduction**
   a. Write the summary equation for the aerobic respiration of glucose. *Underline* the reactants and *circle* the products.

   b. Indicate the products of anaerobic cellular respiration of glucose in:

   Animals _____ Yeast _____

   c. Indicate the number of ATP molecules produced per glucose molecule respired by:

   Aerobic respiration _____ Anaerobic respiration _____

   d. What molecule provides immediate energy for cellular work? _____

   e. Some energy released by cellular respiration is captured in high-energy phosphate bonds. What happens to these high-energy phosphates? _____

   _____

   _____

   What happens to released energy that is not captured? _____

   _____

   _____

   f. What regulates cellular respiration so that energy is released gradually? _____

   _____

2. **Respiration and Carbon Dioxide Production**
   a. Record the results of your experiments in the table below.

   Carbon Dioxide Production in Animals and Germinating Seeds

Tube	Contents Plus Bromthymol Blue	Color of Bromthymol Blue After Test Interval	CO_2 Concentration	
			Increase	No Change
Exp. 1: Humans				
1	Exhaled air	_____	_____	_____
2	Atmospheric air	_____	_____	_____
Exp. 2: Germinating Peas and Crickets				
1	Germinating peas	_____	_____	_____
2	Live crickets	_____	_____	_____
3	Atmospheric air	_____	_____	_____

b. What is the purpose of tube 2 in Experiment 1 and tube 3 in Experiment 2? ______________
____

c. Do the results support the hypothesis? ______________________________
d. State a conclusion from the results of Experiment 1. ______________________
____
____

e. State a conclusion from the results of Experiment 2. ______________________
____
____

3. Respiration and Heat Production

a. Which bottle serves as the control? ______________________________
b. Record the temperature in each vacuum bottle.

Bottle 1 ____________ °C Bottle 2 ____________ °C Bottle 3 ____________ °C

c. Do the results support your hypothesis? ______________________________
d. State a conclusion from your results. ______________________________
____
____

4. Temperature and Respiration Rate: Peas and Crickets

a. Record the data collected for the rate of respiration in germinating pea seeds and crickets at 10°C, room temperature, and 40 °C. *Add your room temperature to the chart.

Oxygen Consumption in Pea Seeds and Crickets

Organism	*Temp. (°C)*	*ml O_2/3 min 5 Replicates*					*Average ml O_2/3 min*	*Total Mass (g)*	*Average ml O_2/hr/g*
		1	*2*	*3*	*4*	*5*			
Peas	10								
	*								
	40								
Crickets	10								
	*								
	40								
Control	10								
	*								
	40								

b. Do the results for peas support the hypothesis? ______________________
State a conclusion from the results. ______________________________
____

c. Do the results for crickets support the hypothesis? ______________________
State a conclusion from the results. ______________________________
____

d. Which organism has the greater average rate of aerobic respiration per gram of body mass? _____

5. Temperature and Respiration Rate: Frog and Mouse

a. Record the data collected for the rate of respiration in a frog and a mouse at 10 °C, room temperature, and 40 °C. *Add your room temperature to the chart.

Oxygen Consumption in a Frog and a Mouse

Organism	Temp. (°C)	ml O_2/3 min 5 Replicates					Average ml O_2/3 min	Total Mass (g)	Average ml O_2/hr/g
		1	2	3	4	5			
Frog	10								
	*								
	40								
Mouse	10								
	*								
	40								

b. What is the independent variable? _____

c. Do the results support the hypothesis? _____

d. Plot the average respiration rate (ml O_2/hr/g) of each organism tested in the graph below. Select and record a different symbol for each organism so your graph may be easily read.

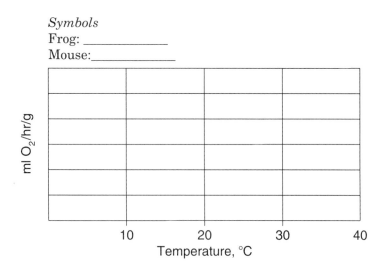

Symbols
Frog: _____
Mouse: _____

e. State a conclusion about the effect of temperature on the rate of aerobic respiration in the frog.

_____

_____

_____

_____

_____

_____

f. State a conclusion about the effect of temperature on the rate of aerobic respiration in the mouse. _____

_____

_____

_____

_____

_____

g. Explain the different responses to 10 °C by the frog and the mouse.

_____

_____

_____

_____

_____

_____

h. What factors or conditions other than temperature may alter the rate of cellular respiration?

_____

_____

_____

_____

LABORATORY REPORT 9

Cell Division

1. Introduction

Write the type of cell division that matches each phrase.

1. Occurs in prokaryote cells   _____

2. Occurs in haploid eukaryote cells   _____

3. Occurs in diploid eukaryote cells   _____

4. Forms cells with identical genetic composition   _____

5. Forms cells with half the chromosome number of
   the parent cell   _____

2. Mitotic Division

a. Write the term that matches each phrase.

1. Compose a replicated chromosome   _____

2. Division of the cytoplasm   _____

3. 5–10% of the cell cycle   _____

4. 90–95% of the cell cycle   _____

5. Stage of interphase where chromosomes replicate   _____

6. Members of a chromosome pair   _____

7. Diploid chromosome number in humans   _____

8. Haploid chromosome number in humans   _____

b. Write the name of the mitotic stage that matches each phrase.

1. Nuclear membrane and nucleolus disappear   _____

2. Spindle is formed   _____

3. Separation of sister chromatids   _____

4. Daughter cells are usually formed   _____

5. Rod-shaped chromosomes are first visible   _____

6. Chromosomes line up on equatorial plane   _____

7. New nuclei are formed   _____

8. Daughter chromosomes migrate to opposite poles
   of the cell   _____

c. Draw whitefish blastula cells in the stages noted below as they appeared on your slide. Label pertinent structures.

Interphase	Prophase	Metaphase

Early Anaphase	Late Anaphase	Telophase

d. Draw onion root tip cells in the following stages as they appeared on your slide. Label pertinent structures.

Prophase	Anaphase	Telophase

e. Describe how mitotic cell division in plants differs from that in animal cells. _____
   _____
   _____

f. In the onion root tip, do the daughter cells occupy the same column of cells as the parent cell? _____ Is it the same for all cells in the root tip? _____

g. Describe the process of mitotic cell division in your own words without using the names of the phases. _____
   _____
   _____
   _____
   _____

3. Meiotic Division

a. Considering meiosis in humans, indicate the:
  1. Number of chromosomes in the parent cell _____
  2. Number of chromatids in daughter cells formed
     by first division _____
  3. Number of chromosomes in daughter cells formed
     by first division _____
  4. Number of chromosomes in daughter cells of
     second division _____
  5. Number of haploid cells formed by meiotic
     division of parent cell _____

b. Diagram the arrangement of the chromosomes in mitosis and meiosis as they would appear in
   the phases noted below where the diploid (2n) chromosome number is 4.

   Mitosis

   | **Metaphase** | **Anaphase** |

   Meiosis I

   | **Metaphase** | **Anaphase** |

   Meiosis II

   | **Metaphase** | **Anaphase** |

   c. Indicate the number of chromosomes in human cells formed by:

      Mitotic cell division _____   Meiotic cell division _____

   d. Summarize in your own words the process of meiosis without using the names of the phases.

_____

_____

_____

_____

_____

_____

   e. Meiotic division is the process leading to the formation of specific cells in animals and plants. Name these cells.

      Animals _____   Plants _____

   f. Indicate the number of chromosomes in a human:

      Egg cell _____   Sperm _____

   g. (True/False)

      The zygote formed by the union of haploid gametes receives:

        A random number of chromosomes from each

          parent   _____

        Equal numbers of chromosomes from each parent   _____

        One member of each chromosome pair from each

          parent   _____

        Both members of each chromosome pair from each

          parent   _____

4. Mini-Practicum

Your instructor has set up cells in various stages of mitotic division under several microscopes. Identify (a) the mitotic stage and (b) the cellular structure indicated by the pointer. Record your answers in the table below.

Microscope No.	Mitotic Stage	Cell Structure
1		
2		
3		
4		
5		

LABORATORY REPORT 10

Student _____

Lab Instructor _____

Heredity

1. **Fundamentals**

   Write the term that matches each phrase.

   a. Traits passed from parents to progeny _____

   b. Part of DNA coding for a specific protein _____

   c. Contain homologous genes _____

   d. Alternate forms of a gene _____

   e. Alleles of a gene pair are identical _____

   f. Alleles of a gene pair are different _____

   g. Observable form of a trait _____

   h. Genetic composition determining a trait _____

   i. Allele expressed in heterozygote _____

   j. Allele not expressed in heterozygote _____

2. **Monohybrid Crosses with Dominance**

   a. In corn plants, plant height is controlled by one gene with two alleles: tall (T) and dwarf (t). How many alleles that control the trait for height are present in each cell nucleus of a corn plant? _____

   How many alleles that control the trait for height are present in the nucleus of each gamete formed by a corn plant? _____

   b. Indicate the genotypes of these corn plants.

   homozygous tall _____ heterozygous tall _____ dwarf _____

   c. Indicate the genotypes of possible gametes of these corn plants.

   homozygous tall _____ heterozygous tall _____ dwarf _____

   d. Determine the predicted phenotype ratio in progeny of a cross between two monohybrid tall corn plants.

   Parent phenotypes _____ _____

   Parent genotypes _____ _____

   Gametes _____ _____ _____ _____

   Punnett square _____ _____

   Genotype ratio _____

   Phenotype ratio _____

e. Examine the corn seedlings that are progeny of a cross of two monohybrid tall corn plants. Record the number of:

Tall plants _____   Dwarf plants _____

Determine the observed phenotype ratio of tall plants to dwarf plants by dividing the number of plants of each type by the number of dwarf plants. Round your answers to whole numbers.

_____ Tall ÷ _____ dwarf = _____   _____ Dwarf ÷ _____ dwarf = 1

Record the observed phenotype ratio: _____ Tall : 1 dwarf

Is the observed ratio different from the predicted ratio? _____

If so, explain why this may have happened. _____

_____

f. Is it possible to observe differences between homozygous and heterozygous tall corn plants?

_____

How can you determine their genotypes? _____

g. In a test cross, a tall corn plant is crossed with a _____ corn plant. Indicate the genotype of the tall corn plant when:

all the progeny are tall.

half the progeny are tall.

h. Indicate the possible genotypes of parent pea plants in crosses yielding the following ratios.

Phenotype Ratio	**Parent Genotypes**
1. All white flowers	_____ × _____
2. All purple flowers (3 possibilities)	_____ × _____
	_____ × _____
	_____ × _____

i. Mary has freckles, but her husband Dick does not. Mary's father has freckles but her mother does not. What is the probability that Mary and Dick's child will have freckles? _____

	Mary	**Dick**
Parent phenotypes	_____	_____
Parent genotypes	_____	_____
Gametes	_____ _____	_____ _____

Punnett square

Genotype ratio   _____

Phenotype ratio   _____

j. Newlyweds Bill and Sue are nonfreckled. Since each had one parent who had freckles, they wonder what the probability is of their children having freckles. What would you tell them?

_____

_____

_____

k. Judy and Tom are wondering what kind of hairline their soon-to-be-born baby will have. Judy has a straight hairline but Tom has a widow's peak. What is Judy's genotype? _____ What are the two possible genotypes for Tom? _____ What is the probability of their baby having a widow's peak if Tom is:

      homozygous? _____   heterozygous? _____

l. The table below lists several human traits that are determined by a simple dominant/recessive mode of inheritance. Record your phenotypes and the frequency of the phenotypes among members of your class.

Human Dominant/Recessive Traits Determined by a Single Gene*

Trait	Phenotype	Your Phenotype	Number in Class with Phenotype
Handedness	Right handed*		
	Left handed		
Ear lobes	Free*		
	Attached		
Freckles	Freckled*		
	Nonfreckled		
Hair line	Widow's peak*		
	Straight		
Little finger	Bent*		
	Straight		
Tongue roll	Yes*		
	No		
Rh factor	Rh^+*		
	Rh^-		

*Indicates dominant traits.

m. For each trait, is the dominant phenotype always more abundant among class members?

_____

n. Assuming your class to be representative of the general population, what can you conclude about the relationship between dominant and recessive phenotypes and their frequencies in the population? _____

3. Incomplete Dominance and Codominance

a. What is the genotype of a pink-flowering snapdragon? _____

b. Indicate the predicted genotype and phenotype ratios in progeny from crossing two pink-flowering snapdragons.

Genotype ratio _____

Phenotype ratio _____

   c. What type of cross would you use to produce progeny that are 100% pink-flowering?

_____

   d. What is the probability that a child with sickle-cell anemia will be born to parents that are each heterozygous for sickle-cell? _____

4. Multiple Alleles: ABO Blood Types

   a. Indicate the expected genotype and phenotype ratios for these matings:

     1. $I^A I^B \times ii$

       Genotype ratio: _____

       Phenotype ratio: _____

     2. $I^A i \times I^B i$

       Genotype ratio: _____

       Phenotype ratio: _____

   b. (True/False)

     _____ A type O child may have two type A parents.

     _____ A type O child may have one type AB parent.

     _____ A type B child may have two type AB parents.

     _____ A type AB child may have one type O parent.

   c. Ann (type A, Rh^+ blood) is suing Joe (type AB, Rh^- blood) for child support claiming that he is the father of her child (type B, Rh^+ blood). On the basis of blood type, is it possible that Joe is the father? _____ What ABO blood type(s) would Joe have to possess to *prove* that he is *not* the father? _____

5. Dihybrid Crosses with Dominance

   a. Determine the predicted phenotype ratio in children of dihybrid parents that are both righthanded and possess free earlobes.

Parent phenotypes   _____   _____

Parent genotypes   _____   _____

Gametes   ____   ____   ____   ____   ____   ____   ____   ____

Punnett square

	____	____	____	____
____				
____				
____				
____				

Phenotype ratio   _____ right-handed free earlobes   _____ left-handed free earlobes

                _____ right-handed attached earlobes   _____ left-handed attached earlobes

b. Determine the predicted phenotype ratio in children of a woman who is dihybrid for right-handed and free earlobes and a man who is left-handed with attached earlobes.

Parent phenotypes   _____   _____

Parent genotypes   _____   _____

Gametes   ____ ____ ____ ____   ____ ____ ____ ____

Punnett square

Phenotype ratio   _____ right-handed   _____ left-handed
                     free earlobes   free earlobes
            _____ right-handed   _____ left-handed
                 attached earlobes   attached earlobes

6. Linked Genes

a. Indicate the types of gametes produced by an AaBb dihybrid when the:

genes are not linked. _____ _____

genes are linked. _____

b. Determine the predicted genotype and phenotype ratios in children of a noncarrier woman with normal vision and a color-blind man.

Parent phenotypes   _____   _____

Parent genotypes   _____   _____

Gametes   ____ ____   ____ ____

Punnett square

Genotype ratio   _____

Phenotype ratio   _____

c. From which parent does a color-blind son inherit the gene for color blindness? _____

7. Pedigree Analysis

Examine the pedigrees shown below, and determine whether the inherited trait (■ or ●) is dominant, recessive, or sex-linked recessive; □ = male; ○ = female.

a.

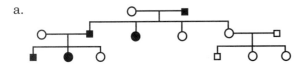

_____

b.

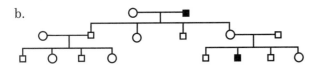

_____

c.

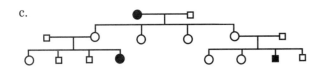

_____

8. Polygenic Inheritance

a. Is it likely that the length of a person's little finger is determined by polygenes? _____

Explain your response. _____

_____

b. List several human traits that seem to be determined by polygenes. _____

_____

9. Chi-Square Analysis

a. In the example of chi-square determination in Table 10.6, what is the hypothesis being tested?

_____

_____

Would the hypothesis be supported by a chi square of 3.952? _____

b. Record your results and do a chi-square analysis.

Chi-Square Analysis of Progeny from a Monohybrid Cross

Phenotype	Actual Results	Expected Results	Deviation (d)	(d^2)	(d^2/e)
Purple kernels	_____	_____	_____	_____	___/___ = ___
White kernels	_____	_____	_____	_____	___/___ = ___

$$\Sigma(d^2/e) = \text{\_\_\_\_}$$
$$\chi^2 = \text{\_\_\_\_}$$

p falls between _____ and _____ .

Is the predicted ratio supported by chi square? _____

LABORATORY REPORT 11

Molecular and Chromosomal Genetics

1. DNA

a. Write the term that matches each phrase.
   1. Molecule containing genetic information _____
   2. Sugar in DNA nucleotides _____
   3. Base that pairs with adenine _____
   4. Base that pairs with cytosine _____
   5. Chemical bonds joining complementary nitrogen
      bases _____
   6. Number of nucleotide strands in DNA _____
   7. Two molecules forming sides of the DNA "ladder" _____

b. In DNA replication, what determines the sequence of nucleotides in the new strands of nucleotides that are formed? _____
   _____
   _____

c. After determining the pairing pattern of the nitrogenous bases in Figure 11.2, add the missing bases to this hypothetical strand of DNA.

A = adenine, C = cytosine, G = guanine, T = thymine

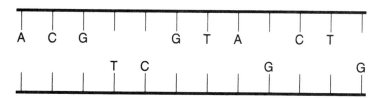

d. At the left is a hypothetical segment of DNA. After replication it forms two strands as shown at the right. Add the missing bases in the replicated strands.

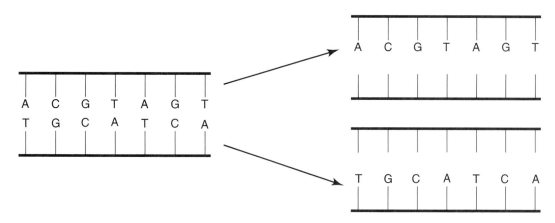

2. RNA Synthesis

a. Write the term that matches each meaning.

1. Sugar in an RNA nucleotide   _____
2. Number of nucleotide strands in RNA   _____
3. RNA base that pairs with adenine   _____
4. Template for RNA synthesis   _____

b. The figure below shows a portion of DNA molecule whose strands have separated to synthesize mRNA. The bases are shown for the strand that is *not* serving as the template. Add the bases of the DNA strand serving as the template and the bases of the newly formed RNA molecule.

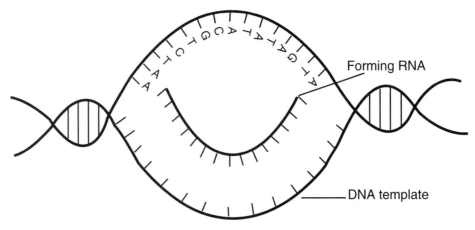

3. Protein Synthesis

a. Write the term that matches each meaning.

1. Carries the genetic code to ribosomes   _____
2. Formed of an mRNA base triplet
3. Specifies a particular amino acid   _____
4. Sites of protein synthesis   _____
5. Carries amino acids to ribosome   _____
6. tRNA triplet that pairs with codon   _____
7. Codon that starts polypeptide synthesis   _____
8. Codons that stop polypeptide synthesis   _____

b. Indicate the DNA base triplet and tRNA anticodons for the mRNA codons in Figure 11.4.

Amino Acid	DNA Base Triplet	Anticodon
Tryptophan	_____	_____
Lysine	_____	_____
Valine	_____	_____
Methionine	_____	_____

c. Using Table 11.1, indicate the possible sequences of mRNA codons that code for the polypeptides below.

methionine - lysine - histidine - tryptophan - glutamine - tryptophan - (stop)

——————— ——— ———————— ————————— —————————— —————————— ———

——— ———————— ————————— —————————— ———

—————

d. Molecular biologists use probes of mRNA to locate specific gene loci on DNA segments. Use the codon sequences determined in item 3c as probes to locate the gene (complementary base sequence) in the segment of template (single-strand) DNA shown below. *Circle* the located gene and indicate the probe that was successful in locating it.

-A-G-T⊦T-C-T⊦T-A-C⊦C-C-T⊦G-A-A⊦C-G-G⊦C-A-T⊦T-C-A⊦
-G-A-C⊦A-T-T⊦C-C-G⊦C-T-A⊦G-A-C⊦C-A-T⊦T-A-C⊦T-T-C⊦
-G-T-A⊦A-C-C⊦G-T-T⊦A-C-C⊦A-T-C⊦G-T-A⊦T-C-A-

Successful probe ————————————————————————————————————

e. Consider the following hypothetical gene. Indicate the mRNA codons that it forms and the sequence of amino acids produced in the polypeptide chain.

-T-A-C⊦T-T-A⊦G-A-A⊦A-T-A⊦C-C-G⊦A-A-G⊦A-C-T-

mRNA codons

——

Amino acid sequence

——

f. For the following mutations, show the effect by indicating the mRNA codons and the amino acid sequence in the polypeptide.

1. Base substitution

-T-A-C⊦T-T-A⊦G-A-G⊦A-T-A⊦C-C-G⊦A-A-G⊦A-C-T-

mRNA codons

——

Amino acid sequence

——

2. Triplet addition

-T-A-C⊦T-T-A⊦G-G-T⊦G-A-A⊦A-T-A⊦C-C-G⊦A-A-G⊦A-C-T

mRNA codons

——

Amino acid sequence

——

3. Base addition

-T-A-C┤T-T-A┤G-A-A┤A-T-C-A┤C-C-G┤A-A-G┤A-C-T

mRNA codons

_____

Amino acid sequence

_____

4. Base deletion (site indicated by arrow)

↓

-T-A-C┤T-T-A┤G-A-A┤A-T┤C-C-G┤A-A-G┤A-C-T

mRNA codons

_____

Amino acid sequence

_____

g. What seems to determine the magnitude of a mutation's effect? _____

_____

_____

_____

h. Summarize how DNA controls cellular functions. _____

_____

_____

_____

_____

_____

4. Chromosomal Abnormalities

a. Prepare karyotypes for patients A, B, and C on the following three pages.

b. Indicate the sex and chromosomal abnormality, if any, for each patient whose karyotype you analyzed.

Patient	Sex	Abnormality
A	_____	_____
B	_____	_____
C	_____	_____

c. Amniocentesis is used to obtain fetal cells from the amniotic fluid for chromosomal analysis. If you were a prospective parent of a fetus with a chromosomal abnormality, what factors would you consider in deciding whether to abort the fetus? _____

_____

_____

_____

_____

Human Karyotype Analysis Form
Patient A

A

1	2	3

B

4	5

C

6	7	8	9	10	11	12	X

D

13	14	15

E

16	17	18

F

19	20

G

21	22	Y

Sex of subject _____   Number of chromosomes _____

Chromosomal disorder _____

Human Karyotype Analysis Form
Patient B

A

B

| 1 | 2 | 3 | 4 | 5 |

C

| 6 | 7 | 8 | 9 | 10 | 11 | 12 | X |

D

E

| 13 | 14 | 15 | 16 | 17 | 18 |

F

G

| 19 | 20 | 21 | 22 | Y |

Sex of subject _____   Number of chromosomes _____
Chromosomal disorder _____

Human Karyotype Analysis Form
Patient C

A

B

| 1 | 2 | 3 | 4 | 5 |

C

| 6 | 7 | 8 | 9 | 10 | 11 | 12 | X |

D

E

| 13 | 14 | 15 | 16 | 17 | 18 |

F

G

| 19 | 20 | 21 | 22 | Y |

Sex of subject _____   Number of chromosomes _____

Chromosomal disorder _____

LABORATORY REPORT 12

DNA Fingerprinting

1. INTRODUCTION

   a. Write the term that matches the phrase.

    1. DNA formed of DNA fragments from two different organism _____

    2. Enzymes cutting DNA at specific sites to form restriction fragments _____

    3. Technique used to separate restriction fragments in this exercise _____

    4. A person's unique RFLP pattern _____

   b. What is the source of restriction endonucleases? _____

   c. Consider three restriction endonucleases: 1 recognizes a four-nucleotide site; 2 recognizes a five-nucleotide site; and 3 recognizes a six-nucleotide site. Indicate which restriction endonuclease will produce:

    1. The greatest number of restriction fragments. _____

    2. The fewest number of restriction fragments. _____

   d. What determines the length of restriction fragments? _____

   e. What are tandemly arranged repeats? _____

   f. How do tandemly arranged repeats affect the lengths of restriction fragments? _____

   g. A single-stranded sticky end with a nucleotide sequence of -A-G-C-T can combine with a complementary sticky end with a nucleotide sequence of _____

   h. In agarose gel electrophoresis of DNA restriction fragments, the restriction fragments:

    1. Migrate toward which pole? _____

    2. Possess, at a neutral pH, an electrical charge that is _____

    3. Are separated according to their _____

   i. If DNA fingerprinting can positively identify an individual, why is it better than regular fingerprints? _____

   j. What are some possible uses of DNA fingerprinting? _____

k. Some persons advocate determining and recording the DNA fingerprint of each newborn baby. What are some pros and cons of such a practice?

Pro. _____

_____

_____

Con. _____

_____

_____

e. DNA fingerprints reflect genetic variability. List a few genetic mechanisms that increase genetic variability. _____

_____

2. READING THE GEL

a. Draw the bands of restriction fragments as they appear on your gel. Circle the band(s) of the smallest restriction fragments.

b. According to your data, which suspect was present at the crime scene?. _____

c. Why was it important to use two restriction endonucleases in determining the DNA fingerprints of the samples? _____

_____

d. Does the use of two or more restriction endonucleases increase the validity of DNA fingerprinting? _____ Explain. _____

_____

_____

Student _____

LABORATORY REPORT 13   Lab Instructor _____

Organization of the Human Body

1. Organs and Organ Systems

Match the organ systems to the phrases below.

a. Cardiovascular   e. Lymphatic   i. Reproductive, female
b. Digestive   f. Muscular   j. Respiratory
c. Endocrine   g. Nervous   k. Skeletal
d. Integumentary   h. Reproductive, male   l. Urinary

Organs

_____ Pancreas, ovaries, testes   _____ Brain and spinal cord
_____ Stomach, liver, intestines   _____ Epidermis and dermis
_____ Bones, cartilages, ligaments   _____ Heart, blood vessels
_____ Spleen, lymph nodes, tonsils   _____ Kidneys, urinary bladder
_____ Trachea, bronchi, lungs   _____ Uterus, oviducts
_____ Thyroid and adrenal glands   _____ Esophagus, liver
_____ Testes, vas deferens, urethra   _____ Urethra, ureters
_____ Sensory receptors, nerves   _____ Pancreas, salivary glands

Functions

_____ Chemical coordination of body functions
_____ Transports materials throughout the body
_____ Coordination of body functions by neural impulses
_____ Protects body from ultraviolet radiation
_____ Contractions enable body movements
_____ Converts nonabsorbable food into absorbable nutrients
_____ Removes metabolic wastes and excess minerals from the body
_____ Exchange of gases between atmospheric air and blood
_____ Collects, cleans, and recycles extracellular fluid
_____ Supporting framework of the body
_____ Prevents excessive evaporative water loss from body surface
_____ Produces urine and removes it from the body

2. Body Cavities

a. List the labels for Figure 13.1.

1. _____   7. _____
2. _____   8. _____
3. _____   9. _____
4. _____   10. _____
5. _____   11. _____
6. _____   12. _____

b. Match the organ with the body cavity in which it is located.

1. Abdominal   3. Pelvic   5. Thoracic

2. Cranial   4. Spinal   6. None

_____ Brain	_____ Liver	_____ Bronchi
_____ Urinary bladder	_____ Heart	_____ Stomach
_____ Adrenal gland	_____ Pancreas	_____ Rectum
_____ Spinal cord	_____ Lungs	_____ Thyroid gland
_____ Testes	_____ Spleen	_____ Uterus
_____ Esophagus	_____ Kidney	_____ Trachea
_____ Duodenum	_____ Colon	_____ Ovaries

3. Epithelial Tissues

a. Draw the following epithelial tissues as they appear on your slides net 100×.

Simple Columnar   **Simple Cuboidal**

**Pseudostratified
Ciliated Columnar**   **Stratified Squamous**

b. Match the epithelial tissue with the phrase.

1. Simple squamous   3. Simple ciliated columnar   5. Pseudostratified ciliated columnar

2. Simple cuboidal   4. Simple nonciliated columnar   6. Stratified squamous

_____ Forms kidney tubules	_____ Lines digestive tract
_____ Lines upper respiratory passages	_____ Lines oviducts
_____ Contains goblet cells	_____ Lines blood vessels
_____ Epidermis of skin	_____ Lines body cavities

4. Connective Tissues

a. Draw the following connective tissues as they appear on your slides at 100×.

| **Loose Fibrous Conn. Tissue** | **Adipose Tissue** | **Dense Fibrous Conn. Tissue** |

| **Hyaline Cartilage** | **Fibrocartilage** | **Bone** |

b. Match the connective tissue with each phrase.

1. Loose fibrous	4. Hyaline cartilage	7. Bone
2. Adipose	5. Elastic cartilage	8. Blood
3. Dense fibrous	6. Fibrocartilage	

_____ Forms ligaments and tendons   _____ Forms cartilage of the nose

_____ Supports trachea and larynx   _____ Attaches skin to muscles

_____ Fat storage tissue   _____ Supports external ear

_____ Forms intervertebral discs   _____ Has a liquid matrix

_____ Forms dermis of skin   _____ Covers ends of long bones

5. Muscle Tissue

a. Draw the following muscle tissues as they appear on your slides. Indicate the magnification for each drawing.

| **Smooth Muscle** | **Skeletal Muscle** | **Cardiac Muscle** |

b. Match the muscle tissue with the phrase.

1. Skeletal   2. Cardiac   3. Smooth

_____ Multinucleated cells   _____ Striations present

_____ Striations absent   _____ Voluntary

_____ Involuntary   _____ Wall of heart

_____ Wall of stomach   _____ Attached to bones

6. Nerve Tissue

a. Draw one or two neuron cell bodies with their neuron processes as they appear on your slide. Indicate the magnification used.

b. Name the two types of cells composing nerve tissue.

_____   _____

Which cells conduct neural impulses? _____

Which cells are supportive cells? _____

c. Which neuron process conducts impulses:

toward the cell body? _____

away from the cell body? _____

7. Mini-Practicum

Your instructor has set up selected tissues under several microscopes. Identify the tissues indicated by the pointers. Record your answers in the table below.

Microscope No.	Tissue	Microscope No.	Tissue
1		5	
2		6	
3		7	
4		8	

Circulation of Blood

1. The Heart

a. List the labels for Figure 15.1.

1. _____	10. _____
2. _____	11. _____
3. _____	12. _____
4. _____	13. _____
5. _____	14. _____
6. _____	15. _____
7. _____	16. _____
8. _____	17. _____
9. _____	

b. Write the term that matches the phrase.

1. Returns blood to right atrium   _____

_____

2. Returns blood to left atrium   _____
3. Separates the ventricles   _____
4. Pumps blood into pulmonary trunk   _____
5. Pumps blood into aorta   _____
6. Prevents backflow of blood into the right atrium   _____
7. Prevents backflow of blood into the left atrium   _____
8. Prevents backflow of blood into the right ventricle   _____
9. Prevents backflow of blood into the left ventricle   _____
10. Fibers restraining AV valve cusps   _____
11. Heart muscle   _____
12. Valves whose closure produces the first heart sound
13. Valves whose closure produces the second heart sound

c. Describe how blood pressure changes in the ventricles open and close the heart valves.

_____

_____

_____

_____

_____

d. The heart cycle is the period from one ventricular contraction to the next ventricular contraction. Based on the heart sounds, does the heart relax between contractions? _____ Explain. _____

_____

2. Dissection of a Sheep Heart

a. What arteries carry blood to supply the heart with nutrients?

_____

b. Which has a thicker wall, the aorta or anterior vena cava?

_____

c. Do atria or ventricles compose most of the heart? _____

d. What vessels open into the left atrium? _____

e. What vessels open into the right atrium? _____

f. *Circle* the terms that describe the chordae tendineae.

                elastic   nonelastic   thick   thin   flexible   stiff

g. Describe the arrangement and appearance of a semilunar valve. _____

_____

_____

h. Which ventricle has the thicker wall? _____

Explain the basis of this condition. _____

_____

_____

3. Pattern of Circulation

a. Write the name of the type of blood vessels described.

1. Carry blood toward the heart   _____

2. Carry blood away from the heart   _____

3. Smallest arteries   _____

4. Smallest veins   _____

5. Smallest blood vessels   _____

b. Trace the pathway of blood from the heart to the lungs and back to the heart. Right ventricle

⟶ _____ ⟶ _____

⟶ lungs ⟶ _____ ⟶ left atrium.

c. Trace the pathway of blood from the heart to the intestine and back to the heart. Left ventricle

⟶ _____ ⟶ _____

⟶ intestine ⟶ _____ ⟶ liver ⟶

_____ ⟶ _____ ⟶ right atrium

d. Describe the flow of blood through a capillary in a frog's foot.

_____

_____

e. Describe the control of blood flow through a capillary. _____

_____

_____

f. By what process are materials exchanged between capillary blood and body cells? _____

_____

g. Draw from your slides cross sections of these vessels showing the relative diameter of the vessel interior and thickness of the wall.

Normal	**Normal**	**Atherosclerotic**
Vein	**Artery**	**Artery**

h. Explain the hazards of atherosclerosis. _____

_____

_____

_____

_____

i. What arteries carry deoxygenated blood? _____

What veins carry oxygenated blood? _____

j. List the labels for Figure 15.5.

1. _____   4. _____   7. _____

2. _____   5. _____   8. _____

3. _____   6. _____   9. _____

4. Blood Pressure

a. Explain the meaning of systolic and diastolic blood pressure.

Systolic _____

_____

Diastolic _____

_____

b. Explain the cause of the pulse. _____

_____

_____

c. Record your pulse rate (beats/min).

At rest _____ /min   Immediately after exercise _____ /min

How long did it take for your pulse rate to return to its resting rate? _____
_____

d. Record the average and range of the resting pulse rates, by gender, among your class members.

Males: Average _____ /min   Range _____ /min to _____ /min

Females: Average _____ /min   Range _____ /min to _____ /min

Total: Average _____ /min   Range _____ /min to _____ /min

Does there seem to be a gender difference in resting pulse rates? _____

Explain. _____
_____
_____

e. Record your blood pressure:

At rest   _____ mm Hg systolic   _____ mm Hg diastolic

After exercise   _____ mm Hg systolic   _____ mm Hg diastolic

3 min later   _____ mm Hg systolic   _____ mm Hg diastolic

What do you think causes an increase in blood pressure during exercise? _____
_____
_____

f. Explain how the sphygmomanometer works in the determination of blood pressure. _____
_____
_____
_____

5. Mini-Practicum

Your instructor has set up a sheep heart with numbered pins indicating certain structures. Write the names of the indicated structures in the table below.

Pin No.	Structure	Pin No.	Structure
1		6	
2		7	
3		8	
4		9	
5		10	

LABORATORY REPORT 16

Blood

1. Blood Cells

a. Use colored pencils to draw 1 or 2 blood cells of these types as they appear on your slide.

Erythrocytes	**Lymphocytes**	**Monocytes**

Neutrophils	**Eosinophils**	**Basophils**

b. Draw sickled erythrocytes as they appear on your slide.

c. Write the term that matches the phrase.
   1. Basic function of blood _____
   2. Liquid portion of blood _____
   3. Blood cells lacking a nucleus _____
   4. Blood cells possessing a nucleus _____
   5. Tiny cell fragments in blood _____
   6. Blood cells containing hemoglobin _____
   7. Most abundant blood cells _____
   8. Most abundant leukocyte _____
   9. Small WBCs with little cytoplasm _____
   10. Phagocytic leukocytes _____
   _____
   11. Transport oxygen and carbon dioxide _____
   12. Produce antibodies _____
   13. Start the clotting process _____

2. Differential White Cell Count

a. Record your tabulation of 100 WBCs and their percentages in normal blood.

Leukocyte	Tabulation	Percentage
Neutrophils		
Eosinophils		
Basophils		
Lymphocytes		
Monocytes		

b. Record your tabulation of 100 WBCs and their percentages in blood of a patient with infectious mononucleosis.

Leukocyte	Tabulation	Percentage
Neutrophils		
Eosinophils		
Basophils		
Lymphocytes		
Monocytes		

c. Indicate the WBCs with higher than normal percentages during these conditions.

1. Antigen-antibody reactions　　　_____
2. Parasitic worm infestation　　　_____
3. Acute bacterial infection　　　_____
4. Chronic infection　　　_____
5. Viral infection　　　_____

3. Blood Typing

a. Indicate your ABO blood group. _____ Rh type _____

b. Indicate compatibility (C) and incompatibility (I) of possible blood transfusions shown in the chart below.

Blood Type and Antigen of Donor	Blood Type (and Antibodies) of Recipient			
	O *(a, b)*	*A* *(b)*	*B* *(a)*	*AB* *(none)*
O				
A				
B				
AB				

c. Which ABO blood type may receive blood from the other three types in emergencies? _____ This is the **universal recipient**.

d. Which ABO blood type may donate blood to the other three types in emergencies? _____ This is the **universal donor**.

e. Considering *both* ABO and Rh blood types, indicate:

   Universal donor _____ Universal recipient _____

f. Infants suffering from erythroblastosis fetalis may require a massive blood transfusion. What blood type or types should be given to a baby with a blood type of A,Rh^+? _____ Explain your answer. _____

   _____

   _____

4. Mini-Practicum

Your instructor has set up selected blood cells under several microscopes. Identify the blood cells indicated by the pointer. Record your answers in the table below.

Microscope No.	Blood Cell	Microscope No.	Blood Cell
1		5	
2		6	
3		7	
4		8	

LABORATORY REPORT 17

Gas Exchange

1. Respiratory System

a. List the labels for Figure 17.1.

1. _____   7. _____
2. _____   8. _____
3. _____   9. _____
4. _____   10. _____
5. _____   11. _____
6. _____

b. List in sequence the air passages that carry air into the lungs.

1. Nostrils _____   6. _____
2. _____   7. _____
3. _____   8. _____
4. Glottis _____   9. _____
5. _____   10. Alveoli _____

c. Write the terms that match the phrases.

1. Site of gas exchange   _____
2. Contains vocal cords   _____
3. Warms and filters inhaled air   _____
4. Separates nasal and oral cavities   _____
5. Opening into larynx   _____
6. Air passageways entering lungs   _____
7. The windpipe   _____
8. The throat   _____
9. Closes glottis when swallowing   _____
10. Tiny air sacs in lungs   _____

d. Draw the following from your slide of trachea, x.s.

Cartilaginous Ring (40×)

Pseudostratified Ciliated Columnar Epithelium (400×)

e. Describe how the ciliated epithelium lining respiratory passages remove particles from inhaled air. _____

_____

_____

f. What is the function of the cartilaginous rings in trachea and bronchi? _____

_____

_____

g. Are the cartilaginous walls of the larynx hard or soft? _____

h. Describe any movement of the larynx during swallowing. _____

_____

_____

i. Draw the microscopic appearance of lung tissue at 40×.

Normal Lung Tissue　　　　　　　　　**Emphysematous Lung Tissue**

j. By what process are oxygen and carbon dioxide exchanged between air and capillary blood in the lungs? _____

k. What is the value of a large respiratory surface area for gas exchange?

_____

_____

l. How is the respiratory surface area reduced in an emphysematous lung? _____

_____

_____

2. Breathing Mechanics

a. Comparing the breathing mechanics model with the body, indicate the body parts simulated by the:

bell jar _____ rubber sheet _____

glass tubing _____ balloons _____

b. When the rubber sheet of the breathing mechanics model is pulled down:

the volume of the bell jar is _____ and the air

pressure in the bell jar is _____ , which causes air

to flow into the lungs.

c. When the rubber sheet of the breathing mechanics model is pushed up:

the volume of the bell jar is _____ and the air

pressure in the bell jar is _____ , which causes air

to flow out of the lungs.

d. During inspiration, atmospheric air pressure is _____ than
    air pressure in the lungs.
    During expiration, atmospheric air pressure is _____ than
    air pressure in the lungs.

3. Lung Volumes

a. Calculate your tidal volume (TV).
    Spirometer reading after 5 normal breathing cycles. _____
    Spirometer reading divided by 5 equals TV. _____ ml

b. Calculate your vital capacity (VC).
    Spirometer reading for each replicate: 1 _____ ml; 2 _____ ml; 3 _____ ml
    Sum of the replicates _____ ml   Sum divided by 3 equals _____ ml
    Is your vital capacity within the normal range? _____

c. Calculate your expiratory reserve volume (ERV).
    Spirometer reading minus 1,000 ml for each replicate:

    1 _____ ml − 1,000 ml = _____ ml
    2 _____ ml − 1,000 ml = _____ ml
    3 _____ ml − 1,000 ml = _____ ml

    Sum of the replicates _____ ml   Sum divided by 3 equals _____ ml

d. Calculate your inspiratory reserve volume (IRV) using the data from the above tests.
    VC − (TV + ERV) = IRV
    VC _____ ml − (TV _____ ml + ERV _____ ml) = IRV _____ ml

e. Calculate your respiratory minute volume (RMV).
    TV _____ ml × breathing cycles/minute = RMV _____ ml/min.

f. Indicate by a check which of the following that you think will change during heavy breathing
    following exercise.

    _____ Tidal volume   _____ Expiratory reserve volume
    _____ Respiratory minute volume   _____ Inspiratory reserve volume
    _____ Breathing cycles/minute   _____ Vital capacity

g. Test your hypotheses by running in place for 3 min and measuring those items that you have
    checked. Were your hypotheses confirmed? _____
    Explain. _____
    _____
    _____

h. Emphysema patients have difficulty expelling air from their lungs because their residual vol-
    ume is increased. What effect would this have on their:
    Vital capacity? _____
    Expiratory reserve volume? _____
    Concentration of oxygen in the lungs? _____
    Effectiveness of gas exchange? _____

Student _____

LABORATORY REPORT 18

Lab Instructor _____

Digestion

1. The Digestive System

a. Write the term that matches the phrase.

1. Produces bile _____

2. Digestive fluid formed by stomach mucosa _____

3. End product of protein digestion _____

4. Where carbohydrate digestion begins _____

5. Site of decay of nondigestible materials _____

6. Absorption of nutrients is completed _____

7. Wave-like contractions of digestive tract _____

8. Separate oral and nasal cavities _____

9. Carries food from pharynx to stomach _____

10. Where digestion of food is completed _____

11. Break food into smaller pieces _____

12. Pushes food into pharynx in swallowing _____

13. Reabsorbs water from nondigestible material _____

b. List the labels for Figure 18.1.

1. _____   4. _____   7. _____

2. _____   5. _____   8. _____

3. _____   6. _____

c. List the labels for Figure 18.2.

1. _____   4. _____   7. _____

2. _____   5. _____   8. _____

3. _____   6. _____

d. List the labels for Figure 18.3.

1. _____   7. _____   13. _____

2. _____   8. _____   14. _____

3. _____   9. _____   15. _____

4. _____   10. _____   16. _____

5. _____   11. _____   17. _____

6. _____   12. _____   18. _____

2. Histology of the Small Intestine

a. List the four layers of the intestine from inside out.

1. _____   3. _____

2. _____   4. _____

b. In what way do villi affect the surface area of the intestinal mucosa? _____

_____

Why is this advantageous? _____

_____

c. What is the function of the smooth muscle tissues in the intestinal wall? _____

_____

_____

d. Draw the following as they appear on your slide at these *total magnifications.*

2 or 3 Villi (40×)   **Columnar Epithelium (400×)**

3. Digestion of Starch

a. Record your results by placing an "S" for a positive test for starch or an "M" for a positive test for maltose in the appropriate squares.

	Tube Numbers and pH																	
	pH 5						pH 8						pH 11					
Temp.	1	1B	2	2B	3	3B	4	4B	5	5B	6	6B	7	7B	8	8B	9	9B
5 °C																		
37 °C																		
70 °C																		

b. List the control tubes expected to give starch tests that are:

positive _____   negative _____

What should be done if these results are not attained? _____

_____

c. Indicate the optimum pH and temperature, among those tested, for amylase activity.

pH _____   Temp. _____ °C

d. How do some pH values and temperatures inactivate amylase? _____

_____

e. Pancreatic amylase is a component of pancreatic juice that is released into the small intestine. What can you hypothesize about the pH of the digestive fluids in the small intestine?

_____

LABORATORY REPORT 19

Neural Control

1. Neurons

a. Write the term that matches the phrase.
   1. Nerve cell _____
   2. Neuron carrying impulses to CNS _____
   3. Neuron carrying impulses from CNS _____
   4. Composed of cranial and spinal nerves _____
   5. Composed of brain and spinal cord _____
   6. Cells providing support for nerve cells _____
   7. Part of neuron containing nucleus _____
   8. Process carrying impulses from cell body _____
   9. Process carrying impulses toward cell body _____
   10. Neuron with all parts located in the CNS _____
   11. Junction of axon tip and adjacent neuron _____
   12. Secretes neurotransmitter _____
   13. Receives neurotransmitter _____
   14. Forms myelin sheath in PNS neurons _____
   15. Forms myelin sheath in CNS neurons _____
   16. Necessary for regrowth of neuron processes _____
   17. Spaces between Schwann cells _____
   18. A stimulatory neurotransmitter _____
   19. An inhibitory neurotransmitter _____
   20. Cells located between neurons in the CNS _____

b. Explain the mechanism of one-way transmission of impulses at a synapse. _____

_____

_____

c. Draw a giant multipolar neuron and a nerve, x.s., as observed on the prepared slides. Label pertinent parts.

Neuron Cell Body   **Nerve, x.s.**

2. The Brain

a. Write the term that matches the phrase.

1. Fissure between cerebral hemispheres   _____

2. Cerebral lobe containing visual center   _____

3. Outer portion of cerebrum   _____

4. Small ridges on surface of cerebrum   _____

5. Cerebral lobe containing hearing center   _____

6. Protective membranes covering CNS   _____

7. Remnant of a third eye   _____

8. Endocrine gland attached to hypothalamus   _____

b. List the labels for Figure 19.4.

1. _____   4. _____   6. _____

2. _____   5. _____   7. _____

3. _____

c. List the labels for Figure 19.7.

1. _____   5. _____   9. _____

2. _____   6. _____   10. _____

3. _____   7. _____   11. _____

4. _____   8. _____   12. _____

d. Matching

1. Cerebrum   2. Cerebellum   3. Thalamus   4. Hypothalamus   5. Medulla oblongata

6. Corpus callosum   7. Ventricles

_____ Uncritical awareness of sensations   _____ Muscular coordination

_____ Contains cerebrospinal fluid   _____ Controls water balance

_____ Critical interpretation of sensations   _____ Largest part of mammalian brain

_____ Regulates heart rate   _____ Controls body temperature

_____ Connects cerebral hemispheres   _____ Intelligence, will

_____ Breathing control center

e. Is the relative size of the following larger in humans or sheep?

Cerebrum _____   Cerebellum _____

f. Matching

1. White matter   2. Gray matter

_____ Exterior of cerebrum   _____ Corpus callosum

_____ Myelinated fibers   _____ Nonmyelinated fibers

_____ Neuron cell bodies   _____ Exterior of the pons

3. Spinal Cord

a. List the labels for Figure 19.8.

1. _____   5. _____   8. _____

2. _____   6. _____   9. _____

3. _____   7. _____   10. _____

4. _____

b. Write the term that matches the phrase.
   1. Neuron carrying impulses to effector   _____
   2. Neuron receiving impulses from receptor   _____
   3. Neuron in dorsal root   _____
   4. Neuron in ventral root   _____
   5. Location of sensory neuron cell body   _____
   6. Connects sensory and motor neurons   _____

c. Describe the effect of an injury that cuts the:

   Dorsal root of a spinal nerve _____

   _____

   Ventral root of a spinal nerve _____

   _____

   Spinal cord _____

   _____

d. Diagram from your slide (1) a cat spinal cord, x.s., labeling gray and white matter, ventral fissure, and central canal, and (2) a motor neuron cell body and associated processes, labeling the nucleus and neuron processes.

       Cat Spinal Cord, x.s.   **Motor Neuron Cell Body**

4. Human Reflexes

a. Rate the strength of your patellar response as strong (3), moderate (2), weak (1), or none (0):

   Left leg: without diversion _____   with diversion _____

   Right leg: without diversion _____   with diversion _____

   Explain any differences in response. _____

   _____

   _____

   _____

b. If you touch a hot stove, you will reflexively jerk back your hand a split second before you feel the pain. Explain the delayed pain response. _____

   _____

   _____

   _____

c. Describe the effect on the size of the pupil when:

   The subject is looking into a darkened area. _____

   The subject is looking into a bright light. _____

d. When a penlight is shined into the left eye, what response is seen:

In the left eye? _____

In the right eye? _____

Explain your observations. _____

_____

_____

5. Reaction Time

a. Record the 5 replicates of your reaction time in milliseconds (msec).

_____   _____   _____   _____   _____

Calculate your average reaction time. _____

b. Record the average reaction time (msec) for each class member by gender.

Males			**Females**		
1. _____	7. _____	13. _____	1. _____	7. _____	13. _____
2. _____	8. _____	14. _____	2. _____	8. _____	14. _____
3. _____	9. _____	15. _____	3. _____	9. _____	15. _____
4. _____	10. _____	16. _____	4. _____	10. _____	16. _____
5. _____	11. _____	17. _____	5. _____	11. _____	17. _____
6. _____	12. _____	18. _____	6. _____	12. _____	18. _____

c. For each gender and the total class, indicate the range of values and calculate the average of the average reaction times.

Males: Range _____ msec to _____ msec   Average _____ msec

Females: Range _____ msec to _____ msec   Average _____ msec

Class: Range _____ msec to _____ msec   Average _____ msec

d. Do the results support the null hypothesis? _____

Does there seem to be a significant gender difference in reaction times? _____

Explain. _____

_____

_____

e. State a conclusion from the results. _____

_____

_____

f. In what types of activities is a faster reaction time advantageous?

_____

g. Indicate your reaction times for 5 replicates after practicing the test for 20 times.

_____   _____   _____   _____   _____

Calculate your average reaction time. _____

Did your reaction time decrease with practice? _____

5. Mini-Practicum

Your instructor has set up a sheep brain with numbered pins indicating certain structures. Write the names of the indicated structures in the spaces below.

1. _____   4. _____   7. _____

2. _____   5. _____   8. _____

3. _____   6. _____   9. _____

Student _____

LABORATORY REPORT 20

Lab Instructor _____

Sensory Perception

1. The Eye

a. List the labels for Figure 20.1.

1. _____ 6. _____ 11. _____
2. _____ 7. _____ 12. _____
3. _____ 8. _____ 13. _____
4. _____ 9. _____ 14. _____
5. _____ 10. _____ 15. _____

b. Write the term that matches each meaning.

1. Outer white fibrous layer _____
2. Fills space in front of lens _____
3. Controls amount of light entering eye _____
4. Transparent window in front of eye _____
5. Layer with photoreceptors _____
6. Controls shape of lens _____
7. Opening in center of iris _____
8. Fills space behind lens _____
9. Site of sharp, direct vision _____
10. Receptors for color vision _____
11. Carries impulses from eye to brain _____
12. Receptors for black and white vision _____
13. Membrane covering front of eye _____
14. Focuses light on retina _____
15. Junction of retina and optic nerve _____

c. *Circle* the terms that describe the sclera.

soft   firm   weak   strong   tough   easy to cut   rigid   flexible

d. *Circle* the terms that describe the vitreous humor.

watery   jelly-like   transparent   murky

e. When looking through the lens across the room, what is different about the image? _____

_____

Does the lens magnify or reduce the print? _____

f. Is the retina thick or thin? _____

Is the retina firmly attached to the choroid layer? _____

What holds the retina in place in an intact eye? _____

2. Visual Tests

a. Record the distance from your eye to Figure 20.3 at which the blind spot is detected.

Left eye _____   Right eye _____

Explain any difference in the distances. _____

_____

b. Record the near-point distance for your eyes.

Left eye: without glasses _____   with glasses _____

Right eye: without glasses _____   with glasses _____

Explain any differences in the distances. _____

_____

How do your near-point values compare with your age? See Table 20.1.

_____

c. Is astigmatism present?

Left eye: without glasses _____   with glasses _____

Right eye: without glasses _____   with glasses _____

d. Record the acuity for each eye.

Left eye: without glasses _20/_____   with glasses _20/_____

Right eye: without glasses _20/_____   with glasses _20/_____

e. Record your responses to the color-blindness test in the table.

Responses to the Ishihara Color-Blindness Test Plates

Plate No.	Subject	Responses*		
		Normal	Red-Green Color Blind	Totally Color Blind
1	_____	12	12	12
2	_____	8	3	0
3	_____	5	2	0
4	_____	29	70	0
5	_____	74	21	0
6	_____	7	0	0
7	_____	45	0	0
8	_____	2	0	0
9	_____	0	2	0
10	_____	16	0	0
11	_____	Traceable	0	0
			Red Cones Absent	Green Cones Absent	
12	_____	35	5	3	0
13	_____	96	6	9	0
14	_____	Can trace two lines	Purple	Red	0

* 0 means the subject sees no pattern in the test plate.

Are you color blind _____   If so, describe your type of color blindness. _____

_____

3. The Ear
a. List the labels for Figure 20.5.

1. _____ 6. _____ 11. _____
2. _____ 7. _____ 12. _____
3. _____ 8. _____ 13. _____
4. _____ 9. _____
5. _____ 10. _____

b. Write the term that matches each meaning.
1. Contains sound receptors   _____
2. Bone in which inner ear is embedded   _____
3. Contains receptors for static equilibrium   _____
4. Connects middle ear to pharynx   _____
5. Interprets impulses as sound   _____
6. Contain receptors for dynamic equilibrium   _____
7. Conduct vibrations across middle ear   _____
8. Membrane separating outer and middle ear   _____

4. Ear Physiology
a. Record the average distance for the watch-tick test.
Left ear _____   Right ear _____   Class average _____

b. State your conclusion from the Rinne test.
Left ear _____
Right ear _____

c. Indicate the subject's degree of wavering in the static equilibrium test. Use the terms *slight, moderate,* and *great* to indicate variations.
Standing on both feet with arms at side
Eyes open _____   Eyes closed _____
Standing on both feet with arms extended
Eyes open _____   Eyes closed _____

d. How important are visual clues in maintaining balance? _____
_____

5. Skin Receptors
a. Indicate the minimal distances that provided a two-point sensation.
Forearm _____   Back of neck _____
Palm of hand _____   Fingertip _____
What is the value of this distribution? _____
_____
_____

b. Indicate the adaptation time when using
A single coin _____   Several coins _____
What is the value of adaptation to stimuli? _____
_____
_____

c. What is the physiological basis for a boy looking for his hat only to discover that he is wearing it? _____
_____

d. Indicate the smallest temperature differential that you could detect by immersing your finger in water. _____ °C

Indicate the range and average of the class.   Range _____ °C   Average _____ °C

e. Was the sensation stronger or weaker when your entire hand was immersed in the water? _____ Explain the sensation. _____

_____

_____

f. Did the sensation change with time when your hands were immersed in

Ice water? _____ Warm water? _____

Explain. _____

_____

_____

g. Describe the sensation when both hands were placed in tap water after immersion in ice and warm water, respectively. _____

_____

How do you account for this? _____

_____

_____

h. What is the physiological basis for a girl believing that the temperature of water in a swimming pool is "fine" after testing it with her foot but complaining that it is "freezing cold" after jumping in? _____

_____

6. Taste

a. For each of the tastes tested, do the locations of your taste receptors agree with those in Figure 20.7?

Sweet _____   Sour _____   Salt _____

Explain any differences. _____

_____

_____

b. Are taste receptors sensitive to dry sugar granules? _____

Explain the role of saliva in taste sensations. _____

_____

c. Indicate the role of the following in sensory perception.

Receptors _____

_____

Brain _____

_____

Nerves _____

_____

Student _____

Lab Instructor _____

LABORATORY REPORT 21

Support and Movement

1. The Skeleton

a. List the labels for Figure 21.1.

1. _____   9. _____   16. _____
2. _____   10. _____   17. _____
3. _____   11. _____   18. _____
4. _____   12. _____   19. _____
5. _____   13. _____   20. _____
6. _____   14. _____   21. _____
7. _____   15. _____   22. _____
8. _____

b. Matching

1. Clavicle   2. Humerus   3. Femur   4. Coxal bones   5. Scapula   6. Ulna and radius
7. Tibia and fibula   8. Mandible   9. Skull   10. Sacrum   11. Sternum   12. Vertebrae
13. Tarsals   14. Carpals   15. Phalanges

_____ Ankle bones   _____ Form backbone
_____ Bones of lower arm   _____ Bone of upper arm
_____ Finger bones   _____ Bones of lower leg
_____ Shoulder blade   _____ Collarbone
_____ Wrist bones   _____ Breastbone
_____ Thighbone   _____ Lower jawbone
_____ Five fused vertebrae   _____ Hipbones

c. Write the terms that match the phrases.

1. Vertebrae of the neck   _____
2. Small bone in the lower leg   _____
3. Number of thoracic vertebrae   _____
4. Part of skull enclosing brain   _____
5. Forms the kneecap   _____
6. Fibrocartilage between vertebrae   _____
7. Movable bone of the skull   _____
8. Bones encasing spinal cord   _____
9. Bones protecting heart and lungs   _____

d. Write the name of the bone(s) that articulates with the:

1. Scapula and sternum   _____
2. Coxal bone and tibia   _____
3. Scapula and ulna   _____
4. Sternum and thoracic vertebrae   _____
5. Carpals and phalanges   _____

e. Matching
1. Immovable   2. Slightly movable   3. Hinge   4. Gliding   5. Pivot   6. Ball and Socket

_____ Vertebrae 1 and 2	_____ Humerus/ulna
_____ Humerus/scapula	_____ Finger joints
_____ Joints between vertebrae	_____ Joints of cranial bones
_____ Femur/coxal bone	_____ Wrist and ankle
_____ Femur/tibia	_____ Lower jaw/skull

2. **Sexual Differences of the Pelvis**

a. Using Table 21.1, record your observations of a male and female pelvis and of the "unknown" pelvis.

Characteristic	Male	Female	Unknown
General structure			
Acetabula			
Pubic angle			
Sacrum			
Coccyx			
Pelvic brim			
Ischial spines			
Pelvic ratio			

b. What is the gender of the "unknown" pelvis? _____

c. What is the advantage of the adaptations of the female pelvis? _____

_____

3. **Fetal Skeleton**

a. What portions of the:

Long bones are formed of bone? _____

Cranial bones are formed of bone? _____

b. Carefully using the tape measure, determine the part of the fetal skeleton that has the largest circumference. _____

c. What is the advantage of the incomplete ossification of the cranial bones prior to birth? _____

_____

_____

4. Macroscopic Bone Structure

a. List the labels for Figure 21.3.

1. _____ 4. _____ 7. _____

2. _____ 5. _____ 8. _____

3. _____ 6. _____

b. Write the term that matches the phrase.

1. Location of yellow marrow   _____

2. Fills spaces in cancellous bone   _____

3. Type of bone forming the diaphysis   _____

4. Type of bone forming the epiphyses   _____

5. Fibrous membrane covering bone   _____

6. Forms blood cells   _____

7. Cartilage between diaphysis and epiphyses   _____

8. Protects articular surface of the bone   _____

9. Site of growth in length   _____

c. Distinguish between the epiphyseal disc and epiphyseal line. _____

_____

_____

d. What advantage is provided by the articular cartilage? _____

_____

e. Why is adequate dietary calcium necessary for the development of strong bones? _____

_____

f. Describe the appearance, feel, and function of the articular cartilage.

1. Appearance _____

2. Feel _____

3. Function _____

g. Circle the following terms that describe the ligaments.

pliable   stiff   weak   strong   elastic   nonelastic

5. Skeletal Muscles

a. Indicate the type of lever described.

1. CF lies between F and R   _____

2. F lies between CF and R   _____

3. R lies between CF and F   _____

b. Which type of lever has the greatest mechanical advantage and can move the largest resistance with the least contraction force? _____

Explain. _____

_____

_____

c. Indicate the action of the antagonists of muscles that produce:

Flexion _____   Adduction _____

d. The stationary end of a contracting muscle is the: _____

The moving end of a contracting muscle is the: _____

e. Indicate the type of movement and lever class for each illustration in Figure 21.6.

Illustration	Movement	Lever Class
(a)	_____	_____
(b)	_____	_____
(c)	_____	_____
(d)	_____	_____
(e)	_____	_____
(f)	_____	_____
(g)	_____	_____

f. Does the body seem to be adapted for great strength or speed of movement? _____
   Explain. _____
   _____
   _____

6. Ultrastructure and Contraction

a. Draw the appearance of striations of the fibers before and after adding ATP, K$^+$, and Mg^{++}.

Before   **After**

b. Record the length of the fibers before and after adding ATP, K$^+$, and Mg^{++}.

Before _____ mm   After _____ mm

c. Write the term that matches the phrase.
   1. Thick myofilament with cross-bridges   _____
   2. Thin myofilaments attached to Z lines   _____
   3. Form boundaries of sarcomere   _____
   4. Myofilaments within the I band   _____

7. Mini-Practicum

Your instructor has set up several "unknown" bones for you to identify. Write the names of the bones in the spaces below.

1. _____	4. _____	7. _____
2. _____	5. _____	8. _____
3. _____	6. _____	9. _____

Student ______________________

LABORATORY REPORT 22

Lab Instructor ______________________

Excretion

1. Nitrogenous Wastes

a. List three forms of nitrogenous wastes excreted by animals.

1. ______________ 2. ______________ 3. ______________

b. Which compound is most toxic? ______________________

2. Urinary System

a. List the labels for Figures 22.1 and 22.2.

Figure 22.1	**Figure 22.2**	
1. ______	1. ______	8. ______
2. ______	2. ______	9. ______
3. ______	3. ______	10. ______
4. ______	4. ______	11. ______
5. ______	5. ______	12. ______
6. ______	6. ______	13. ______
7. ______	7. ______	14. ______
8. ______		

b. Describe the function of each:

Kidney ______________________

Urethra ______________________

Urinary bladder ______________________

Ureter ______________________

c. Diagram the structural relationship between the Bowman's capsule, glomerulus, afferent arteriole, and efferent arteriole.

    d. List in sequence the parts of a nephron.

       1. _____ 3. _____

       2. _____ 4. _____

3. Urine Formation

    a. Explain the value of increased blood pressure in the glomerulus. _____

_____

_____

    b. Describe what happens in each:

       Glomerular filtration _____

_____

       Tubular reabsorption _____

_____

       Tubular secretion _____

_____

    c. Using the data in Table 22.1, what percentage of the blood passing through the kidneys becomes filtrate? _____ filtrate becomes urine? _____

       What happens to the rest of the filtrate? _____

    d. Why is a low volume of urine production important in humans and other terrestrial animals but unimportant in aquatic animals? _____

_____

    e. Considering Table 22.2, explain how the high concentration of urea in urine (column 4) is attained. _____

_____

    f. Does the kidney simply maintain the concentration of urea at a safe level or remove all of it from the blood? _____

      Explain. _____

    g. Comparing filtrate and urine, how many times is urea concentrated in the urine (columns 3 and 4)? _____

      How is that accomplished? _____

    h. Comparing the concentration of mineral salts in columns 1 and 5, very little is removed from the blood. What happens to most of the salts in the filtrate? _____

    i. Explain the protein concentrations in each column:

       Column 2 _____

       Column 5 _____

    j. Does glucose easily pass from the glomerulus into Bowman's capsule? _____

       How do you know this? _____

       What happens to the glucose in the filtrate? _____

       Explain the slightly lower concentration of glucose in column 5 as compared to column 1.

_____

_____

4. Urinalysis

a. Record the results of your analysis of the simulated urine samples in the table below. Use an X to indicate the presence of a component and record the color and specific gravity.

Characteristic	Tubes						
	1	2	3	4	5	6	Your Urine
Glucose							
Bilirubin							
Ketone							
Specific gravity							
Blood							
pH							
Protein							
Nitrite							
Color							

b. Indicate for each urine sample the health condition of the "patient" based on your urinalysis. Select the health conditions from the list that follows.

Tube 1 _____

Tube 2 _____

Tube 3 _____

Tube 4 _____

Tube 5 _____

Tube 6 _____

Your urine _____

Diabetes insipidis

This disorder is caused by a deficiency of antidiuretic hormone, which results in the inability of kidney tubules to reabsorb water. Symptoms are constant thirst, weight loss, weakness, and production of 4–10 liters of urine each day. The urine is very dilute (a low specific gravity) and is nearly colorless.

Diabetes mellitus

This disorder is caused by a deficiency of insulin, which results in a decreased ability for glucose to enter cells. Therefore, glucose accumulates in the blood and fats are used excessively in cellular respiration, producing ketones as a waste product. Symptoms include weakness, fatigue, weight loss, and excessive blood glucose levels. In uncontrolled diabetes, a urine sample contains excess glucose, ketones, and a low pH.

Hepatitis

Hepatitis is usually caused by a viral infection. In type A hepatitis, symptoms may include fatigue, fever, generalized aching, abdominal pain, and jaundice (yellowish tinge to the whites of the eyes due to excessive blood levels of bilirubin). Urine is dark amber in color due to excessive bilirubin.

Glomerulonephritis
In this kidney disease, excessive permeability of Bowman's capsule allows proteins and red blood cells to enter the filtrate. They are present in the urine since they cannot be reabsorbed.

Hemolytic anemia
A number of conditions cause the destruction of red blood cells, releasing hemoglobin into the blood plasma, for example, malaria and incompatible blood transfusions. The urine of such patients contains hemoglobin and may be red-brown or smoky in color.

Normal
See Table 22.3 for the composition of normal urine.

Strenuous exercise and a high-protein diet
Athletes in excellent health may produce urine with an acid pH, presence of a small amount of protein, and an elevated specific gravity due to the presence of the proteins.

Urinary tract infection
Bacterial urinary tract infections may involve the urethra (urethritis), urinary bladder (cystitis), ureters, or kidney. Production of alkaline urine promotes the growth of infectious bacteria. Urethritis often causes a burning pain during urination. Symptoms may include a low-grade fever and discomfort of the affected region. Urine will show a positive test for nitrite because bacteria in the urine convert nitrate, a normal component, into nitrite. Severe infections may also cause the urine to contain blood and pus (leukocytes), producing a cloudy urine.

c. Explain the relationship between the volume of urine, its specific gravity, and its color. _____

_____

_____

_____

_____

_____

_____

5. Mini-Practicum
Your instructor has set up a sectioned sheep kidney and a model of the urinary system with selected structures indicated by numbered pins or tags. Write the names of the structures in the spaces below.

1. _____ 4. _____ 7. _____

2. _____ 5. _____ 8. _____

3. _____ 6. _____ 9. _____

LABORATORY REPORT 23

Reproduction

1. Reproductive Systems

a. List the labels for Figure 23.1.

1. _____   6. _____   11. _____
2. _____   7. _____   12. _____
3. _____   8. _____   13. _____
4. _____   9. _____   14. _____
5. _____   10. _____   15. _____

b. Trace the path of sperm from the seminiferous tubules to the urethra. _____

_____

c. List the labels for Figure 23.2.

1. _____   5. _____   9. _____
2. _____   6. _____   10. _____
3. _____   7. _____   11. _____
4. _____   8. _____

d. Contrast the function of the urethra in males and females.

Males _____

Females _____

e. Spermatozoa of all animals require a fluid medium in which to swim to the egg. How is this provided in humans? _____

_____

_____

f. Matching

1. Testes   2. Ovaries   3. Oviduct   4. Vas deferens   5. Penis   6. Prostate gland
7. Prepuce   8. Vagina   9. Uterus   10. Ejaculatory duct

_____ Secretion that activates sperm   _____ Male copulatory organ
_____ Produces eggs   _____ Carries eggs to uterus
_____ Site of embryo development   _____ Female copulatory organ
_____ Carries sperm to urethra   _____ Removed in circumcision
_____ Produces spermatozoa   _____ Lined with ciliated cells

g. Indicate the (1) site of production and (2) function of these hormones.

Testosterone   1. _____
                2. _____

Estrogen   1. _____
                2. _____

Progesterone   1. _____
                2. _____

2. Gametogenesis

a. Matching

1. Diploid   2. Haploid   3. Formed by first meiotic division
4. Formed by second meitoic division

_____ Spermatogonium   _____ Oogonium

_____ Spermatid   _____ Secondary oocyte

_____ First polar body   _____ Spermatozoa

_____ Secondary spermatocyte   _____ Primary oocyte

b. Draw the appearance of the following from your slides.

Graafian Follicle   **Corpus Luteum**   **Sperm**

3. Birth Control

a. Contraceptives are birth control methods that prevent the union of sperm and a secondary oocyte. Which of the methods in Table 23.1 are *not* contraceptives? _____

b. Which contraceptives provide a barrier to the entrance of sperm into the uterus? _____

c. Which contraceptives are chemicals that kill sperm? _____

d. Which contraceptive uses hormones to prevent ovulation? _____

e. Which birth control method prevents the implantation of an early embryo? _____

f. Explain why the rhythm method is not very effective. _____

g. Does using a condom guarantee that you will not be infected with human immunodeficiency virus (HIV) or pathogens of other sexually transmitted diseases? _____
Explain. _____

h. Induced abortion is the removal of a developing embryo or fetus from the uterus prior to term. Is it better to prevent unwanted pregnancies by abstinence or birth control or to rely on induced abortion? _____
Explain. _____

LABORATORY REPORT 24

Fertilization and Development

1. Introduction

Write the term that matches each phrase.

a. Penetration of egg by a sperm _____

b. Mitotic division of fertilized egg _____

c. Fusion of egg and sperm nuclei _____

d. Solid ball of cells (embryonic stage) _____

e. Hollow ball of cells (embryonic stage) _____

f. Cell formed by fertilization _____

g. Embryonic stage formed by gastrulation _____

h. Color of secretion containing sperm _____

i. Color of secretion containing eggs _____

2. Activation and Early Embryology

a. Write the term that matches each phrase.

1. Chemicals that attract sperm to eggs _____

2. Hemisphere of egg containing most yolk _____

3. Hemisphere of egg penetrated by sperm _____

4. Extension of egg engulfing sperm head _____

5. Prevents penetration by additional sperm _____

6. Time until most eggs are activated _____

b. From your slide, draw a few nonactivated and activated eggs. Show the distribution of sperm around the activated eggs. Label pertinent parts.

Nonactivated Eggs   **Activated Eggs**

c. Explain the distribution of the sperm. _____

_____

d. List the labels for Figure 24.5.

1. _____   4. _____   7. _____

2. _____   5. _____   8. _____

3. _____   6. _____   9. _____

  e. Record the time required to reach each stage.

    Two-cell stage _____   Four-cell stage _____   Eight-cell stage _____

  f. Indicate the most abundant stage after each time period.

    12 hr _____   24 hr _____   48 hr _____

3. Chordate Development

  a. Indicate the plane (polar or equatorial) of the cleavage divisions.

    1. First division   _____

    2. Second division   _____

    3. Third division   _____

  b. Are the planes the same as in the sea urchin? _____

  c. Why are the cells formed in the vegetal hemisphere larger than those of the animal hemisphere?

    _____

    _____

  d. In which hemisphere does gastrulation occur? _____

    Is this the same site as in the sea urchin? _____

  e. Indicate the embryonic tissue that forms these structures.

    1. Embryonic gut   _____

    2. Mesodermal pouches   _____

    3. Notochord   _____

    4. Neural tube   _____

    5. Lining of the coelom   _____

  f. Matching

    1. Allantois   2. Amnion   3. Chorion   4. Yolk sac

    In Reptiles

    _____ Envelops embryo   _____ Envelops yolk

    _____ Embryonic urinary bladder   _____ Outermost membrane

    _____ Gas exchange organ   _____ Provides "private pond" for embryo

    In Humans

    _____ Envelops embryo   _____ Part of placenta

    _____ Outermost membrane   _____ Part of umbilical cord

    _____ Brings embryonic blood   _____ Provides "private pond" for embryo
           vessels to placenta

  g. Contrast the function of the allantois in reptiles and humans.

    Reptiles _____

    _____

    Humans _____

    _____

  h. Contrast the function of the yolk sac in reptiles and humans.

    Reptiles _____

    _____

    Humans _____

    _____

4. Human Development

a. By what process are materials exchanged between embryonic and maternal bloods? _____

_____

b. Describe the function of the placenta and umbilical cord in humans.

Placenta _____

_____

Umbilical cord _____

_____

c. List the ages and distinctive visible developmental features of the fetuses observed.

Age	Characteristics
1	
2	
3	
4	
5	
6	
7	
8	
9	

d. At what age is an embryo implanted in the uterus? _____

e. At what age are germ layers formed? _____

f. At what age is the fetus recognizably human? _____

g. At 8 weeks, what percentage of the fetus' length consists of the head? _____

_____

h. At 8 weeks, are arms, legs, fingers, and toes evident? _____

i. What trend occurs in the head-body proportions during development? _____

_____

_____

_____

j. What is the duration of pregnancy in humans? _____

k. Physicians usually advise pregnant women to avoid all drugs during pregnancy. What is the rationale for such avoidance? _____

_____

_____

_____

l. Why does a fetus assume a "fetal position" in the uterus? _____

_____

_____

_____

LABORATORY REPORT 25

Monerans, Protists, and Fungi

1. Monera

a. Draw a few bacterial cells from your slide showing the three types of shapes in bacterial cells.

Bacillus   **Coccus**   **Spirillum**

b. Can you see the flagella of the motile bacteria? _____

   Can you see the cellular contents? _____

c. Indicate the diameters of the no-growth areas.

Antibiotic	Concentration	Diameter of the No-Growth Area (mm)	
		S. epidermidis	*E. coli B*
Ampicillin	10 ug	_____	_____
Erythromycin	10 ug	_____	_____
Cephalothin	10 ug	_____	_____
Chloromycetin	10 ug	_____	_____

d. Which antibiotic was most effective against both organisms? _____

e. Which species was most susceptible to the antibiotics? _____

f. Do some strands of *Oscillatoria* exhibit movement? _____

g. Draw these cyanobacteria as they appear on your slides.

   Gloeocapsa   ***Oscillatoria***

2. Animal-like Protists

a. Draw the appearance of *Trichonympha* from your slide. Label the nucleus and flagella.

b. *Trichonympha* and a termite live in a mutualistically beneficial relationship. What does *Trichonympha* gain from the relationship? _____

   What does the termite gain from the relationship? _____

   _____

c. Draw the appearance of *Pelomyxa* from your slide. Label a food vacuole, a contractile vacuole, a nucleus, and the plasma membrane.

d. Can *Paramecium* move backward as well as forward? _____

   Is the anterior end pointed or rounded? _____

   Does *Paramecium* rotate when swimming? _____

e. Do the cilia beat in unison or in coordinated groups? _____

   How many contractile vacuoles are present in your specimen? _____

3. Plant-like Protists

a. Is the flagellum of *Euglena* at the anterior or posterior end?

   What gives the color to *Euglena*? _____

   Is *Euglena* flexible or rigid? _____

b. Describe the distribution of *Euglena*. _____

   _____

   Explain the distribution. _____

   _____

c. List the plant-like characteristics of dinoflagellates. _____

   _____

d. Draw 1 or 2 of these organisms as observed on the prepared slides.

   Dinoflagellates   **Diatoms**

e. How many species of diatoms are visible in the slide of diatomaceous earth? _____

4. Fungus-like Protists

a. Which stage of slime molds is the feeding stage? _____

b. What stimulates formation of the reproductive stage? _____

_____

5. Pond Water

Draw 2 or 3 monerans and 4 or 5 protists observed in pond water. Indicate the magnification used and the division or phylum of each.

6. Fungi

a. Write the term that matches the phrase.

1. Filaments composing a fungus _____
2. Dormant reproductive cells _____
3. Nonreproductive body of a fungus _____
4. Reproductive structure containing
   spore-forming hyphae _____

b. Contrast saprotrophic and parasitic modes of nutrition.

Saprotrophic _____

_____

Parasitic _____

_____

c. Describe how fungi obtain nutrients from organic matter. _____

_____

_____

d. Where is the youngest portion of a *Rhizopus* colony located? _____

_____

Indicate the color of: mature sporangia _____

immature sporangia _____ zygospores _____

e. Indicate whether the following are haploid (n) or diploid (2n) in *Rhizopus*.

Hyphae _____ Mitospores _____ Sporangia _____ Zygospore _____

f. What is the advantage of sexual reproduction? _____

_____

g. Draw a few budding yeast cells from your slide. Label a nucleus and a bud.

h. In *Penicillium*, are the youngest conidiospores at the center or near the edge of the colony?

_____

i. Are conidiospores formed with a sporangium? _____

Explain. _____

_____

j. Are conidiospores formed by mitotic or meiotic division? _____

k. Indicate the type of nuclei (n; n + n; 2n) in these cells of *Coprinus*.

1. Cells of hyphae forming a mushroom mycelium   _____

2. Cells of hyphae forming a mushroom fruiting body   _____

3. Basidia after nuclear fusion   _____

4. Basidiospores   _____

l. Are basidiospores formed by mitotic or meiotic division? _____

m. Draw a few basidia with basidiospores from you slide.

7. **Mini-Practicum**

Write the numbers of the "unknown" specimens in the correct spaces.

_____ Bacteria   _____ Euglenoids

_____ Cyanobacteria   _____ Dinoflagellates

_____ Flagellated protozoans   _____ Diatoms

_____ Amoeboid protozoans   _____ Slime molds

_____ Ciliated protozoans   _____ Fungi

LABORATORY REPORT 26

Plants

1. **Algae**
   a. Write the term that matches the phrase.
   1. Type of nutrition in algae _____
   2. Material forming cell wall _____
   3. Nutrient storage in green algae _____
   4. Organelles containing pigments _____
   5. Chlorophylls in green algae _____
   6. Chlorophylls in brown algae _____
   7. Chlorophylls in red algae _____
   b. Is *Chlamydomonas* larger than bacterial cells? _____
   c. Explain the distribution of *Chlamydomonas* in the partially shaded Petri dish. _____
   _____
   _____
   d. What is the shape of a *Volvox* colony? _____
   e. Describe the movement of *Volvox*. _____
   _____
   _____
   What enables this movement? _____
   f. What is the shape of the chloroplast in a *Spirogyra* cell? _____
   g. How many cells wide is each *Spirogyra* filament? _____
   How are new cells added to the filament? _____
   h. In brown algae, what is the function of a:
   Holdfast _____
   Stipe _____
   Blade _____
   Air bladder _____
   i. Considering the habitat of each, explain the robust body of brown algae in contrast to the delicate body of red algae. _____
   _____
   _____
   j. In paired *Spirogyra* filaments, are the zygospores located randomly in either filament or do all zygospores occur in one filament? _____
   _____
   _____
   Explain this distribution. _____
   _____
   _____

k. In alternation of generations:

Name the two adults in the life cycle. _____

What process forms gametes? _____ spores? _____

Which adult forms gametes? _____ spores? _____

2. Moss Plants

a. Write the term that matches the phrase.

1. Dominant generation   _____

2. Main photosynthetic organs   _____

3. Spore producing generation   _____

4. Process producing gametes   _____

5. Process producing spores   _____

6. Organ producing spores   _____

7. Organ producing sperm   _____

8. Organ producing egg   _____

9. Produced by spore germination   _____

10. Anchors gametophyte to soil   _____

b. Explain why the process of sperm transport restricts mosses to moist areas. _____

_____

_____

c. Draw the appearance of these structures from your slides.

Antheridia **with** **Sperm**	**Archegonium** **with** **Egg**	**Sporangium** **with** **Spores**

3. Ferns

a. Write the term that matches the phrase.

1. Type of nutrition in sporophytes   _____

2. Type of nutrition in gametophytes   _____

3. Dominant generation in life cycle   _____

4. Process producing gametes   _____

5. Process producing spores   _____

6. Anchors gametophyte to soil   _____

b. What is formed by the fusion of sperm and egg? _____

What does this cell develop into? _____

c. How many eggs are formed in an archegonium? _____

d. How are ferns better adapted to terrestrial life than mosses? _____
_____

e. In what way are ferns no better adapted to terrestrial life than mosses? _____
_____

f. What happens when sporangia dry out? _____
_____

g. Draw a sporangium from your slide.

4. Gymnosperms

a. Write the term that matches the phrase.

1. Dominant generation _____

2. Cones forming pollen grains _____

3. Cones forming female gametophytes _____

4. Cones containing seeds _____

5. Develops into embryo sporophyte _____

6. Process-forming microspores _____

7. Process-forming gametes _____

b. Draw the following from your slides.

Pollen	**Female Gametophyte with Archegonium**	**Seed Section with Embryo Sporophyte**

c. How are pollen grains and seeds adapted for dispersal by wind?

Pollen _____

Seeds _____

d. Explain the advantage of pollination in the life cycle. _____
_____
_____

e. Explain the advantage of seeds in the life cycle. _____
_____
_____

5. Flowering Plants

a. Write the term that matches the phrase.

1. Cells that become pollen grains   _____

2. Major pollinating agents   _____

3. Develops from functional megaspore   _____

4. Develops from pollen grain   _____

5. Process-forming megaspores   _____

6. Process-forming gametes   _____

7. Composed of a mature, ripened ovary   _____

8. Dominant generation   _____

b. Explain the role of flowers in the life cycle. _____

_____

_____

c. List the labels for Figure 26.8.

1. _____   5. _____   9. _____

2. _____   6. _____   10. _____

3. _____   7. _____   11. _____

4. _____   8. _____

d. Are lily pollen grains adapted for wind transport? _____

e. How many compartments compose the ovary? _____

What is located within each compartment? _____

f. List the name and indicate the type of each numbered fruit.

     a. Dry dehiscent   b. Dry indehiscent   c. Fleshy

	Name	**Type**		**Name**	**Type**
1.	_____	_____	4.	_____	_____
2.	_____	_____	5.	_____	_____
3.	_____	_____	6.	_____	_____

g. What part of a pea flower develops into a pea pod? _____

What part of a pea flower develops into a pea seed? _____

h. Explain why a corn kernel is a fruit and not a seed. _____

_____

i. Where is starch concentrated in the:

Bean seed? _____   Corn fruit? _____

j. Explain the role of fruits in the life cycle. _____

_____

_____

6. Mini-Practicum

a. Match the numbers of the "unknown" plants with the correct group.

_____ Algae   _____ Ferns   _____ Angiosperms

_____ Mosses   _____ Gymnosperms

b. Identify the structures indicated by numbered pins or tags.

1. _____   4. _____   7. _____

2. _____   5. _____   8. _____

3. _____   6. _____   9. _____

LABORATORY REPORT 27

Structure of Flowering Plants

1. External Structure

a. List the labels for Figure 27.2.

1. _____   4. _____   7. _____

2. _____   5. _____   8. _____

3. _____   6. _____   9. _____

b. Write the term that matches the phrase.

1. Site of leaf attachment to a stem   _____

2. Type of leaf venation in dicots   _____

3. A leaf stalk   _____

4. Type of leaf venation in monocots   _____

5. Arrangement of vascular bundles in:

   dicots   _____

   monocots   _____

6. Number of cotyledons in seeds of:

   dicots   _____

   monocots   _____

7. Arrangement of flower parts in:

   dicots   _____

   monocots   _____

8. Type of root systems in:

   dicots   _____

   monocots   _____

c. Compare the *Coleus* and corn seedlings.

Structure	Coleus	Corn
Root type		
Leaf venation		
Arrangement of vascular bundles		

d. Identify as a dicot or monocot.   *Coleus* _____   Corn _____

e. Identify the "unknowns" as dicots or monocots.

1. _____   3. _____   5. _____

2. _____   4. _____   6. _____

2. Roots

a. Write the term that matches the phrase.

1. Root-tip region forming new cells   _____

2. Root-tip region of greatest growth   _____

3. Root-tip region of root hairs   _____

4. Tissue transporting water and minerals   _____

5. Tissue transporting organic nutrients   _____

6. Tissue forming branch roots   _____

7. Tissue of water-impermeable cells   _____

b. List the three functions of roots. _____

_____

c. Draw a young root of a germinated seed with root hairs at 40× and an epidermal cell with its root hair at 100×. Label the oldest and youngest root hairs.

Young Root   **Root Hair**

d. Are the cells of an *Allium* root tip arranged in an orderly pattern or in a nonordered mass? Explain. _____

_____

e. Draw the arrangement of roots in tap root and fibrous root systems.

Tap Root System   **Fibrous Root System**

f. What are adventitious roots? _____

_____

g. Which type of root system is best for:

preventing erosion of surface soil? _____

reaching deep water sources? _____

3. **Stems**
   a. Describe the function of stems. _____

   _____

   _____

   b. Describe the arrangement of vascular bundles in a cross section of a *Zea* stem. _____

   _____

   Are they more abundant near the periphery of the stem? _____

   Explain any mechanical advantage in their arrangement? _____

   _____

   _____

   c. In a *Medicago* vascular bundle, is xylem closer to the epidermis or pith? _____

   Which vascular tissue has larger cells? _____

   has thicker walls? _____ provides greater support? _____

   d. In a young *Quercus* stem, what tissue produces new (secondary) xylem and phloem? _____

   Is phloem formed interior or exterior to this tissue? _____

   What is the outermost tissue in the young stem? _____

   e. In an old *Quercus* stem, is the newest xylem near the pith or vascular cambium? _____

   Is the oldest phloem next to the cortex or vascular cambium? _____

   List the tissues forming the bark. _____

   What tissue forms wood in a tree stem? _____

   f. How can you distinguish spring wood and summer wood in an annual ring? _____

   _____

   Does spring or summer wood provide the greatest growth? _____

   g. What forms the cork cells of the bark? _____

   What is the function of cork cells? _____

   i. Explain why removing a strip of bark completely around a tree stem causes the tree to die.

   _____

   _____

   j. What is the age of the demonstration section of tree stem? _____

4. **Leaves**
   a. Write the term that matches the phrase.

   1. The broad, flat part of a leaf   _____

   2. A leaf stalk   _____

   3. Venation in monocots   _____

   4. Basic venation in dicots   _____

   b. Indicate the type of venation of the numbered "unknown" leaves.

   1. _____   4. _____

   2. _____   5. _____

   3. _____   6. _____

c. Write the term that matches the phrase.
   1. Photosynthetic tissue of a leaf _____
   2. Tiny openings in the epidermis _____
   3. Waterproof coating of epidermis _____
   4. Compose a leaf vein _____
   _____
   _____
   5. Control size of stomata _____
d. Explain the advantage of the palisade mesophyll cells being closely packed together. _____
   _____
   _____

e. Explain the advantage of the many spaces between spongy mesophyll cells. _____
   _____
   _____

f. Draw a stoma showing the arrangement of the guard cells and adjacent epidermal cells. Label the stoma, guard cells, and chloroplasts.

g. What advantage is there in having stomata open during the day? _____
   _____
   _____

h. Draw a portion of a *Zea* (corn) leaf, x.s., from your slide. Label the epidermis, stomata, mesophyll, and vein.

6. **Mini-Practicum**

   Identify the structures indicated by the numbered pins, tags, or pointers of numbered microscopes. Write your answers in the spaces below.

1. _____	5. _____	9. _____
2. _____	6. _____	10. _____
3. _____	7. _____	11. _____
4. _____	8. _____	12. _____

LABORATORY REPORT 28

Simple Animals

1. Sponges
   a. Which type(s) of sponge skeletons are soft and pliable? _____
   b. How does a sponge obtain food? _____
   _____
   c. What creates the water current flowing through a sponge? _____
   _____

2. Cnidarians
   a. Body form exhibited by *Hydra*. _____
   Body form exhibited by jellyfish. _____
   b. In *Obelia*, indicate the body form that reproduces:
   sexually _____ asexually _____
   c. When in a feeding position and waiting for prey, is *Hydra* extended or contracted? _____
   _____ inverted or upright? _____
   d. In Hydra, which tissue layer contains stinging cells? _____
   e. Describe the feeding behavior of *Hydra*. _____
   _____
   _____
   _____
   _____

3. Flatworms
   a. Which end of *Dugesia* is more sensitive to touch? _____
   Is movement of *Dugesia* fast or slow? _____ constant or jerky? _____
   b. Describe the two-step digestive process in coelenterates and flatworms.
   1. _____
   2. _____
   c. What is the advantage of a highly branched gastrovascular cavity? _____
   _____
   d. Does *Dugesia* prefer shade or lighted areas? _____
   e. Draw the appearance of these tapeworm structures from your slides.

   Scolex   **Gravid Proglottid**   **Encysted Larvae**

   f. How can human infestation by beef tapeworms be prevented? _____

_____

   g. Are humans the intermediate or final host of the beef tapeworm? _____

   h. Give the name and infestation site in humans for each fluke observed.

      1. _____

_____

      2. _____

_____

      3. _____

_____

4. Roundworms

   a. Is *Turbatrix* visible to the naked eye? _____

     Does *Turbatrix* move slowly or rapidly? _____

   b. How can human infestation with *Ascaris* be prevented? _____

_____

   c. What is the flat, ribbon-like structure within *Ascaris*? _____

   d. *Ascaris* has a tube-within-a-tube body plan. What forms the "outer tube?" _____

     the "inner tube?" _____

     To what system do most of the internal organs of *Ascaris* belong? _____

_____

   e. How are humans infected with *Trichinella*? _____

_____

     How can infestation be prevented? _____

_____

     Are *Trichinella* larvae visible to the naked eye? _____

5. Review

Match the animal group with the characteristic.

1. Sponges   2. Cnidarians   3. Flatworms   4. Roundworms

_____	Tissue level	_____	Lack contractile fibrils
_____	Bilateral symmetry	_____	Radial symmetry
_____	Organ level	_____	Organ system level
_____	Filter feeders	_____	Tube-within-a-tube body plan
_____	Gastrovascular cavity	_____	Asymmetrical
_____	All adults immobile	_____	Some adults immobile
_____	False coelom	_____	Sac-like body plan
_____	Cellular-tissue level	_____	Some are human parasites

6. Mini-Practicum

Write the phylum to which each specimen belongs in the spaces below.

1. _____	5. _____	9. _____
2. _____	6. _____	10. _____
3. _____	7. _____	11. _____
4. _____	8. _____	12. _____

LABORATORY REPORT 29

Mollusks, Annelids, and Arthropods

1. Mollusks

a. List the four major characteristics of mollusks.

_____   _____

_____   _____

b. Write the term that matches the phrase describing a clam.

1. Covers visceral mass; secretes shell _____

2. Brings water into mantle cavity _____

3. Organs of gas exchange _____

4. Organs collecting food from water _____

5. Used to bury clam in mud or sand _____

c. Matching

1. Monoplacophorans   2. Chitons   3. Snails and slugs   4. Tooth shells

5. Clams and oysters   6. Octopi and squids

_____ Image-forming eyes   _____ Conical shell open at each end

_____ Radula present   _____ Shell of eight plates

_____ Remnants of segmentation   _____ Shell of two valves

_____ All predaceous   _____ Filter feeders

_____ Digging foot   _____ Crawl on flattened foot

_____ Marine only   _____ Foot modified into tentacles

d. Place the numbers of the "unknown" mollusks in the correct spaces.

_____ Monoplacophorans   _____ Snails and slugs

_____ Chitons   _____ Clams, oysters, and mussels

_____ Tooth shells   _____ Octopi and squids

2. Annelids

a. What is the primary distinguishing characteristic of annelids? _____

_____

b. Write the term that matches the phrase regarding earthworms.

1. Name of body segments _____

2. Number of setae per segment _____

3. Secretes the egg case _____

4. Type of circulatory system _____

5. Large white structures near hearts _____

6. Remove metabolic wastes _____

7. Grinds food into tiny pieces _____

8. Number of hearts _____

c. Matching

1. Oligochaetes   2. Polychaetes   3. Leeches

_____ Head with simple eyes   _____ Blood-sucking parasites

_____ Few setae   _____ Parapodia and many setae

_____ Hermaphroditic   _____ Superficial rings on somites

_____ Suckers for attachment   _____ Marine only

d. Place the numbers of the "unknown" annelids in the correct spaces.

_____ Oligochaetes   _____ Polychaetes   _____ Leeches

3. **Arthropods**

a. List the major characteristics of arthropods. _____

_____

_____

b. Write the term that matches the phrase regarding a crayfish.

1. Number of legs   _____

2. Body divisions   _____

3. Number of antennae   _____

4. Type of eyes   _____

5. Type of circulatory system   _____

6. Protects the gills   _____

7. Number of appendages per body segment   _____

8. Large yellowish gland around stomach   _____

c. Write the term that matches the phrase regarding a grasshopper.

1. Number of legs   _____

2. Body part legs are attached to   _____

3. Number of tympani   _____

4. Number of spiracles per abdominal segment _____

5. Number of antennae   _____

d. Matching

1. Arachnids   2. Crustaceans   3. Centipedes   4. Millipedes   5. Insects

_____ Three pairs of legs   _____ Compound eyes only

_____ Two pairs of antennae   _____ Four pairs of legs

_____ Cephalothorax and abdomen   _____ Head, thorax, and abdomen

_____ Elongate, dorsally convex body _____ Body segments fused in pairs

_____ Compound and simple eyes   _____ No antennae

_____ Elongate, flattened body   _____ Five pairs of legs

_____ Wings attached to thorax   _____ Simple eyes only

e. Place the numbers of the "unknown" arthropods in the correct spaces.

_____ Arachnids   _____ Centipedes   _____ Insects

_____ Crustaceans   _____ Millipedes

Echinoderms and Chordates

1. **Echinoderms**
   a. Write the term that matches the phrase.
      1. Type of symmetry in adult   _____
      2. Body surface where mouth is located   _____
      3. Formed from the embryonic blastopore   _____
      4. Major distinctive characteristic   _____
      5. Type of skeleton   _____
   b. Matching
      1. Brittle stars   2. Sea cucumbers   3. Sea lilies   4. Sea stars   5. Sea urchins

      _____ Star-shaped with broad-based   _____ Elongate forms lacking arms and
              arms   spines

      _____ Star-shaped with narrow-based   _____ Globose forms with movable spines;
              arms   mouth with five teeth

      _____ Attached by stalk from aboral   _____ Ambulacral grooves with tube
              surface   feet

      _____ Arms used for locomotion   _____ Ambulacral grooves without tube
                                                              feet

   c. Write the numbers of the "unknown" echinoderms in the correct spaces.
      _____ Brittle stars   _____ Sea cucumbers   _____ Sea lilies
      _____ Sea stars   _____ Sea urchins
2. **Chordates**
   a. List the four distinguishing characteristics of chordates. _____
      _____

      Are they all present in: larval tunicates? _____
      adult tunicates? _____ adult amphioxus? _____
   b. What is the function of gills in tunicates and amphioxus? _____
      _____
   c. What is the function of gills in fish? _____
   d. What trend is evident from lamprey to perch in the number of:
      gill slits? _____   fins? _____
   e. If a shark stops swimming, it will sink, but a perch will not.
      Explain why this occurs. _____
      _____

      _____
   f. Do most fish utilize internal or external fertilization? _____
      internal or external development? _____

g. Draw a cross section of lamprey and shark vertebrae. Label vertebra, nerve cord, and noto-chord, if present.

      Lamprey   **Shark**

h. Do amphibians utilize internal or external fertilization? _____

 internal or external development? _____

i. How are reptiles better adapted to terrestrial life than amphibians? _____

 _____

j. Embryos of reptiles, birds, and mammals develop pharyngeal pouches that never become func-tional gills in the adult. What is the evolutionary significance of this? _____

 _____

k. What are the advantages of homeothermy? _____

 Indicate any disadvantages. _____

 _____

l. What are the advantages of internal embryonic development in mammals? _____

 _____

m. Matching

 1. Jawless fish (modern) 2. Cartilaginous fish 3. Bony fish 4. Amphibians

 5. Reptiles 6. Birds 7. Mammals

_____ Diaphragm and hair	_____ Feathers, horny beak
_____ Notochord in adult	_____ Cartilaginous endoskeleton
_____ Leathery shelled eggs	_____ Intrauterine development
_____ No scales, jaws, or operculum	_____ Lungs, paired limbs, no scales
_____ Epidermal scales	_____ Dermal scales
_____ Operculum, swim bladder	_____ Subterminal mouth, paired fins
_____ Poikilothermic	_____ Homeothermic
_____ Two-chambered heart	_____ Four-chambered heart
_____ Seven pairs of gill slits	_____ Bony endoskeleton
_____ Milk-secreting glands	_____ Claws or nails
_____ Hard-shelled eggs	_____ Lightweight dermal scales
_____ Rib cage	_____ Three-chambered heart

n. Write the numbers of the "unknown" chordates in the correct spaces.

_____ Tunicates	_____ Cartilaginous fish	_____ Reptiles
_____ Lancelets	_____ Bony fish	_____ Birds
_____ Jawless fish	_____ Amphibians	_____ Mammals

LABORATORY REPORT 31

Human Evolution

1. Introduction

a. Why are modern apes not considered ancestral to humans? _____
_____

b. Based on Figure 31.1, and if modern apes and modern humans share a common ancient ancestor, what would you expect to find in hominid ancestors of modern humans regarding these
skull characteristics?

1. Brow ridges. _____
_____

2. Shape of the face. _____
_____

3. Size of canine teeth. _____
_____

4. Chin. _____
_____

5. Position of skull attachment. _____
_____

2. Hominids

a. List the labels for Figure 31.2.

1. _____   4. _____   7. _____
2. _____   5. _____   8. _____
3. _____   6. _____   9. _____

b. Write the term that matches each phrase.

1. Earliest hominid   _____
2. First human   _____
3. Continent where humans evolved   _____
4. First modern humans   _____
5. First human living on three continents   _____

c. What subsequent evolutionary changes seem to have resulted from the bipedal gait of early hominids? _____
_____
_____
_____
_____

3. Skull Analysis

a. Record the cephalic indexes for your class members below and determine the average and range.

Cephalic Indexes of Class Members

1. _____	6. _____	11. _____	16. _____	21. _____	26. _____
2. _____	7. _____	12. _____	17. _____	22. _____	27. _____
3. _____	8. _____	13. _____	18. _____	23. _____	28. _____
4. _____	9. _____	14. _____	19. _____	24. _____	29. _____
5. _____	10. _____	15. _____	20. _____	25. _____	30. _____

Range _____   Average _____

b. Plot the frequencies of the cephalic indexes for your class on the graph below.

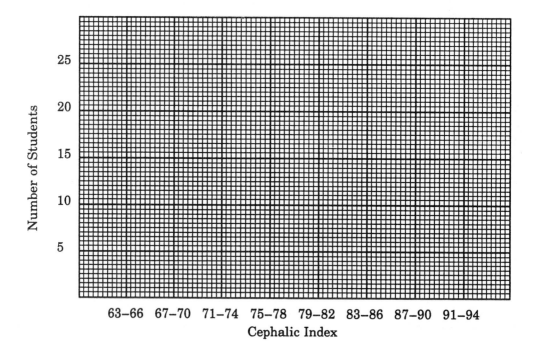

c. Considering the data in item 3b, what difficulties are encountered in trying to make firm conclusions about a species from a single fossil specimen? _____
_____
_____

d. Why are indexes better than simple measurements for comparing fossil specimens? _____
_____
_____

e. Record the results of your skull analyses in the table below.

Skull Comparison

Characteristics	Chimpanzee	A. africanus	H. erectus	H. sapiens neanderthalensis	H. sapiens sapiens (Cro-Magnon)	H. sapiens sapiens (modern)
Cranial breadth						
Cranial length						
Cranial index						
Cranial breadth						
Facial breadth						
Skull proportion index						
Facial projection length						
Skull length						
Facial projection index						
Brow ridges						
Forehead						
Canine length						
Skull and vertebra attachment						
Chin						

f. Which of the fossil hominid skulls has a cranial (cephalic) index that falls within the range of indexes for the class? _____

_____

g. Have there been any major skull changes in humans during the last 40,000 years? _____
Explain your response _____

_____

_____

h. For each hominid skull, plot the skull proportion index and the facial projection index against each other on the graph. Label each plot as to organism.

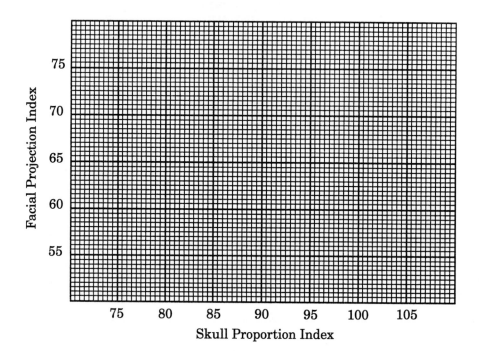

i. Summarize the human evolutionary trends based on your analysis.

Cranium size _____

_____

_____

Face _____

_____

_____

Brow ridges _____

_____

_____

Skull attachment _____

_____

_____

LABORATORY REPORT 32

Ecological Relationships

1. Energy Flow

a. Write the term that matches each meaning.

1. All populations in an area _____
2. All populations and the nonliving environment in an area _____
3. Members of a species in an area _____
4. Ultimate source of energy _____
5. Animals feeding on plants _____
6. Animals feeding on animals _____
7. Convert sunlight into chemical energy of organic molecules _____
8. Obtain nutrient energy from dead organisms and organic wastes _____

b. If only 10% of the energy in one trophic level is passed to the next trophic level, what happens to the remaining 90%? _____
_____ %

c. What percentage of the energy captured by photosynthesis is available to top carnivores?
_____

d. What is the ultimate source of energy? _____

e. List the organisms (by number) in Figure 32.2 according to their roles in energy flow. Note: A carnivore may occupy more than one role.

Producers _____ Herbivores _____

Primary carnivores _____ Secondary carnivores _____

Tertiary carnivores _____

f. If humans were added to the food web in Figure 32.2, what role(s) would they fulfill? _____
_____
_____

g. What would happen to the populations in the food web:
If pesticides significantly reduced the populations of herbivorous insects? _____
_____
_____

If trapping significantly reduced the populations of coyotes and foxes? _____
_____
_____

h. Is it possible to significantly reduce any population in the food web without affecting the other members of the community? _____

Explain. _____

_____

_____

i. An important ecological role filled by certain organisms is not shown in Figure 32.2. What is the role? _____

2. Cycling of Materials

a. Match the steps in the carbon cycle with these processes.

Combustion _____   Consumption _____

Death _____   Photosynthesis _____

Respiration _____

b. Ignoring fossil fuels, what is the nonliving reservoir of carbon?

_____

What group of organisms can use carbon dioxide for their carbon needs?

_____

c. In what form must humans obtain carbon? _____

3. Environmental Pollution

a. Record the number of surviving *Daphnia* from your experiments and the total class below. Calculate the percentage of *Daphnia* surviving for the total class.

1. Acid Pollution

Your group: Number surviving   pH7____   pH6____   pH5____   pH4___ pH3___

Total class: Number surviving   pH7____   pH6____   pH5____   pH4___ pH3___

           Number exposed   pH7____   pH6____   pH5____   pH4___pH3___

           % surviving   pH7____   pH6____   pH5____   pH4___pH3___

2. Pesticide Pollution

Your group: Number surviving 0.000%___ 0.001%___ 0.002%___ 0.003%___ 0.004%___

Total class: Number surviving 0.000%___ 0.001%___ 0.002%___ 0.003%___ 0.004%___

           Number exposed   0.000%___ 0.001%___ 0.002%___ 0.003%___ 0.004%___

           % surviving   0.000%___ 0.001%___ 0.002%___ 0.003%___ 0.004%___

3. Thermal Pollution

Your group: Number surviving   20°C____   25°C____   30°C____35°C___

Total class: Number surviving   20°C____   25°C____   30°C____35°C___

           Number exposed   20°C____   25°C____   30°C____ 35°C___

           % surviving   20°C____   25°C____   30°C____ 35°C___

b. Plot the percentage of surviving *Daphnia* below and determine the LC$_{50}$ for each pollutant. Use Figure 32.4 as a guide.

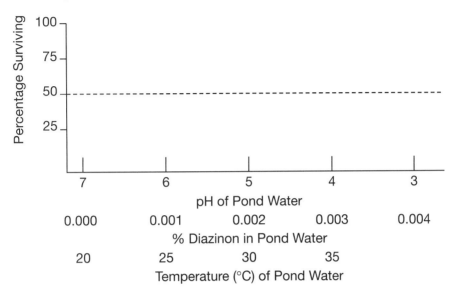

c. Record the estimated LC$_{50}$ for each pollutant.
   Acid pollution _____   Pesticide pollution _____   Thermal pollution _____
d. What is the role of *Daphnia* in an aquatic food web? _____
e. If only the *Daphnia* population of a lake is reduced by pollution, what would you predict to be the effect of *Daphnia* reduction on populations of:
   Microscopic algae? _____
   Small fish? _____
   Larger fish? _____
f. Is it possible that *Daphnia* that survived exposure to these pollutants for 1 hr may:
   Not survive exposure for hours, days, or weeks? _____
   Not survive exposure to two or more pollutants at the same time? _____

4. Behavioral Ecology
a. Describe the normal movement of sowbugs. _____
   _____
   _____
   Does their speed of movement vary? _____
   Explain. _____
   _____
   Do they start and stop frequently? _____
b. What observations would support the null hypothesis? _____
   _____
   _____

c. Record your observations.

Elapsed Time	Number of sowbugs	
	In Shade	In Light
3 min		
6 min		
9 min		
12 min		
15 min		
Ave./3 min		

d. Do your results support the hypothesis? _____
   State a conclusion from your results. _____
   _____
   _____
   _____

e. Record your observations.

Elapsed Time	Number of sowbugs	
	In Moist Area	In Dry Area
3 min		
6 min		
9 min		
12 min		
15 min		
Ave./3 min		

f. Do the results support the hypothesis? _____
   State a conclusion from your results. _____
   _____
   _____
   _____

g. Based on your experiments, what environmental factors characterize the habitat of sowbugs?
   _____

LABORATORY REPORT 33

Population Growth

1. **Introduction**
   Write the term that matches each meaning.
   a. Sum of all limiting factors   _____
   b. Maximum theoretical reproductive capacity   _____
   c. Limiting factors whose effect is unchanged by
      population size   _____
   d. Limiting factors whose effect changes propor-
      tionately with population size   _____

2. **Growth Curves**
   a. Plot the theoretical and realized growth curves below.

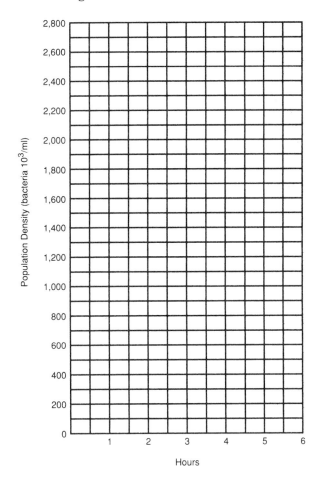

Theoretical and realized growth curves of bacterial populations

b. Indicate the time interval when the growth rate was greatest.

Theoretical growth _____ Realized growth _____

c. In the realized growth curve:

When did cell deaths most nearly equal cell "births"? _____

Why did the realized bacterial population die out? _____

_____

What limiting factors determined the realized population growth pattern? _____

_____

Were density-dependent or density-independent factors involved? _____

_____

d. In item 2a, show the shape the realized growth curve would have taken if additional nutrient broth was added at the fifth hour. Use a red pencil to draw the new curve.

e. Write the term that matches each meaning regarding Figure 33.1.

1. Growth curve with a J shape _____

2. Growth curve with an S shape _____

3. Phase of explosive growth _____

4. Point where limiting factors take effect _____

5. Phase of equilibrium between biotic potential and environmental resistance

6. Where birth equals death _____

7. Where resources per capita are least _____

3. Human Population Growth

a. Plot the human population growth curve below using data in Table 33.3.

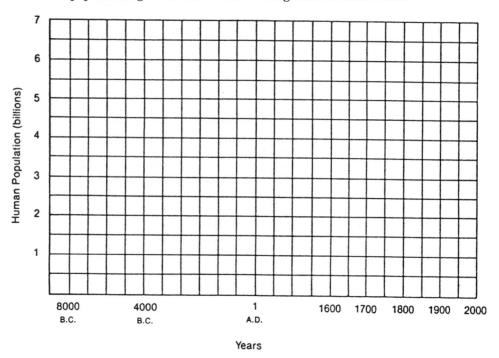

Human population growth curve

b. Does the human population growth curve resemble the theoretical growth curve or the realized growth curve? _____

c. In what phase of the realized growth curve does the human population curve appear to be?

_____

d. What events in human history have enabled the exponential growth of the human population?

_____

_____

e. Are human populations subject to the interaction of biotic potential and environmental resistance like natural populations? _____

_____

f. List any limiting factors controlling the size of natural populations that do not affect human populations. _____

g. Humans are subject to certain limiting factors that are not observed in other animal populations. Name two of these. _____

_____

h. If nothing is done to check the growth of the human population, will deaths ultimately equal births? _____ What will be the shape of the population growth curve? _____

i. Ultimately, there are only two ways to decrease population growth. Name them. _____

_____

Which is preferable? _____

j. Record the growth rate and doubling time as calculated for Table 33.4.

	Growth Rate (%/yr)	**Doubling Time** (yr)
Africa	_____	_____
Asia	_____	_____
Europe	_____	_____
Latin America	_____	_____
North America	_____	_____
Oceania	_____	_____

k. What region has the:
Fastest population growth? _____
Slowest population growth? _____

l. The rapid growth rate in developing countries has resulted from a decreased death rate without a proportionate decrease in birth rate. What has brought this about? _____

_____

_____

m. Are the earth's resources infinite? _____

n. The standard of living is based on available resources per person. If a country's population is increasing at a rate of 2.0% per year, what rate of increase in productivity is required to maintain the same standard of living? _____

o. At the carrying capacity, deaths equal births and resources are just adequate to maintain the existing population size. Would it be advantageous to curtail the human population (zero growth) before it levels off at the carrying capacity due to natural causes? _____ Explain your response. _____

_____

_____

p. Are developing countries likely to improve their standard of living without decreasing their population growth rate? _____ Explain your answer. _____

_____

_____

q. A famine occurs in a certain geographic region, and developed countries provide food to help people faced with starvation. Does the donated food constitute a real or artificial increase in the region's carrying capacity? _____ Explain. _____

_____

_____

What is the long-range solution? _____

_____

_____

_____

_____

r. What are the major obstacles to decreasing the rate of population growth in developing countries? _____

_____

_____

_____

_____

_____

Appendix A

Common Prefixes, Suffixes, and Root Words

Prefixes of Quantity and Size

amphi-, diplo-	both, double, two
bi-, di-	two
centi-	one hundredth
equi-, iso-	equal
haplo-	single, simple
hemi-, semi-	one-half
hex-	six
holo-	whole
quadri-	four
macro-	large
micro-	small
milli-	one thousandths
mono-, uni-	one
multi-, poly-	many
oligo-	few
omni-	all
pento-	five
tri-	three

Prefixes of Direction or Position

ab-, de-, ef-, ex-	away, away from
acro-, apici-	top, highest
ad-, af-	to, toward
antero-, proto-	front
anti-, contra-	against, opposite
archi-, primi-, proto-	first
circum-, peri-	around
dia-, trans-	across

deutero-	second
ecto-, exo-, extra-	outside
endo-, ento-, intra-	inner, within
epi-	upon, over
hypo-, infra-, sub-	under, below
hyper-, supra-	above, over
inter-, meta-	between
medi-, meso-	middle
pre-, pro-	before, in front of
post-, postero-	behind
retro-	backward
ultra-	beyond

Miscellaneous Prefixes

a-, an-, e-	without, lack of
chloro-	green
con-	together, with
contra-	against
dis-	apart, away
dys-	difficult, painful
erythro-	red
eu-	true, good
hetero-	different, other
homo-, homeo-	same, similar
leuco-, leuko-	white
meta-	change, after
neo-	new
necro-	dead, corpse
pseudo-	false

re-	again	chondr, -i, -o	cartilage
sym-, syn-	together, with	chrom, -at, -ato, -o	color
tachy-	quick, rapid	coel, -o	hollow
		crani, -o	skull

Miscellaneous Suffixes

-ac, -al, -alis,	pertaining to	cuti	skin
-an, -ar, -ary		cyst, -i, -o	bag, sac, bladder
-asis, -asia, -esis	condition of	cyt, -e, -o	cell
-blast	bud, sprout	derm, -a, -ato	skin
-cide	killer	entero	intestine
-clast	break down, broken	gastr, -i, -o	stomach
-elle, -il	little, small	gen	to produce
-emia	condition of blood	gyn, -o, gyneco	female
-fer	to bear	hem, -e, -ato	blood
-ia, -ism	condition of	hist, -io, -o	tissue
-ic, -ical, -ine, -ous	pertaining to	hydro	water
-id	member of group	lip, -o	fat
-itis	inflammation	mere	segment, body section
-lysis	loosening, split apart	morph, -i, -o	shape, form
-oid	like, similar to	myo	muscle
-logy	study of	neph, -i, -o	kidney
-oma	tumor	neur, -l, -o	nerve
-osis	a condition, disease	oculo	eye
-pathy	disease	odont	tooth
-phore, -phora	bearer	oo, ovi	egg
-sect, -tome, -tomy	cut	oss, -eo, osteo	bone
-some, -soma	body	oto	ear
-stat	stationary, placed	path, -i, -o	disease
-tropic	change, influence	phag, -o	to eat
-vor	to eat	phyll	leaf
		phyte	plant
		plasm	formative substance

Miscellaneous Root Words and Combining Vowels

andro	man, male	pneumo, -n	lung
arthr, -i, -o	joint	pod, -ia	foot
aut, -o	self	proct, -o	anus
bio	life	soma, -to	body
blast, -i, -o	bud, sprout	stasis, stat, -i, -o	stationary, standing still
brachi, -o	arm	sperm, -a, -ato	seed
branch, -i	gill	stoma, -e, -to	mouth, opening
bronch, -i	windpipe	therm, -o	heat
carcin, -o	cancer	troph	food, nourish
cardi, -a, -o	heart	ur, -ia	urine
carn, -i, -o	flesh	uro, uran	tail
cephal, -i, -o	head	viscer	internal organ
chole	bile	vita	life
		zoo, zoa	animal

Appendix B

Common Metric Units
and Temperature Conversions

Common Metric Units

Category	Symbol	Unit	Value	English Equivalent
Length	km	kilometer	1000 m	0.62 mi
	m	meter*	1 m	39.37 in.
	dm	decimeter	0.1 m	3.94 in.
	cm	centimeter	0.01 m	0.39 in.
	mm	millimeter	0.001 m	0.04 in.
	μm	micrometer	0.000001 m	0.00004 in.
Mass	kg	kilogram	1000 g	2.2 lb
	g	gram*	1 g	0.04 oz
	dg	decigram	0.1 g	0.004 oz
	cg	centigram	0.01 g	0.0004 oz
	mg	milligram	0.001 g	
	μ	microgram	0.000001 g	
Volume	l	liter*	1 l	1.06 qt
	ml	milliliter	0.001 l	0.03 oz
	μ	microliter	0.000001 l	

* Denotes the base unit.

TEMPERATURE CONVERSIONS

Fahrenheit to Celsius

$$°C = \frac{5(°F - 32)}{9}$$

Celsius to Fahrenheit

$$°F = \frac{9 \times °C}{5} + 32$$

Appendix C

Oil-Immersion Technique

The oil-immersion objective enables a magnification of $95\times$ to $100\times$ because the oil prevents loss of light rays and permits the resolution of two points as close as $0.2\,\mu$m (1/100,000 in.).

The working distance of the oil-immersion objective is extremely small, and care must be exercised to avoid damage to the slide or the objective or both. Although focusing with the low-power and high dry objectives before using the oil-immersion objective is not essential, it is the preferred technique.

Use the following procedure:

1. Bring the object into focus with the low-power objective.
2. Center the object in the field and rotate the high-dry objective into position.
3. Refocus using the fine-adjustment knob and readjust the iris diaphragm. Also, recenter the object if necessary.
4. Rotate the revolving nosepiece halfway between the high-dry and oil-immersion objectives to allow sufficient room to add the immersion oil.
5. Place a drop of immersion oil on the slide over the center of the stage aperture. Move quickly from bottle to slide to avoid dripping the oil on the stage or table.
6. Rotate the oil-immersion objective into position. Note that the tip of the objective is immersed in the oil.
7. A slight adjustment with the fine-adjustment knob may be needed to bring the object into focus. Remember that the distance between the slide and objective cannot be decreased much without damaging the slide or objective or both.
8. When finished with the oil-immersion objective, wipe the oil from the objective, slide, and stage.

Appendix D

The Classification of Organisms

This classification of organisms fits quite well with most introductory biology texts. However, it should be noted that biologists are not in total agreement regarding the classification of all organisms, and this classification may differ from your text or the preferences of your instructor. You should keep in mind that the placement of organisms in a classification system is a dynamic, rather than a static, process. A lack of unity is expected in such a process as changes are brought about by new data.

Kingdom MONERA   Unicellular prokaryotes without distinct nuclei and membranebound organelles.

Subkingdom ARCHAEBACTERIA   Bacteria-like prokaryotes living in harsh environments. Includes methanogens, halophiles, and thermophiles.

Subkingdom EUBACTERIA   True bacteria. All other prokaryotes including chemosynthetic, saprotrophic, and parasitic forms.

Kingdom PROTISTA   Mostly unicellular eukaryotes with distinct nuclei and membrane-bound organelles.

Animal-like Protists

Phylum ZOOMASTIGOPHORA*	Flagellated protozoans
Phylum SARCODINA	Amoeboid protozoans
Phylum CILIOPHORA	Ciliated protozoans
Phylum SPOROZOA	Nonmotile, parasitic protozoans

* Botanists use the term *division* for this category, while zoologists use the term *phylum*.

Plantlike Protists

Division BACILLARIOPHYTA	Diatoms and golden-brown algae
Division EUGLENOPHYTA	Flagellated algae lacking cell walls
Division DINOFLAGELLATA	Fire algae (dinoflagellates)

Fungus-like Protists

Division MYXOMYCOTA	Plasmodial slime molds
Division ACRASIOMYCOTA	Cellular slime molds
Division OOMYCOTA	Water molds

Kingdom FUNGI   Saprotrophic, eukaryotic heterotrophs typically with multinucleated cells. Body usually multicellular and composed of hyphae.

Division ZYGOMYCOTA	Conjugating or algal fungi
Division ASCOMYCOTA	Sac fungi
Division BASIDIOMYCOTA	Club fungi
Division DEUTEROMYCOTA	Imperfect fungi

Kingdom PLANTAE   Multicellular eukaryotes with rigid cell walls; primarily photosynthetic autotrophs.

Complex Algae

Division RHODOPHYTA	Red algae
Division PHAEOPHYTA	Brown algae
Division CHLOROPHYTA	Green algae

Nonvascular Land Plants

Division BRYOPHYTA	Mosses, liverworts, and hornworts

Vascular Land Plants Without Seeds

Division PSILOPHYTA	Psilopsids
Division LYCOPHYTA	Club mosses
Division SPHENOPHYTA	Horsetails
Division PTEROPHYTA	Ferns

Vascular Plants with Seeds

Gymnosperms

Division CONIFEROPHTYA	Conifers
Division CYCADOPHYTA	Cycads
Division GINKGOPHYTA	Ginkgos
Division GNETOPHYTA	Gnetophytes

Angiosperms

Division ANTHOPHYTA	Flowering plants
Class MONOCOTYLEDONES	Monocots (e.g., grasses, lilies)
Class DICOTYLEDONES	Dicots (e.g., beans, oaks, snapdragons)

Kingdom ANIMALIA   Multicellular eukaryotic heterotrophs; cells without cell walls or chlorophyll; primarily motile.

Phylum PORIFERA   Sponges

Radial Protostomes
Phylum CNIDARIA   Cnidarians (diploblastic)
   Class HYDROZOA   Hydroids
   Class SCYPHOZOA   True jellyfish
   Class ANTHOZOA   Sea anemones and corals

Bilateral Protostomes
Phylum PLATYHELMINTHES   Flatworms (triploblastic)
   Class TURBELLARIA   Free-living flatworms
   Class TREMATODA   Flukes (parasitic)
   Class CESTODA   Tapeworms (parasitic)
Phylum NEMATODA   Roundworms
Phylum ROTIFERA   Rotifers
Phylum MOLLUSCA   Mollusks; soft bodied and usually with a shell
   Class POLYPLACOPHORA   Chitons
   Class MONOPLACOPHORA   Neopilina (remnants of segmentation)
   Class SCAPHOPODA   Tooth shells
   Class GASTROPODA   Snails and slugs
   Class BIVALVIA   Clams and mussels
   Class CEPHALOPODA   Squids and octopi
Phylum ANNELIDA   Segmented worms
   Class POLYCHAETA   Sandworms
   Class OLIGOCHAETA   Earthworms
   Class HIRUDINEA   Leeches
Phylum ONYCHOPHORA   Peripatus
Phylum ARTHROPODA   Animals with an exoskeleton and jointed appendages
   Class CRUSTACEA   Crustaceans; crabs, crayfish, barnacles
   Class ARACHNIDA   Scorpions, spiders, ticks, mites
   Class CHILOPODA   Centipedes
   Class DIPLOPIDA   Millipedes
   Class INSECTA   Insects

Deuterostomes
Phylum ECHINODERMATA   Spiny-skinned, radially symmetrical animals
   Class CRINOIDEA   Sea lilies, feather stars
   Class ASTEROIDEA   Sea stars
   Class OPHIUROIDEA   Brittle stars
   Class ECHINOIDEA   Sea urchins
   Class HOLOTHUROIDEA   Sea cucumbers
Phylum HEMICHORDATA   Acorn worms
Phylum CHORDATA   Chordates
   Subphylum UROCHORDATA   Tunicates
   Subphylum CEPHALOCHORDATA Lancelets
   Subphylum VERTEBRATA   Vertebrates
     Class AGNATHA   Jawless fishes
     Class CHONDRICHTHYES Cartilaginous fishes

Class OSTEICHTHYES	Bony fishes
Class AMPHIBIA	Salamanders, frogs, toads
Class REPTILIA	Lizards, snakes, crocodiles, turtles
Class AVES	Birds
Class MAMMALIA	Mammals
Subclass PROTOTHERIA	Egg-laying mammals
Subclass METATHERIA	Marsupial mammals
Subclass EUTHERIA	Placental mammals